CARDIOVASCULAR TOXICITY AND THERAPEUTIC MODALITIES TARGETING CARDIO-ONCOLOGY

CARDIOVASCULAR TOXICITY AND THERAPEUTIC MODALITIES TARGETING CARDIO-ONCOLOGY
From Basic Research to Advanced Study

Edited by

Dr. Imteyaz Qamar
Assistant Professor (Senior Level), School of Biotechnology, Gautam Buddha University, Uttar Pradesh, India

Dr. Pawan Kumar Maurya
Professor, Department of Biochemistry, Coordinator, Department of Vocational Studies and Skill Development, Central University of Haryana, Mahendergarh, India

ELSEVIER

ACADEMIC PRESS
An imprint of Elsevier

Academic Press is an imprint of Elsevier
125 London Wall, London EC2Y 5AS, United Kingdom
525 B Street, Suite 1650, San Diego, CA 92101, United States
50 Hampshire Street, 5th Floor, Cambridge, MA 02139, United States
The Boulevard, Langford Lane, Kidlington, Oxford OX5 1GB, United Kingdom

ISBN 978-0-323-90461-2

For information on all Academic Press publications
visit our website at https://www.elsevier.com/books-and-journals

Publisher: Stacy Masucci
Acquisitions Editor: Katie Chan
Editorial Project Manager: Franchezca A. Cabural
Production Project Manager: Omer Mukthar
Cover Designer: Victoria Pearson Esser

Typeset by STRAIVE, India

Contents

4. Stroke risk assessment and atrial fibrillation (AF) in cancer patients

Saima Wajid and Sana Kauser

5. Biochemistry of biomarkers in cardiotoxicity induced in chemotherapeutic drugs

Swaraj Mohanty and Yasmin Ahmad

6. Cardiotoxic effects of tyrosine kinase inhibitors directed against VEGFR

Savneet Kaur

7. Critical insights into cardiotoxicity of anthracyclines

Hanumappa Ananda and Imteyaz Qamar

8. Noninvasive cardiovascular imaging techniques

Bipul Kumar Sinha, Uditi, and Aman Rathore

9. Cardiovascular risk induced by radiotherapy

Mari Uyeda

10. Cardiovascular complication from cancer therapy

Navin Kumar, Swagata Chatterjee, and Bhavna Solanki

11. Novel anticancer drugs related to cardiotoxicity

Pallabi Banerjee and Imteyaz Qamar

12. Principles of cardiovascular rehabilitation

Preeti Sharma, Poornima Pandey, Tikam Chand Dakal, and Narendra Kumar Sharma

13. Hematopoietic stem cell transplantation: Sources of cells, cryopreservation, and applications

Kumar Pranay, Deepak Kumar Jha, Rakesh Bhaskar, and Mukesh Kumar Gupta

Index

Contributors

Yasmin Ahmad Peptide and Proteomics Division, Defence Institute of Physiology & Allied Sciences (DIPAS), Defence Research & Development Organization (DRDO), Timarpur, Delhi, India

Hanumappa Ananda Department of Clinical Embryology, Kasturba Medical College, Manipal, Manipal Academy of Higher Education, Manipal, Karnataka, India

Pallabi Banerjee School of Biotechnology, Gautam Buddha University, Greater Noida, Uttar Pradesh, India

Rakesh Bhaskar Department of Biotechnology and Medical Engineering, National Institute of Technology Rourkela, Rourkela, Odisha, India

Swagata Chatterjee School of Biotechnology, Gautam Buddha University, Greater Noida, Uttar Pradesh, India

Tikam Chand Dakal Genome & Computational Biology Lab, Department of Biotechnology, Mohanlal Sukhadia University, Udaipur, Rajasthan, India

Mukesh Kumar Gupta Department of Biotechnology and Medical Engineering, National Institute of Technology Rourkela, Rourkela, Odisha, India

Riya Gupta Amity Institute of Biotechnology, Amity University Uttar Pradesh, Noida, India

Deepak Kumar Jha P.C. Vigyan Mahavidyala, Chapra, Bihar, India

Savneet Kaur Department of Molecular and Cellular Medicine, Institute of Liver and Biliary Sciences (ILBS), New Delhi, India

Sana Kauser Department of Biotechnology, School of Chemical and Life Sciences, Jamia Hamdard, New Delhi, India

Saima Naz Khan Laboratory for Molecular Reproduction and Genetics, All India Institute of Medical Sciences, New Delhi, India

Navin Kumar School of Biotechnology, Gautam Buddha University, Greater Noida, Uttar Pradesh, India

Pawan Kumar Maurya Department of Biochemistry, Central University of Haryana, Mahendragarh, India

Swaraj Mohanty Peptide and Proteomics Division, Defence Institute of Physiology & Allied Sciences (DIPAS), Defence Research & Development Organization (DRDO), Timarpur, Delhi, India

Md. Kausar Neyaz Bio-Services, New Delhi, India

Poornima Pandey Department of Bioscience and Biotechnology, Banasthali Vidyapith, Banasthali Tonk, Rajasthan, India

Kumar Pranay Department of Biochemistry, Indira Gandhi Institute of Medical Sciences, Patna, Bihar, India

Imteyaz Qamar School of Biotechnology, Gautam Buddha University, Greater Noida, Uttar Pradesh, India

Aman Rathore School of Biotechnology, Gautam Buddha University, Greater Noida, Uttar Pradesh, India

Narendra Kumar Sharma Department of Bioscience and Biotechnology, Banasthali Vidyapith, Banasthali Tonk, Rajasthan, India

Preeti Sharma Department of Bioscience and Biotechnology, Banasthali Vidyapith, Banasthali Tonk, Rajasthan, India

Bipul Kumar Sinha Fortis Escorts Heart Institute & Research Centre, New Delhi, Delhi, India

Bhavna Solanki School of Biotechnology, Gautam Buddha University, Greater Noida, Uttar Pradesh, India

Uditi School of Biotechnology, Gautam Buddha University, Greater Noida, Uttar Pradesh, India

Mari Uyeda AC Camargo Cancer Centre, São Paulo, Brazil

G. Vidya Department of Zoology, St. Joseph's College (Autonomous) Irinjalakuda, Irinjalakuda, Kerala, India

Saima Wajid Department of Biotechnology, School of Chemical and Life Sciences, Jamia Hamdard, New Delhi, India

Preface

In recent years, cardio-oncology has emerged as an important subfield in cardiology aimed at reducing cardiovascular morbidity and mortality in cancer survivors and improving their quality of life. The long-term cancer survival rate has increased significantly with the advent of new, more potent antineoplastic therapies. Recent studies have shown that specific cancer treatments are associated with an increased rate of cardiovascular illness, and in the past 20 years, we have noticed an increase in death from heart failure, ischemic heart disease, arrhythmias, and hypertension among cancer survivors. Specifically, cardio-oncology is concerned with preventing, diagnosing, and treating cardiotoxicity caused by chemotherapy and/or radiotherapy.

Short- and long-term cardiovascular toxicity from anticancer therapy can cause heart failure, as well as exacerbating and/or unraveling preexisting heart disease. Cancer treatment and cardiovascular disease treatment are aimed at improving patients' quality of life and survival. Radiotherapy, cytotoxic chemotherapy, molecularly targeted inhibitors of the immune system, and antibodies designed to target immune checkpoints are among the current therapies employed to treat advanced cancers. Radiation therapy and chemotherapy, for example, have the potential to negatively impact the cardiovascular system, and significant experience has been gained in both cases. In this book, we have discussed the general aspects related to antineoplastic therapy, including action mechanisms of the most commonly used anticancer agents, cardiotoxic effects of tyrosine kinase inhibitors directed against VEGFR, classification of cardiotoxicity, and mechanisms underlying anthracycline-induced cardiotoxicity. Various anticancer agents are being developed daily, and we have summarized the different varieties of antineoplastic drugs, including traditional, new targeted, and immuno-oncological drugs, as well as their cardiovascular toxicity. There are a number of classic clinical cardiovascular risk factors that are known to predict cardiotoxicity, and along with cardiac biomarkers and imaging techniques, these factors can be used to diagnose cardiotoxicity early on in the subclinical/asymptomatic stage, allowing preventive measures to be taken. Therapeutic advances have also been made in the symptomatic stage, whose outcome is outlined. Several advanced technologies are available for isolating resident cells directly from the heart or for conditioning noncardiac cells to achieve a disease-specific phenotype for optimal outcome. As regenerative principles become more prevalent in clinical practice, rehabilitation providers across the patient continuum will be increasingly involved. Thus, there are numerous opportunities for regenerative rehabilitation as a new multidisciplinary field.

In view of such clinical and healthcare importance of cardiovascular health monitoring, this book has been proposed. The book starts with a brief introduction about the cardiovascular system and the cardiovascular changes associated with aging as a background for consideration of the management of hypertension in an elderly population. Then it focuses on the advances in noninvasive cardiac imaging methods to monitor chemotherapy-induced cardiotoxicity, early detection of impairment of contractile myocardial function, and novel anticancerous drugs related to cardiotoxicity, which are very important for developing onsite clinical applications that can be used in the prevention of diseases.

One of the greatest advantages of this book is the fact that it covers common cardiovascular morbidities, including arrhythmias and heart failure comprehensively. In addition to addressing the fundamental biology and pathophysiology of cardiotoxicity, the book also provides insight into the emergence of new clinical trials, for example, the development and use of QT/QTc monitoring protocols within the context of the most recent clinical algorithms utilized by practitioners.

After numerous deliberations, we came up with the idea to explore the possibility of developing a book on cardiovascular toxicity to partially (if not completely) fill this void. Finally, we decided to develop a book by inviting chapters from experts in the field who have relevant research experience and an understanding of the intricacies of the subject. We had in mind a book that would help to alleviate most of the worries of both students and instructors. We discussed, argued, and disagreed until we came up with the thought that a resource book would be a reasonable format, as it could provide sufficient information and literature for instructors to teach the subject, while providing students with ample information to gain better insight about the subject. Once we formulated these thoughts to develop a resource book, the ball started rolling, and we identified various experts and convinced them to contribute chapters.

This is our maiden effort to produce a book that will serve as a major reference for students and cardiologists involved in general cardiovascular practice, therapeutic modalities targeting cardio-oncology, or advanced cardiovascular imaging. This book will also be of value for clinical oncologists and mid-level providers taking care of patients with various cancers. We hope that we will get support from the readers of this book. We are open to criticism, suggestions, and recommendations that can help to improve the content and presentation of the book. Your suggestions and criticisms will give us an opportunity to explore other aspects of oxidative stress in our future ventures and endeavors.

CHAPTER

1

Overview of changes in the cardiovascular system

Riya Gupta[a] *and Pawan Kumar Maurya*[b]

[a]Amity Institute of Biotechnology, Amity University Uttar Pradesh, Noida, India, [b]Department of Biochemistry, Central University of Haryana, Mahendragarh, India

1 Introduction

The cardiovascular system is made up of the heart and the circulatory system [1,2]. The heart works like a pump and transports blood to all the organs, tissues, and cells of the body. The heart beats over 100,000 times a day pumping blood to 60,000 miles of blood vessels [3]. The heart is just like any other tissue and needs a continuous supply of oxygen and nutrient [4]. Heart receives its blood from the coronary arteries, which arise from the aorta roots. The heart consists of three layers: the endocardium (inner layer), the epicardium (middle layer), and the myocardium (outer layer) [5]. The outermost layer is the pericardium, which is the protective membrane of the heart [3]. As blood moves throughout the body, it delivers oxygen and nutrients to the cell and, on the other hand, removes carbon dioxide and waste products accumulated in those cells [4]. Blood moves from the heart to the rest of the body, through the complex networks of arteries, arterioles, and capillaries. Blood is carried back to the heart through venules and veins [6].

The heart is a discrete organ. It consists of four chambers in humans. The right side of the heart has the right atrium and the right ventricle, which receives blood from the periphery and carries it to the lungs via the pulmonary artery. This blood is deoxygenated and is revived in the lungs with oxygen [7]. Once this happens, the re-oxygenated blood is carried to the left side of the heart. The pulmonary vein takes the blood from the lungs to the left atrium, from where it enters the left ventricle and gets pumped to the aortic arch for distribution to the entire body. [6]

Cardiovascular Toxicity and Therapeutic Modalities Targeting Cardio-oncology
https://doi.org/10.1016/B978-0-323-90461-2.00010-9

2 Circulatory system

The heart is one of the first organs developed during the embryonic development. [8] It is a powerful pump which circulates blood, nutrients, and oxygen to the entire body. It was considered the most important organ in Greek medicine. Heart disease is one of the leading causes of death around the world, which makes it extremely necessary to understand the genetic, molecular, and environmental basis of the cardiovascular development and diseases [9].

There are two circulatory pathways that arise from the heart.

- The pulmonary circulatory pathway—Is a short loop that starts from the heart, goes to the lungs, and comes back to the heart again.
- The systemic circulatory pathway—This is the pathway that carries blood to parts of the body except the lungs and comes back to the heart again.

2.1 Pulmonary circulation

Pulmonary artery is the main artery coming from the heart. It is a high-flow, low-pressure circuit. The low pressure prevents the fluid from moving out of the pulmonary vessels into the interstitial space and allows the right ventricle to operate at a minimal energy cost [10]. The pulmonary artery divides into two main branches, which brings blood to the lungs from the heart. The blood picks up oxygen and releases carbon dioxide in the lungs. This oxygenated blood then returns to the heart through the pulmonary veins [11].

The oxygen uptake (VO_2) and removal of carbon dioxide (VCO_2) is increased during exercise increases and reaches to almost 20-fold above the resting values. This increase in gas exchange increases the cardiac output up to sixfold. The sixfold increase in cardiac output increases the convective transport demands of these gases to and from muscles. Thus, exercise causes stress on pulmonary circulation, taking an entire cardiac output while keeping the lungs dry and right ventricles compensated [10].

2.2 Systemic circulation

A functional blood supply is provided to all parts of the body through systemic circulation. Blood returning to the heart picks up oxygen from the lungs, which can then be delivered throughout the body [8]. Systemic circulation also picks up all the waste and carbon dioxide from the body parts and in return provides cells with oxygen and nutrients. The aorta is the main artery leaving the heart with oxygenated blood. The aorta branches off, through arteries and to capillaries to all the tissues and muscles where it delivers oxygenated blood [11]. The deoxygenated blood returns through a system of veins from the tissue capillaries to the right atrium of the heart [12].

The neural control of systemic circulation contains feedback mechanisms, which operate both in short and in long terms through the autonomic nervous system, primarily the sympathetic nervous system. Short-term changes in sympathetic activity are triggered either by reflex mechanisms, which have peripheral receptors, or by a centrally generated response. On the other hand, short-term changes in sympathetic activity are triggered by reflex mechanisms, which involve peripheral receptors or using a centrally generated response. On the other hand, long-term changes are evoked through the modulation of sympathetic nervous system by humoral factors and central mechanisms that involve the hypothalamus [13].

3 Blood vessels

Blood vessels are the channels through which the blood is distributed throughout the body. All the vessels make two closed systems, which start and end at the heart. The pulmonary system takes blood from the right ventricle to the lungs and then brings it back to the left atrium [8]. On the other hand, the systemic system carries blood from the left ventricle of the heart and delivers it to the tissues all around the body, which then eventually returns to the right atrium of the heart. Based on the function and structure, blood vessels can be classified as veins, arteries, or capillaries [14].

3.1 Veins

Veins carry blood to the heart. After passing through the capillaries, the blood enters the smallest of the veins, called venules [15]. From the venules, the blood enters the veins and progressively flows to the larger veins till it reaches the heart. The pulmonary veins in the pulmonary system carry the blood from the lungs to the left atrium of the heart. This blood is highly oxygenated as it is coming out of the lungs. The systemic veins transport deoxygenated blood from the body to the right atrium of the heart. This blood has reduced oxygen due to metabolic activities in the tissue cells [15].

There are three layers in the veins. The innermost layer of the vein is called the tunica intima (or the tunica interna). This layer is made up of simple squamous epithelium, which is surrounded by connective tissue basement membrane with elastic fibers. The middle layer is called the tunica media, and this is primarily made up of smooth muscle [15]. The vessel changes its diameter to regulate the flow of the blood and blood pressure. The outermost layer of the vessel that is attached to the surrounding tissue is the tunica externa or tunica adventitia. This layer contains varying amounts of elastic and collagenous fibers. The connective tissue of this layer is dense and is adjacent to the tunica media, but changes to loose connective tissue near the periphery [16].

In the veins, the smooth muscle and connective tissue layers are thin, which makes the walls thin and thus allows more blood to flow through. This is because the blood in the veins has lower pressure as compared to the arteries in which the walls are thicker. This allows the veins to hold more blood. At any given time, almost 70% of the blood volume is in the veins. Medium and large veins have venous valves. These help keep the blood flow toward the heart. Venous valves are important, especially in the arms and legs, to prevent the backflow of blood against gravity [14].

3.2 Arteries

Arteries carry the blood away from the heart. Pulmonary arteries are responsible for the transport of less oxygenated blood from the right ventricle toward the lungs. On the other hand, systemic arteries transport oxygenated blood from the left ventricle to the tissues. Blood is carried from the vesicles into the large, elastic arteries, which branch into smaller and smaller arteries. The smallest microscopic arteries are called arterioles. Arterioles regulate blood flow into the tissue capillaries. At any given time, 10% of the total blood is in the arterial system. The arteries have the same three walls as the veins. However, the tunica interna and tunica media are thicker in the arteries, and therefore more flexible [14].

Arteries undergo structural changes as they age. Changes include gradual thickening of arterial walls, increase in conduit artery diameter, and changes in wall content (i.e., less elastin, advanced glycation endproducts); these changes occur in both the central and peripheral arteries. The effect of this is on artery function as they increase arterial stiffness. Most importantly, arterial function changes directly affect cardiovascular events, which are independent of age and other cardiovascular risk factors. Thus, the measurement of arterial function and health can predict one's biological and chronological age [17].

3.3 Capillaries

Capillaries are the most abundant and the smallest of the blood vessels. They connect the vessels and carry blood away from the heart (arteries) and return the blood back to the heart (veins). Capillaries perform the function of exchange of materials between the blood and the tissue cells.

Distribution of capillaries varies with the body tissues and depends on the metabolic activity of the location [18].

The skeletal muscle, liver, and kidneys have an extensive network of capillaries due to their high metabolic activity and thus require an abundant supply of nutrients and oxygen. In contrast, tissues like the connective tissues have a comparatively less supply of capillaries. Lens, cornea, and the epidermis of skin all completely lack any supply of the capillary network. At any given time, 5% percent of the total blood volume is present in the capillaries. And the other 10% is in the lungs. Smooth muscle cells of the arterioles regulate the transportation of blood flow in the capillaries [14].

4 Heart valves

The heart consists of two types of valves. These help the blood flow in the right direction. The valves present between the atria and the ventricles are known as atrioventricular valves (they are also called the cuspid valves), and the bases of the large vessels that leave the ventricles are known as semilunar valves.

The right atrioventricular valves is called the tricuspid valve, and the left atrioventricular valve is called the bicuspid or mitral valve. The valve present between the right ventricle and the pulmonary trunk is the pulmonary semilunar valve. The valve present between the left ventricle and the aorta is called the aortic semilunar valve. Atrioventricular valves close when the ventricles contract to prevent blood from flowing back into the atria. Semilunar valves prevent the blood from flowing back into the ventricles when they relax [19].

5 Physiology of heart

Heart's conduction system consists of several components. The first component is the sinoatrial node, which initiates the impulse at the rate of 70 to 80 times per minute [20]. This is called the pacemaker of the heart because it establishes the rhythm of the heartbeat. Other parts of the conduction system include the atrioventricular node, atrioventricular bundle, bundle branches, and conduction myofibers. These components coordinate the relaxation and contraction of the heart chambers [14].

6 Cardiac cycle

The cardiac cycle describes all the events that occur during a single heartbeat. It is a stepwise process of how the heart works, altering from contraction to relaxation of the myocardium walls in the heart chambers, which is coordinated by the conduction system during a single heart beat. Systole is the contraction phase, and diastole is the relaxation phase of the cardiac cycle. In the normal heart rate, one cardiac cycle is 0.8 s. [14] Dr. Carl Wiggers made a detailed diagram of all the specific steps of the cardiac cycle in his publication in 1921. This diagram came to be known as the Wiggers diagram. It is still the most commonly used diagram to teach cardiac cycle today [21]. A paper published by the Wright brothers in 2020 introduced the Wright table, which is a novel tool for teaching and learning the cardiac cycle. This table supplements the 100-yr-old Wiggers diagram, which is usually difficult to learn. The Wright table is a compact representation of the information showing (1) the pressures and flow change over time of each component and (2) the heart working as a pump, filling, and emptying the ventricles [21].

7 Heart sounds

Heart sounds are produced from the blood flowing through the heart chambers and the opening and closing of the heart valves during the cardiac cycle. [20] The vibrations of heart chambers from the blood flow create audible sounds: the more turbulent the blood flow, the stronger the sound due to more vibrations being created. These sounds are known as lub-dub. As the sounds are created from the blood flow, the turbulence of the blood flow can be measured using the same variables that measure the flow of all liquids. These variables include density, viscosity, velocity, and the diameter of the column through which the fluid is traveling [22].

A typical medical examination uses a stethoscope as a cornerstone of physical medical examinations and a valuable first-line tool to evaluate a patient. If the heart sounds are unusual, they are called murmurs [14]. There are other pathological sounds which are very characteristic that have major pathophysiological consequences. These lesions are heard during systole, diastole, or continuously throughout the cardiac cycle [22].

8 Heart rate

Sinoatrial node alone produces a constant heart rhythm [23]. The atrioventricular node regulates the heart rate by increasing or decreasing it as per the bodies need. Changes in the heart rate are mediated by the cardiac center present in the medullar oblongata in the brain. This center contains both sympathetic and parasympathetic components, which adjust the heart rates according to the changing needs of the body. Heart rate can be affected by peripheral factors like emotions, body temperatures, or ion concentrations. These are mediated through the cardiac center [14].

9 Blood

Blood is the main component of the vascular system. Blood consists of 55% fluids (plasma) and 45% solids (cells). The plasma consists mainly of water, nutrients, hormones, proteins, dissolved waste products, and antibodies [23]. Blood and the bone marrow consist of multiple cells of different kinds. These cells provide functions like hemostasis, oxygen transport, immune defense, and phagocytosis, which are independent of each other and provided by their specialized cells. [24]

The three main categories of blood cells are:

Erythrocytes
Erythrocytes are also called red blood cells. These are small, red, disk-shaped cells [25]. Erythrocytes consist of hemoglobin that combines with oxygen in the lungs and is then transported to the cells of the body. Hemoglobin then brings all the carbon dioxide waste back to the lungs. Erythrocytes are formed in the bone marrow [3].

Leukocytes
Leukocytes are white blood cells. They help in fighting bacteria and infections. If a tissue gets damaged or there is an infection, the number of leukocytes in the blood increases. Leukocytes are formed in the small ends of bones. These can be classed as granular leukocytes (eosinophils, neutrophils, and basophils), and there are three types of nongranular (monocytes, T-cell lymphocytes, and B-cell lymphocytes) [3].

Thrombocytes
Thrombocytes are also called platelets, and they aid in the formation of blood clots by releasing various protein substances. Whenever there is an injury, thrombocytes disintegrate and cause a chemical reaction with the proteins of plasma, forming a threadlike structure called fibrin. Fibrin catches other blood cells, which forms clots and prevents further loss of blood and helps in the healing process [3].

10 Effects on the cardiovascular system due to various physical/environmental factors

10.1 Effects due to aging

Aging has been vastly studied over the last half-century; significant advances have also been made in understanding the same. Interventions to reduce aging such as calorie restriction, pharmacological agents, exercise training, and others have all shown significant alterations in the cardiovascular system as well as other parts.

The cardiovascular system goes through structural as well as functional changes through the ages. As people undergo aging, the arterial veins become stiffer, which leads to higher systolic blood pressure and pulse pressure and also leads to a higher risk of systolic hypertension; this predisposes to LV hypertrophy and maintains a normal LVEF. Due to reduced diastolic filling rate, the heart becomes older and more reliant on late diastolic filling from the left atrium. These may lead to myocardial ischemia and heart failures in older individuals. Mutated proteins accumulate over time, which may lead to ATTRwt in the elderly. Other characteristics of advancing age are posture, exercise, and afterload stress due to blunted CV. Recent studies in caloric restriction, antioxidants, and exercise training all show promise of CV aging. Further research in this field will help in reducing incidence of clinical CV diseases that occur in older adults. [26]

10.2 Effects due to exercise

Exercise causes various changes in the heart at the tissue level. Cardiac hypertrophy is one of the most dramatic changes that occur. This is when the heart mass increases by an evident amount. In humans, exercise can increase the left ventricular mass by 20% or more. This is thought to result primarily from cellular hypertrophy of the cardiomyocytes themselves, and this would lead to an increase in the thickness of the cardiac wall, or in series (eccentric hypertrophy), which leads to the increase in the chamber volume. Hypertrophy is also influenced by the type of exercise, like concentric hypertrophy is induced by increased pressure and endurance exercises. It results in eccentric hypertrophy.

Exercise triggers the release of IGF-1 and neuregulin in tissues. In the heart tissues, it results in increased autocrine and paracrine signaling. Increased IGF-1 is a strong driver for physiological hypertrophy, leading to larger hearts with greater number of cardiomyocytes and cardiac specific overexpression of IGF-1's tyrosine kinase receptor. IGF-1R results in the increased chamber weight, increased myocyte size, and improved systolic function without histological signs of pathology. Overexpression of IGF-1 also protects the mice from a broad range of cardiac diseases, including diabetic cardiomyopathy, angiotensin II-mediated oxidative stress, chronic ischemia, and ischemia-reperfusion injury. Neuroglin-1 produces a similar effect [27].

10.3 Effects due to anesthetics

Anesthetic agents have some effects on the heart. This was first demonstrated by Embley in 1902 in a dog in which after the administration of chloroform, dog's heart rate lead to progressive slowing and reached the point of complete stoppage. However, the administration of atropine would protect the animals against it. Anesthetics may also directly affect the heart's muscular ventricle wall. [28]

11 Conclusion

Heart plays a very important role in maintaining various metabolic and physiological changes in the body. It is affected by various endogenous as well as exogenous materials. Maintaining healthy heart is important in order to maintain a proper circulation of blood.

References

[1] S.S. Virani, A. Alonso, E.J. Benjamin, et al., Heart disease and stroke statistics—2020 update: a report from the American Heart Association, Circulation (2020) E139–E596, https://doi.org/10.1161/CIR.0000000000000757.

[2] J.M. Nerbonne, R.S. Kass, Molecular physiology of cardiac repolarization, Physiol. Rev. 85 (4) (2005) 1205–1253, https://doi.org/10.1152/PHYSREV.00002.2005.

[3] Texas Heart Institute, Anatomy of the Heart and Cardiovascular System | Texas Heart Institute. https://www.texasheart.org/heart-health/heart-information-center/topics/anatomy-of-the-heart-and-cardiovascular-system/. (Accessed 27 March 2021).

[4] G. Liebau, Heart beat and blood movement, Z. Gesamte Exp. Med. 125 (5) (1955) 482–498. https://pubmed.ncbi.nlm.nih.gov/13247998/. (Accessed 24 August 2021).

[5] G.A. MacGowan, J.D. Parikh, K.G. Hollingsworth, Diffusion tensor magnetic resonance imaging of the heart: looking into the layers of the myocardium, J. Am. Coll. Cardiol. 69 (6) (2017) 677–678, https://doi.org/10.1016/J.JACC.2016.10.080.

[6] Cardiovascular System, Cardiovascular System, 2021. https://sphweb.bumc.bu.edu/otlt/MPH-Modules/PH/PH709_Heart/PH709_Heart2.html. (Accessed 27 March 2021).

[7] J.S. Pober, G. Tellides, Participation of blood vessel cells in human adaptive immune responses, Trends Immunol. 33 (1) (2012) 49–57, https://doi.org/10.1016/J.IT.2011.09.006.

[8] A. Aspelund, M.R. Robciuc, S. Karaman, et al., Lymphatic system in cardiovascular medicine, Circ. Res. 118 (3) (2016) 515–530, https://doi.org/10.1161/CIRCRESAHA.115.306544.

[9] A. Granger, N. Emambokus, Cell metabolism ♥ cardiovascular biology, Cell Metab. 21 (2) (2015) 151, https://doi.org/10.1016/J.CMET.2015.01.018.

[10] R. Naeije, N. Chesler, Pulmonary circulation at exercise, Compr. Physiol. 2 (1) (2012) 711, https://doi.org/10.1002/CPHY.C100091.

[11] Nemours KidsHealth, Heart and Circulatory System (for Teens)—Nemours KidsHealth, 2021. https://kidshealth.org/en/teens/heart.html. (Accessed 27 March 2021).

[12] SEER Training, Circulatory Pathways | SEER Training. https://training.seer.cancer.gov/anatomy/cardiovascular/blood/pathways.html. (Accessed 28 August 2021).

[13] R.H. Anderson, E.J. Baker, A.N. Redington, M.L. Rigby, D.J. Penny, G. Wernovsky, Paediatric cardiology, Paediatr. Cardiol. (2010), https://doi.org/10.1016/B978-0-7020-3064-2.X0001-X.

[14] SEER Training, Classification & Structure of Blood Vessels | SEER Training, 2021. https://training.seer.cancer.gov/anatomy/cardiovascular/blood/classification.html. (Accessed 27 March 2021).

[15] M. Loukas, S. Bilinsky, et al., Cardiac veins: a review of the literature, Clin. Anat. 22 (1) (2009) 129–145, https://doi.org/10.1002/CA.20745.

[16] S. Hochauf, R. Sternitzky, S.M. Schellong, Structure and function of the peripheral venous system, Herz 32 (1) (2007) 3–9, https://doi.org/10.1007/S00059-007-2951-X.

[17] D.H.J. Thijssen, S.E. Carter, D.J. Green, Arterial structure and function in vascular ageing: are you as old as your arteries? J. Physiol. 594 (8) (2016) 2275, https://doi.org/10.1113/JP270597.

[18] I.S. Checkman, Capillaries of living systems: natural nanomechanisms of the functioning, Likars'ka Sprav. 3 (2013) 3–10. https://pubmed.ncbi.nlm.nih.gov/25016740/. (Accessed 24 August 2021).

[19] SEER Training, Structure of the Heart | SEER Training, 2021. https://training.seer.cancer.gov/anatomy/cardiovascular/heart/structure.html. (Accessed 27 March 2021).

[20] K. Verhoeff, J.R. Mitchell, Cardiopulmonary physiology: why the heart and lungs are inextricably linked, Adv. Physiol. Educ. 41 (3) (2017) 348–353.

[21] B.E. Wright, G.L. Watson, N.J. Selfridge, The Wright table of the cardiac cycle: a stand-alone supplement to the Wiggers diagram, Adv. Physiol. Educ. 44 (4) (2020) 554–563, https://doi.org/10.1152/ADVAN.00141.2019.

[22] S. Dornbush, A.E. Turnquest, Physiology, Heart Sounds—PubMed, 2021. https://pubmed.ncbi.nlm.nih.gov/31082054/. (Accessed 29 August 2021).

[23] G.G. Berntson, J.T. Bigger Jr., D.L. Eckberg, et al., Heart rate variability: origins, methods, and interpretive caveats, Psychophysiology 34 (6) (1997) 623–648, https://doi.org/10.1111/J.1469-8986.1997.TB02140.X.

[24] C. Finch, L. Harker, J. Cook, Kinetics of the formed elements of human blood, Blood 50 (4) (1977) 699–707, https://doi.org/10.1182/BLOOD.V50.4.699.699.

[25] J.M. Higgins, Red blood cell population dynamics, Clin. Lab. Med. 35 (1) (2015) 43–57, https://doi.org/10.1016/J.CLL.2014.10.002.

[26] N.S.V. Singam, C. Fine, J.L. Fleg, Cardiac changes associated with vascular aging, Clin. Cardiol. 43 (2) (2020) 92–98, https://doi.org/10.1002/clc.23313.

[27] C. Platt, N. Houstis, A. Rosenzweig, Using exercise to measure and modify cardiac function, Cell Metab. 21 (2) (2015) 227–236, https://doi.org/10.1016/j.cmet.2015.01.014.

[28] A.R. Hunter, The pharmacological action of anaesthetic drugs on the cardiovascular system, Br. J. Anaesth. 28 (1956), https://doi.org/10.1093/bja/28.12.544.

CHAPTER

2

Cardiovascular disease in cancer survivors: Risk and management

Saima Naz Khan[a] *and Md. Kausar Neyaz*[b]

[a]Laboratory for Molecular Reproduction and Genetics, All India Institute of Medical Sciences, New Delhi, India, [b]Bio-Services, New Delhi, India

1 Introduction

Although advances in medical science and therapeutic strategy have led to an improvement in the survival of patients with cancer, complications that increase morbidity and mortality have also been widely observed [1,2]. Improvement in treatment modalities, including radiotherapy and systemic therapies, has led to better prognosis for patients with malignancies. According to new research published in the *European Heart Journal*, more than one in 10 cancer patients does not die from their cancer but from heart and blood vessel problems instead. For some cancers, such as breast, prostate, endometrial, and thyroid cancer, around half will die from cardiovascular diseases (CVDs) [3]. In the general population, CVDs are major causes of morbidity [4] and mortality, accounting for 30%–50% of all deaths in most developed countries. Because of this high background rate, even a minor increase in the risk of CVD will have an important impact on morbidity and mortality.

Cardiovascular diseases normally include coronary heart disease (CHD), rheumatic heart disease (RHD), strokes, cardiomyopathy, and other heart diseases [5]. According to WHO, 16.7 million people globally die of cardiovascular diseases each year in 2002 that represent about 1/3 of all deaths globally [6].

Several studies from the past were suggestive that CVD was once considered to be the disease of developed countries, but now it is affecting other low- and middle-income countries, where it is taking a toll for nearly 30% of all deaths [7]. With every passing year, researchers estimate that by the end of 2030, the noncommunicable diseases will account for

Cardiovascular Toxicity and Therapeutic Modalities Targeting Cardio-oncology
https://doi.org/10.1016/B978-0-323-90461-2.00003-1

more than three-quarters of global deaths. CVD alone will be responsible for more deaths in developing countries than infectious diseases [8]. Not only genetics or environment acts as a contributor, but also the therapeutic interventions given to the cancer patients are strongly contributory toward cardiovascular diseases. Certain cancer therapies, including radiation therapy and some types of chemotherapies, are associated with increased risk of cardiovascular disease (CVD) [9] as these therapies have direct toxic effects on the heart, along with the effects of the cancer itself on the blood and vascular system. Additional risk factors common to both are obesity and dietary factors as previous studies have also shown that many major cancers have the same long-term, cumulative risk factors (e.g., smoking, obesity, physical inactivity, low consumption of fruit and vegetables, sedentary lifestyle). Not only this, but the stress of major illness, and its treatments, may aggravate CVD risk, and certain cancer treatments can present specific cardiovascular hazards, such as weight gain (obesity) and metabolic disturbances. The known link between obesity and cancer suggests an important interaction between the host lipid metabolism and tumorigenesis. Researchers have demonstrated that tumor-induced hyperlipidemia encompasses a feed-forward loop that reprograms hepatic lipoprotein homeostasis in part by providing LDL-cholesterol to support tumor growth [10].

The literature widely suggests that comorbidities such as obesity, abnormal lipid profiles, and insulin resistance are often associated with CVD. Insulin is a key hormone that functions as a regulator of cellular metabolism in many tissues in the human body. During insulin resistance, several metabolic alterations induce the development of cardiovascular disease. For instance, insulin resistance can induce an imbalance in glucose metabolism that generates chronic hyperglycemia, which in turn triggers oxidative stress and causes an inflammatory response that leads to cell damage. Notably, obesity associated with an abnormal lipid profile at a younger age has been strongly correlated with insulin resistance [11,12].

As cancer patients and survivors of cancer have a greater burden of cardiovascular disease, this much of the elevated cardiovascular risk in these individuals is likely attributable to hypertension. Treatment with chemotherapy is an independent risk factor for hypertension due to direct effects of many agents on endothelial function, sympathetic activity, and renin-angiotensin system activity as well as nephrotoxicity [13].

Preventive measures include adequate screening, the use of medications such as ACE inhibitors/angiotensin receptor blockers and/or beta blockers, statin therapy, and aspirin in persons who warrant these medications, as well as therapeutic lifestyle modifications such as exercise/physical activity, weight loss, and appropriate diet for a healthy lifestyle. Periodic follow-up with a good primary care physician who understands the risks associated with cancer therapy is important, and referral to oncocardiology for further management of cardiovascular risk in these survivors is based on a patient's cardiovascular risk level and the type, amount, and duration of cancer therapies received during the patient's lifetime [14].

2 Section 1: Underlying risk factors: Genetic and environmental

2.1 Common risk factors

Cardiovascular diseases and cancer represent the most frequent causes for mortality and morbidity in industrialized countries [15]. It has been well studied that both diseases share common risk factors such as lifestyle and age and many patients are affected by both disease types. Studies are suggestive that there is a close relationship between CVD and cancer. Nevertheless, for decades, in clinical routine cardiovascular diseases and cancer have been viewed separately and interdisciplinary treatment concepts were rarely considered.

However, with the development of highly effective antitumor therapies, the prognosis and long-term survival of cancer patients improved, leading to an increased incidence of cardiovascular problems in these patients [16].

There are several reports suggestive of the fact that some risk factors seem to be cancer specific, such as cancer-causing viral infections [17], or specific for cardiomyopathies, such as high blood pressure and cardiomyopathy-causing gene variants, and many other risk factors are common for both disease types, such as smoking, metabolic syndrome (hypertension, dyslipidemia, insulin resistance, hyperinsulinemia, glucose intolerance, and obesity), irradiation, age, air pollution, and environmental toxins.

Several cardiovascular diseases such as heart failure, arrhythmias, atherosclerosis, and thrombosis are the result of many anticancer drugs, e.g., anthracyclines (such as epirubicin [18], daunorubicin [19], doxorubicin [20], or idarubicin [21]), antibodies (trastuzumab [22], bevacizumab [23]), or small molecules (such as dasatinib, sunitinib [24], and sorafenib [25]), as these drugs have cardiotoxic side effects. Several studies have documented that there are various chemotherapeutic agents that are contributory toward cardiomyopathy with both systolic and diastolic dysfunction and could eventually progress to clinical heart failure; for example, anthracyclines are particularly known to increase the risk of cardiomyopathy with systolic dysfunction and heart failure by generation of free radicals, leading to cardiac cell damage [18].

These cardiovascular diseases can occur acutely during anticancer treatment or even years after the initial anticancer treatment. Far more than half of all childhood cancer survivors develop treatment-related chronic health issues later in life, of which cardiovascular complications make up a major fraction [26]. With aging, the immune system also gets affected and proceeds to immunosenescence, a term describing the gradual deterioration of the immune system during aging. Compromised immune system is one of the leading causes of increased susceptibility to infections, cancer, and cardiovascular diseases [27]. Several documented experimental studies has revealed that tumors can directly impact on the heart, e.g., by modifying cardiac metabolism resulting in impairment of cardiac function.

Furthermore, mouse model study revealed that tumors can directly impact the heart, which is achieved by modifying cardiac metabolism further resulting in impairment of cardiac function. Reduction of systemic insulin levels in melanoma tumor-bearing mice due to consumption of large amounts of glucose by the tumor tissue interfered with cardiac glucose uptake and was associated with cardiac atrophy and dysfunction [28]. Common conditions such as obesity, hyperglycemia, hypertension, and hypertriglyceridemia induce inflammation [29–32], and this may, in part, explain why CVD and cancer share several risk factors. Apart from several cancer therapy, age, prior cardiac dysfunction, coronary disease, hypertension, smoking, and obesity are well-described risk factors for anthracycline-related cardiotoxicity. It has been documented in several epidemiological studies that prevalent heart failure increases the risk of developing cancer [33–36]. In other words, CVD and cancer are closely linked and associated because of their shared mode of expression and risk factors.

2.2 Genetic risk factors

Genetic background plays an important role in disease predisposition for cancer and heart failure as well as for (cardio)toxic response to antitumor therapy.

Besides the other common risk factors such as environmental factors, poor lifestyle, and social habits, genetic background plays an important role in the pathogenesis of both the diseases. Several studies demonstrated that genetics might not only contribute to the pathogenesis of each disease type, but also connect cardiovascular (CV) disease and cancer. Both CV disease and cancer have numerous genetic risk factors, but not many are shared, and still there are several inherited (germ line) genetic mutations and acquired (somatic) genetic mutations that play an important role in CVD and cancer.

Cancer can affect cardiac function directly while antitumor therapies may have acute- and/or late-onset cardiotoxic effects and vice versa as recent studies suggest that heart failure might promote tumorigenesis and tumor progression [34].

In both cancer and cardiovascular diseases, genetic predisposition played an important role in the disease onset and development. Several studies are supportive of the fact that genetic variants classically associated with cardiomyopathies increase the risk for toxic side effects on the cardiovascular system. Genetic variants associated with increased cancer risk are frequent in patients with peripartum cardiomyopathy complicated by cancer, pointing to a common genetic predisposition for both diseases [37]. Furthermore, recent studies discovered interesting genetic interactions between the two diseases; for example, heart failure was shown to promote transition of pretumor stages and tumor growth [38]. In addition, there are other supportive documented evidences which are suggestive that the occurence of cancer predisposition syndrome mutations (CPS) are more frequent in patients with peripartum cardiomyopathy (PPCM) as compared with age-matched women in the healthy population [39].

Most of the cancers arise from somatically acquired changes in the DNA [40,41]. This does not mean, however, that all of the somatic abnormalities present in a cancer genome are involved in the development of the cancer. In fact, it is likely that some or most of these somatic abnormalities make no contribution at all. Therefore, the DNA sequence of a cancer cell genome (and of most normal cell genomes) has acquired a set of mutations. Such mutations are termed somatic mutations to differentiate them from germ line mutations, which are inherited from parents [42,43].

Many scientific studies have pointed toward the role of somatic mutations, such as CHIP mutations, which start accumulation during the aging process and are known to increase the risk of hematological neoplasia. Such studies diverted the attention and simultaneously became the focus of other scientific groups working in the field of oncocardiology. Not only contributory toward the increased risk of hematological neoplasia, there are several studies suggestive of the fact that CHIP mutations also accelerate the development of cardiovascular diseases such as atherosclerosis, coronary heart disease, and ischemic stroke, and worsened the outcome of heart failure patients [44].

For example, in the Wnt/β-catenin pathway, germ line mutations play an important role. The Wnt/β-catenin signaling pathway is involved in cell proliferation, polarity, migration, and cell fate determination [45]. Aberrant Wnt signaling plays an important role in the pathogenesis of atherosclerosis and also in several cancers [46,47]. Furthermore, mutations in one of the binding proteins of Wnt, LRP6, are involved in both coronary artery disease and different forms of cancer [46]. More acquired mutations that strengthen the genetic relation between heart failure and cancer are, for example, DYRK1B. DYRK1B is overexpressed in many types of tumors, and a gain-of-function mutation in the DYRK1B gene is linked to CV disease [48,49]. This genetic overlap might represent novel biological shared pathways between cancer development and heart failure.

There is evidence that patients with mutations associated with cardiomyopathies have more sensitive responses to the cardiotoxic side effects of anticancer treatment [37]. It has been document that the implication of genetic alteration in cancer etiology, prognosis, drug response, and safety have significant genetic variations among patients with breast cancer [50].

There is evidence that patients with mutations associated with cardiomyopathies have more sensitive responses to the cardiotoxic side effects of anticancer treatments. Genome-wide association studies (GWAS) identified variants in genes, belonging to the cardiac remodeling pathway, which influence left ventricular functional changes after anthracycline exposure [51].

Okwuosa et al. have also documented that anthracyclines are associated with long-term risk of heart failure even over 20 years after administration of the chemotherapy, particularly when followed by administration of trastuzumab; radiation therapy can cause significant coronary artery disease (CAD), autonomic dysfunction, valvular heart disease, pericardial diseases, tachyarrhythmia, or bradycardia years after radiation therapy; and cisplatin leads to long-term increased risk of hypertension, metabolic syndrome, and cardiovascular events, including MI and stroke [52]. Genetic variants associated with dilated cardiomyopathy, especially Titin-truncating variants that increase the risk of cancer therapy-induced cardiomyopathies, are considered to be the most prominent variants [53], and therefore, genetic testing to assess the individual risk of chemotherapy cytotoxicity has been suggested [54]. Along this line, in the study conducted by Ware and Mazaika et al., pathogenic Titin variants are frequent in patients with peripartum cardiomyopathy (PPCM) who have a 16-fold higher risk of cancer. This relationship underlines the complex and close connection between both the diseases [55].

Apart from these evidences, mutations in DNA damage response/ repair system (DDR) genes are also very important and equally contributory toward the development of cancer and other diseases because mutations to these genes result in loss of function, downregulation, or even upregulation of respective proteins which can predispose individuals to the above-mentioned illness [56]. In addition, DDR mutations may also increase the risk of cardiotoxicity from antitumor therapies, as seen, e.g., in breast cancer patients with BRCA1 and BRCA2 mutations who have a higher risk of cardiac toxicity following anthracycline-based chemotherapy [57]. Another set of genes that are the most frequently mutated in the cancer genome are the kinases, making them attractive therapeutic targets for drug design.

Targeted anticancer drugs were initially considered to be magic bullets as these were thought to affect tumors but not normal tissue in which kinases were not constitutively active. Thus, the hope of targeted therapy was one of high efficacy, with minimal adverse effects. However, unexpected reports of cardiotoxicity from approved targeted drugs suggest that these agents are not fully safe [58]. In fact, as long as a kinase maintained expression in the heart, targeting that kinase in cancer could, theoretically, cause cardiotoxicity.

All the alterations to the cancer genome lead to the irreversible changes in the DNA sequence or structure, and in the number of sequences, genes, or chromosomes. These changes can simultaneously affect hundreds of genes and regulatory transcripts such as promoter regions. Once the promoter region is affected, it acts as a precursor to the epigenetic modifications. Epigenetic modifications of DNA or histones by methylation, acetylation, and other mechanisms have also become recognized as key mediators of cancer phenotype. Epigenetic modifications result in activation and/or inhibition of various biological events involved in cancer and other diseases.

The cardio-oncologist needs to understand cancer and its biology. Central to cancer management is the understanding of cancer as the cumulative phenotypic consequence of acquired genetic and epigenetic alterations in cancer cells and its possible association responsible for cardiovascular diseases. Genetic background plays an important role in disease predisposition for cancer and CVD along with cardiotoxic response to antitumor therapy. A broader understanding of genetic architecture of cancer and CVD along with pretherapeutic pharmacogenetic testing is required, which may provide important information for optimal antitumor therapy with minimal side effects (cardiovascular) to ensure best possible and safe outcome for cancer patients.

2.3 Environmental risk factors

Cardiovascular disease (CVD) and cancer are the largest contributors to the burden of chronic disease at the global level. The burden of these diseases is rising because of the aging of the population, and increasing exposure to established carcinogenic chemicals, viruses, and radiations, along with adoption of sedentary unhealthy lifestyle and poor social habits such as smoking, alcohol intake, obesity, and reduced or no physical activity. A trail of scientific evidence is suggestive of the fact that CVD and cancer possess several similarities and possible interactions, including a number of similar risk factors, indicating a shared biology. Although inflammation appears to be a major unifying factor in the etiology and progression of these diseases [59,60], additional risk factors are equally contributory. Current biological mechanisms of cancer and CVD suggest that both the diseases are originated from both environment and genetics, meaning that there are multiple external factors combined with internal genetic changes which lead to cancer and CVD.

Tobacco use, physical inactivity, obesity, exposure to environmental pollutants, and a poor diet not only contribute to cancer but also accelerate the other known risk factors for CVD, such as hypertension, diabetes mellitus, and the development of atherosclerosis [61]. The heart and vascular system are highly vulnerable to a variety of environmental agents. As per the WHO research, 35% of deaths caused by cancer worldwide are due to potentially modifiable risk factors resulting from lifestyle, including smoking and alcohol consumption, in countries with low, medium, and high incomes, infections, parasites, exposure to ultraviolet light and tanning using devices that emit ultraviolet radiation, environmental tobacco smoking, dietary factors, hormone replacement therapy, and exposure to ionizing radiation [62–64].

2.3.1 *Inflammation in CVD and cancer*

Inflammation is an ancient evolved process, which involves activation, recruitment, and action of cell of innate and adaptive immunity. Inflammation is considered to be a common factor among a variety of diseases, including both CVD and cancer [63,64]. Several traditional illnesses associated with CVD are obesity, hyperglycemia, hypertension, and hypertriglyceridemia-induced inflammation [64]. Atherosclerosis was once considered as a lipid storage disease as it begins with the accumulation of LDL, although it is now known that inflammation mediates all of its stages, from initiation to progression and, ultimately, thrombosis. Hypertension, smoking, dyslipidemia, and insulin resistance are the conditions that all appear to trigger atherosclerosis [29,64].

Considering the role of inflammation in promoting carcinogenesis and tumor progression, then it is very well established that inflammation predisposes to the development of cancer and accelerates all stages of tumorigenesis. Previous studies have highlighted the role of inflammation in host defense against pathogens [65], but during the last couple of decades, the contribution of the immune system and inflammation to cancer development, progression, and therapy has regained enormous interest. Extensive factual evidences have shown that inflammation within the tumor microenvironment has effects that promote malignant transformation of cells, carcinogenesis, and its progression [66]. Moreover, as tumors grow, their survival depends on the release of chemicals that signal immune cells to the tumor.

Relationship between oxidative stress (OS) and inflammation has been documented by many authors. Evidences indicated that oxidative stress plays a pathogenic role in chronic inflammatory diseases. Inflammation is a natural defense mechanism against pathogens, and it is associated with many pathogenic diseases such as microbial and viral infections, exposure to allergens, radiation and toxic chemicals, autoimmune and chronic diseases, obesity, consumption of alcohol, tobacco use, and a high-calorie diet. Many of chronic diseases linked with higher production of reactive oxygen species (ROS) result in oxidative stress and a variety of protein oxidations [67]. Chronic inflammation, found in conditions common to both CVD and cancer (e.g., diabetes mellitus, hypertension, obesity), also induces oxidative stress [68].

Main targets of oxidative stress are proteins, lipids, and DNA/RNA, and modifications in these molecules may increase the chances of mutagenesis. ROS overproduction over a prolonged period of time can cause damage to the cellular structure and functions and may induce somatic mutations and preneoplastic and neoplastic transformations. If not removed quickly, then excessive production of ROS in cells and tissues may be deleterious [69]. Indeed, excessive ROS production may cause irreversible damage to cells, resulting in cell death by the necrotic and apoptotic processes [70].

2.3.2 *Air pollution*

Emissions from motor vehicles, industrial processes, power generation, the household combustion of solid fuel are the sources that pollute the ambient air across the world. The mixtures of ambient air pollution invariably contain specific chemicals known to be carcinogenic to humans.

Recent estimates suggest that the disease burden due to air pollution is substantial. Exposure to ambient fine particles ($PM_{2.5}$) was recently estimated to have contributed 3.2 million premature deaths worldwide in 2010, due largely to cardiovascular disease, and 223,000 deaths from lung cancer. More than half of the lung cancer deaths attributable to ambient $PM_{2.5}$ were estimated to have been in China and other East Asian countries [71].

In 2014, the WHO identified air pollution as the largest single environmental-health risk [72]. Furthermore, a pooled analysis published in 2013 estimated the excess risk of cardiovascular mortality as 11% per $10\,\mu g/m^3$ rise in the levels of $PM_{2.5}$, with stronger associations linked to mortality from ischemic heart disease (IHD) [73]. In 2012, 3.7 million deaths in people aged < 60 years were because of outdoor air pollution, and 80% of these deaths were the result of cardiovascular or cerebrovascular disease [72]. The extent of exposures to indoor and outdoor air pollution is well recognized. There has long been concern that airborne carcinogens contribute to the global burden of cancer, especially of the lung, which receives the most substantial inhaled doses.

Acute and long-term exposures to pollutants are associated with the risks of chronic heart disease (CHD), heart failure (HF), and stroke, as well as with preclinical risk factors such as atherosclerosis, endothelial dysfunction, and hypertension. Research consistently demonstrates the role of environmental exposures in the development and progression of the most pervasive CVDs and cancer.

2.3.3 Tobacco use and smoking

Tobacco use is the most preventable cause of death; rather, it is the most preventable cause of cardiovascular diseases and cancer.

Tobacco smoking produces a number of irritants, carcinogens, proinflammatory stimuli, and oxidizing agents. These processes stimulate abnormal signaling pathways found within smoking-related cancer and CVDs [73]. Smoking-induced CVDs comprise a large number of conditions and are considered to be the leading cause of death globally, accounting for an estimated 17.3 million to 17.5 million deaths yearly [74]. Smoking promotes thrombosis [75], elevates the levels of platelet-activating substance [76], and furthermore leads to more binding of platelets to white blood cells, a process that is both proinflammatory and prothrombotic and changes the structure of platelets to make them more susceptible to aggregation.

Cigarette smoking enhances oxidation of plasma LDL-cholesterol, the "bad cholesterol" that is proatherogenic and known to impair endothelial function [77]. Smoking is also associated with increased insulin resistance and hyperinsulinemia, which often co-occurs with derangement in lipid metabolism [78]. Further devastating effects of smoking are documented in large, well-conducted population studies that markers of inflammation, including white blood cells, fibrinogens, interleukin-6, and other proteins, are elevated in smokers [79,80].

Tobacco use is closely associated with several cardiovascular disease risk factors, such as dyslipidemia, hypertension, and diabetes. Smoking is not only associated with CVDs but is also responsible for coronary heart disease, hearth failure, and carotid and cerebrovascular diseases.

Likewise, tobacco use and smoking are extensively contributory toward cancer. It has been documented that each year, more than 180,000 Americans die from cancer as a result of smoking. More than 85% of all lung cancers and 30% of all deaths are caused by smoking. Secondhand smoke can also increase the risk of cancer by 5%. Smokers are more likely than nonsmokers to develop different types of cancers, such as lung, larynx, mouth, esophagus, bladder, kidney, throat, stomach, pancreas, cervix, and acute myeloid leukemia.

Among 4000 chemicals that have been identified in tobacco smoke, at least 400 are known to be harmful to human health, such as hydrogen cyanide, carbon monoxide, and ammonia. Almost 40 (10%) of these toxic chemicals are shown to be carcinogenic, such as arsenic, benzene, chromium, ethylene oxide, nickel, polonium, vinyl chloride, formaldehyde, benzo[α]pyrene, and toluene [81].

Poisons in tobacco smoke can damage or change a cell's DNA. DNA is the cell's "instruction manual" that controls a cell's normal growth and function. When DNA is damaged, a cell can begin growing out of control and create a cancer tumor.

2.3.4 Alcohol consumption

Alcohol has been classified by the World Health Organization (WHO) as group 1 carcinogen. Alcohol consumption also has an established role in association with CVDs and several cancers such as oral cavity cancer, laryngeal, esophageal, liver, colon, and rectal cancers, and, in women, breast cancer. It is estimated that alcohol is responsible for ~ 2.5 million deaths each year and for 4.5% of the global burden of disease and injury [82].

Genetic susceptibility plays an important part in alcohol-related cancer, and evidence suggests that the effect of alcohol is modulated by polymorphisms in genes encoding enzymes for ethanol metabolism, folate metabolism, and DNA repair [83]. Likewise, data from numerous epidemiologic studies over the past two decades have revealed that alcohol use has complex effects on cardiovascular health and complex associations with conditions such as hypertension, coronary heart disease, stroke, peripheral arterial disease, and cardiomyopathy, although these associations are strongly modulated by the dose and pattern of alcohol consumption. Low-to-moderate daily alcohol consumption is associated with a reduced risk of CV disease and mortality, whereas greater amounts of alcohol consumption have been linked to an increased risk.

In patients suffering from cardiovascular disease, meta-analyzed data (8 observational studies) of 16,351 patients found moderate alcohol consumption to be associated with reduced CVD, whereas with healthy individuals, there has been an extensively documented J-shaped dose-effect curve, with excessive alcohol use leading to increased cardiovascular events and all-cause mortality [84,85].

The mechanisms by which alcohol consumption exerts its carcinogenic effect are various and not fully understood. Polymorphisms of the genes that encode enzymes for ethanol metabolism affect the ethanol/acetaldehyde oxidizing capacity and are responsible for the limited action of the enzyme that converts acetaldehyde to acetate that is not toxic to the body [86–88]. Therefore, higher exposure to acetaldehyde, the first metabolite of ethanol, (which is extremely harmful to the body), will induce a defect in alcohol dehydrogenase gene and the tumor suppressor gene (BRCA) inactivation that can lead to the liver and upper aerodigestive tract and even breast cancers [89]. Certain other factors such as alcohol-related increase in estrogen and androgen levels in women might promote the development of the breast cancer [90], and also the alcohol-related immunodeficiency and immunosuppression also facilitate carcinogenesis of several organs [91]. Alcohol also acts as a solvent for tobacco carcinogens and leads to the production of reactive oxygen species and reactive nitrogen species [89]. Therefore, conclusively, alcohol abuse leads to iron accumulation, retinoic acid impairment, acetaldehyde genotoxicity, increased estrogen concentration, free radical production, and folate metabolic alterations which are enough to cause major damage to the human body.

2.3.5 Diet

People having sedentary lifestyle including poor diet with reduced physical activity are considered to be at higher risk of several diseases. Obese people will be at higher risk of coronary heart disease, stroke, high blood pressure, diabetes, and cancers (esophagus, breast, uterus, colon, rectum, and prostate).

Accumulating data relating diet to CVDs have predominantly been generated from high-income countries (HICs), but over 80% of CVD deaths occur in low- and middle-income countries (LMICs). It has been forecasted that noncommunicable diseases are going to increase substantially in LMICs because of lifestyle transitions. This transition is associated with increasing urbanization, economic development, and globalization. According to the Global Burden of Disease study, diet is considered to be a major factor behind the rise in hypertension, diabetes, obesity, and other CVD components [71].

Poor-quality diets are high in refined grains and added sugars, salt, unhealthy fats, and animal-source foods, and low in whole grains, fruits, vegetables, legumes, fish, and nuts.

The role of dietary composition in cancer risk too ranges from known carcinogens in food sources (e.g., aflatoxins, nitrosamines) to dietary components impacting obesity, hypertension, hyperlipidemia, and chronic inflammatory patterns that mediate cancer risk. Several reports by AICR/WCRF highlighted associations between diet/food and several types of cancers, red/processed meat with colorectal cancer, aflatoxins with liver cancer, β-carotenes with lung cancer risk, cantonese-style salted fish with nasopharyngeal cancer, salty foods with stomach cancer, glycemic load with endometrial cancer, maté with esophageal cancer, and arsenic in drinking water and skin cancer [92–95]. Obese people may produce high amount of insulin-like growth factor-1 (IGF-1) in their blood, which may lead to certain cancers. Also, fat tissue produces high quantity of estrogen, which is linked to an increased risk of cancers. Leptin and adiponectin are produced by fat cells that can promote and inhibit cell proliferation, respectively [96,97]. Talking about aflatoxins, then, these are natural mycotoxins and are carcinogenic. These aflatoxins can intercalate with chromosome DNA base pairs through its epoxide moiety. These changes can cause genetic mutation in p53 gene (prevent cell cycle progression during mutagenic insult to the DNA) [98,99].

Common biological pathways between diet, CVD, and cancer have been described. Genetic mutations in the folate metabolism pathway, in conjunction with inadequate folate intake, are associated with an increased risk of both CVD and colorectal cancer [100]. On the one hand, conjugated linoleic acids have shown antiatherosclerotic and anticarcinogenic effects in experimental models [101]; on the other hand, meat-dominated dietary patterns have been associated with several cancers, especially colorectal cancer [102]. Increased dietary fat is one of the leading cause for the accumulation of carcinogenic bile acids, and several genotoxic substances (e.g., nitrosamines) that can act directly on DNA and can cause mutations (point mutations, deletions, and insertions) [103].

Diet/food is considered to be a modifiable risk factor for CVD and several cancers, and therefore, changing one's eating habits and restricting unhealthy food can significantly reduce the disease burden.

Having a healthy diet and being physically active may help to reduce cancer risks. Doctors suggest a healthy diet that includes plenty of foods that are high in fiber and low in fat, vitamins, minerals, whole-grain breads, cereals, fruits, and vegetables every day [104]. It has been observed and hypothesized that antioxidant vitamins may protect against cancer caused due to oxidative DNA damage. Taking an antioxidant- and polyphenol-rich diet may help, by preventing oxidative DNA damage, and would simultaneously prevent progression of cancer because they have a promising effect on several metabolic pathways, including mitogen-activated protein kinases, phosphatidylinositol 3-kinases, IGF-1, nuclear factor-κB, and reactive oxygen species [105].

Weight control equally plays an important role in preventing CVDs and several cancers. The reduction of adipose tissue through physical activity lowers the production of circulating sex and metabolic hormones, insulin, leptin, and inflammatory markers, several of which are potentially carcinogenic [106].

2.3.6 Organic and inorganic chemicals

A number of chemical substances revealed to be dangerous to our health, when present at high concentration in our body. These chemical substances are known as carcinogens, which can induce cancer, by damaging the genome or through disruption of cellular metabolic processes. Carcinogens usually have an insidious toxic effect rather than an acute toxic effect. There are several chemical toxins causing CVDs and promoting cancer. People who have certain jobs such as painting, construction, pesticide, and petroleum workers have an increased risk of cancer. Many studies have shown that exposure to asbestos, benzene, benzidine, cadmium, nickel, arsenic, radon, and vinyl chloride in the workplace can cause cancer. Apart from the above-mentioned toxins, cigarette smoke, polycyclic aromatic hydrocarbons (PAHs), nitrosamines (dimethylnitrosamine), aflatoxins (prominent DNA-damaging agents), chemotherapy (e.g., cytostatic agents—melphalan, nitrosourea, etoposide), and dioxins (industrial toxins) are considered to have a vital role in the development and progression of cancer and lead to associated CV illness [107].

Metals such as lead, cadmium, and arsenic have been listed among the WHO's top 10 environmental chemicals of concern and are considered to mediate the development and progression of CVD.

Arsenic is highly toxic to human health. Inorganic and most toxic forms of arsenic (arsenate and arsenite) are found in soils, crops, and water, particularly in groundwater from deep wells, often used as drinking water.

Arsenic has been associated with heritable changes in gene expression, such as DNA hypermethylation of p16 and p53 tumor suppressor genes, histone modifications, and RNA interference. FDA in the year 2000 has approved arsenic trioxide to be used for acute promyelocytic leukemia (APL) and psoriasis [108].

One of the previous prospective cohort studies has reported that long-term exposure to low-to-moderate arsenic levels is associated with CVD incidence and cardiovascular mortality [109].

In addition to clinical CVD outcomes, arsenic exposure is associated with established CVD risk factors such as hypertension [110], electrocardiographic abnormalities [111–113], diabetes [114,115], and kidney disease [116,117].

Likewise, cadmium is another toxin, which can damage vascular tissues, induce endothelial dysfunction, and promote atherosclerosis by oxidative mechanisms [118] and can also contribute to atherosclerosis indirectly through hypertension and renal effects.

Other potential mechanisms for the action of cadmium include calcium channel agonism [119], disruption of endothelial-barrier functions [120], induction of necrosis [121], and inhibition of vasodilator substances such as NO [122].

Lead as a toxin too causes a lot of metabolic disturbances as it promotes oxidative stress, inflammation, endothelial dysfunction, and proliferation of vascular smooth muscle cells and fibroblasts and inhibits heart-rate variability [123,124]. Lead might induce epigenetic modifications. Studies have been conducted in the United States, China, and Spain to evaluate the association between exposure to lead and DNA methylation, generally discovering a trend toward hypomethylation [125–127].

Research consistently demonstrates the role of environmental exposures in the development and progression of the most pervasive CVDs and progressive cancers. An understanding of the factors that influence susceptibility to exposure and to the effects of exposure to all the environmental toxins (organic and inorganic) is especially important, to facilitate the prevention and treatment of CVD and several cancers.

3 Section 2: Health behaviors: Overweight and obesity, hyperlipidemia, hypertension, insulin resistance, and metabolic syndrome

Survival rate among cancer patients has gradually increased, but with long-term life-threatening complications. Cardiovascular morbidity and mortality have particularly been highlighted for cancer survivors [128,129]. The encouraging increase in overall survival is accompanied by increasing numbers of cancer survivors whose prognosis and quality of life are hampered by the potentially harmful long-term and late side effects of their treatments. Several studies have documented that in survivors of childhood cancer and adult-onset cancer, secondary malignancies and cardiovascular diseases are common severe late effects that can affect life expectancy and quality of life.

The Childhood Cancer Survivor Study assessed that long-term cancer survivors were 8.2 times more likely to die of cardiac-related events than the standard population [128] and were more likely to develop coronary artery disease and cerebrovascular events as compared to healthy siblings [130]. Cardiovascular disease can be the result of direct cardiovascular damage caused by cancer treatment or of accelerated atherosclerosis due to cancer treatment-related cardiovascular risk factors [131]. During the last three decades, several studies have emphasized that these cardiovascular risks tend to cluster together and are known as metabolic syndrome (MS).

3.1 Metabolic syndrome

The metabolic syndrome is defined as a constellation of cardiovascular risk factors (hypertension, dyslipidemia, obesity, and insulin resistance) and is highly associated with cardiovascular events and death [132,133]. This syndrome is associated with inflammatory and prothrombotic features and might be an important link between cancer treatment, cardiovascular toxicity, and accelerated atherosclerosis in cancer survivors [134]. At least 25% of the population in the Americas, Europe, and India has MS, and persons with the metabolic syndrome are at essentially twice the risk for cardiovascular disease compared with those without the syndrome [135]. Accumulating evidences are suggestive that patients with MS are at increased risk of developing a CV event or type 2 diabetes mellitus [136,137]. Early detection is always a key for better treatment or prevention, whatever possible, to reduce cardiovascular morbidity [136].

The etiology of MS is multifactorial and differs between treatment type, cancer diagnosis, and patient's characteristics. Surgery, radiotherapy, chemotherapy, and hormonal therapy have been shown to induce MS, probably due to different and sometimes overlapping mechanisms. Occurrence of MS in cancer survivors is an open gateway for cardiovascular damage, as it might contribute to earlier development of atherosclerosis and related diseases in predisposed individuals. Signs of cardiovascular damage may be present before cardiovascular disease becomes evident.

3.2 Overweight and obesity

Obesity, or excess adiposity, is just one component of a constellation of cardiovascular disease risk factors experienced by cancer survivors.

The high prevalence of weight gain and sedentary lifestyle in cancer population [138] is a contributing factor to the higher occurrence of MS and cardiovascular morbidity among cancer patients and survivors. With the fact that obesity is rapidly taking over smoking as the most preventable cause of cancer in the United States [139], it is credible that obesity is part of a vicious circle of cancer treatment-related fatigue [140], impaired physical function, discomfort, reduced physical activity, and weight gain [141]. The most recently proposed criteria for metabolic syndrome emphasize the role of central obesity [142] and reflect the hypothesis that visceral adipose tissue is causally associated with the syndrome [143]. In contrast to subcutaneous fat, visceral adipose tissue might directly lead to metabolic syndrome, because of its hyperlipolytic state and the contribution of excess free fatty acids to insulin resistance.

Obesity, cancer, and CVD have a complex relationship mediated by several factors such as diet, body fat distribution, physical activity, hormones (sex hormones, insulin and insulin-like growth factor [IGF] signaling, and adipokines), chronic inflammatory burden, and oxidative stress. Moreover, adipose tissue is increasingly considered an endocrine organ that in obese people releases an excess of proinflammatory cytokines, such as tumor necrosis factor-α (TNF-α) and interleukin-6 (IL-6), shows a low release of the anti-inflammatory and insulin-sensitizing adipokine adiponectin, and produces plasminogen-activator inhibitor 1 (PAI-1) [144,145], leptin, angiotensinogen, resistin (linking obesity to diabetes mellitus), and CRP, several of which have antiapoptotic and proangiogenic properties that not only help sustain fat storage, but also have tumorigenic effects at other sites.

These factors might contribute to associated conditions such as insulin resistance and a proinflammatory and prothrombotic state.

According to the lipid overflow-ectopic fat model, the inability of dysfunctional subcutaneous adipose tissue to store excess energy, which is caused by excess caloric intake, a sedentary lifestyle, or both, leads to ectopic fat deposition in visceral adipose tissue and in organs such as the liver, skeletal muscle, and heart [144].

3.3 Insulin resistance

Insulin is an anabolic hormone and coordinates glucose either to oxidation or to storage in the body. There are several functions; for example, it decreases hepatic glucose production and increases glucose uptake in fat and muscle tissue. Furthermore, it promotes growth of cells by stimulating lipogenesis and inhibiting lipolysis, increasing protein synthesis, and inhibiting protein breakdown. Several studies have documented that insulin has tumorigenic effects on preneoplastic cells that have insulin receptors, and insulin-like growth factors (IGFs) have also been implicated in tumorigenesis [146,147]. In another study, glucose intake has also been shown to stimulate cancer growth via the inflammatory cascade, the 12-lipoxygenase pathway, in mice fed sucrose-enriched diet [148].

Insulin resistance (IR) is a pathological condition characterized by a decrease in the efficiency of insulin signaling for blood sugar regulation. Insulin resistance is a major component of metabolic syndrome, which is a group of risk factors that generally occur together and increase the risk for various diseases (e.g., type 2 diabetes mellitus) [149,150]. IR occurs as a physiological adaptive response to several stresses such as pregnancy, fasting, and exercise and is also seen in chronic diseases such as type 2 diabetes and cancer cachexia [151,152]. Malignant tumors have witnessed the presence of chronic insulin resistance, but not benign tumors, and are hypothesized to develop in cancer cachexia due to chronic exposure of proinflammatory cytokines, TNF-α, IL-6, and insulin growth factor binding protein [153], which results in insulin resistance [154]. The literature widely suggests that free fatty acids are the predominant substrate used in the adult myocardium for ATP production; however, the cardiac metabolic network is highly flexible and can use other substrates, such as glucose, lactate, or amino acids. During insulin resistance, several metabolic alterations induce the development of cardiovascular disease.

An imbalance in glucose metabolism is induced due to IR, which in turn generates chronic hyperglycemia, further leading to oxidative stress and causing an inflammatory response that leads to cell damage. As already mentioned, IR can also alter systemic lipid metabolism, which then leads to the development of dyslipidemia. The occurrence of dyslipidemia is facilitated by the well-known lipid triad, which is high levels of plasma triglycerides, low levels of high-density lipoprotein, and the appearance of small dense low-density lipoproteins. This triad, along with endothelial dysfunction, which can also be induced by aberrant insulin signaling, contributes to atherosclerotic plaque formation [155].

Interestingly, a strong correlation between insulin resistance and risk to develop CVD has been established [156]. Accumulating evidences have established the fact that several molecular mechanisms contribute to the association between insulin resistance and CVD [157–160]. These mechanisms include the role of insulin resistance in atherosclerosis development, vascular function, hypertension, and macrophage accumulation [160].

As the physiological effect of insulin varies dramatically from tissue to tissue, and this complexity is crucial to understand the adverse impact of insulin resistance/compensatory hyperinsulinemia in human disease. Overall, IR contributes to generate cancers and CVD via several independent and associated pathways. Currently, there are no special guidelines for treatment of the metabolic syndrome in survivors of cancer and treatment is the same as in the general population. However, when a candidate etiological factor can be identified, treatment of this factor might intervene at a central point and reduce multiple components of the metabolic syndrome. Future research is needed to understand the precise mechanism between insulin resistance and its progression toward cancer and CVD with a focus on new therapy development.

3.4 Hyperlipidemia

Obesity, among the metabolic syndromes is highly associated with hyperlipidemia. The known link between obesity and cancer suggests an important interaction between the host metabolism and tumorigenesis. Hyperlipidemia is a condition wherein the level of fat and cholesterol in your body is too high.

Metabolic reprogramming of lipid metabolism present in a wide range of cancers plays a critical role in the pathogenesis of human malignancies and cancer-associated cachexia. Aberrant lipoprotein profiles are often observed in cancer patients. Low plasma levels of total cholesterol (TC) and low-density lipid-cholesterol (LDL-C), as well as high triglyceride (TG) concentrations, are generally correlated with a high risk for various types of hematological and solid tumors [161,162]. Excess lipid supply and adipose expansion lead to the onset of insulin resistance and chronic inflammation which can distort the normal balance between cell proliferation and differentiation and apoptosis, which eventually contribute to obesity-induced cancer incidence [163].

Rapidly proliferating tumor cells generally require high amounts of fatty acids (FAs) and cholesterol. Numerous studies have confirmed hyperactivation of de novo lipogenesis in various types of neoplasia. Targeting lipogenesis by inhibiting fatty acid synthase (FASN), stearoyl-CoA desaturase-1 (SCD1), ATP-citrate lyase (ACLY), or sterol regulatory element-binding protein-1 (SREBP1) was effective in tumor suppression [164].

Cancer cells can also rapidly convert exogenous fatty acids to lipids required for proliferation and protumorigenic lipid signaling [165].

Apart from its role in the development and progression of cancer, serum lipids have a well-known association with coronary artery diseases [166]. Atherogenesis begins when excess lipoproteins such as low-density lipoprotein accumulate in the subendothelial space, are oxidatively modified, and are taken up selectively by macrophages and monocytes. The cholesterol metabolite 27-hydroxycholesterol is similar in both structure and action to estradiol (an estrogen) and has been recently implicated in breast cancer [167]. This forms the basis for an intriguing relationship between CVD and cancer through cholesterol and its metabolism.

3.5 Hypertension

Cancer patients and survivors of cancer have a greater burden of cardiovascular disease compared with the general population [168]. Accordingly, hypertension is the foremost modifiable risk factor of adverse cardiovascular outcomes among cancer patients. Much of the elevated cardiovascular risk in these individuals is likely attributable to hypertension, because individuals with cancer have a particularly high incidence of hypertension following cancer diagnosis. Treatment with chemotherapy is an independent risk factor for hypertension due to direct effects of many agents on endothelial function, sympathetic activity, and renin-angiotensin system activity, as well as nephrotoxicity.

Hypertension is the most common severe adverse event in patients with cancer receiving chemotherapy [169]. One retrospective study analyzed the incidence of new-onset hypertension in a population of 25,090 adults with solid malignancies in the United States and found that approximately one-third developed hypertension during the follow-up [170]. Cancer survivors equally share the burden of hypertension as studies are suggestive that patients who have a history of cancer have a high prevalence of hypertension compared with the general population. It was observed in the Childhood Cancer Survivor Study that hypertension was more common in > 10,000 adults who have survived childhood cancer versus > 3000 siblings [168].

Diagnosis and management of hypertension in cancer patients requires accurate blood pressure measurement and consideration of potential confounding factors, such as adjuvant treatments and acute pain, that can temporarily elevate blood pressure readings. Apart from keeping a check on blood pressure at home, selection of antihypertensive agents in cancer patients should account for treatment-specific morbidities and target organ damage.

4 Section 3: Management of risk factors

The intimate connections and the almost obligatory consistency in risk factors suggest that these conditions may have much more in common than previously believed. Several studies have highlighted the presence of the extensive overlap in risk factors and disease prevention for CVD and cancer which further suggests that these seemingly diverse diseases have some common basic molecular pathways or networks.

Chronic inflammation may have a considerable role because it contributes to both diseases and occurs in conditions such as obesity, diabetes mellitus, hypertension, and dyslipidemia. Controlling CVD risk factors can help reduce the risk of cancer and vice versa. There is an urgent need to improve the health status of the population to reduce the prevalence of disease. Further understanding of the delicate interaction between CVD and cancer may lead to better prevention, earlier detection, and safer treatment strategies.

Furthermore, the knowledge of the etiology of the metabolic syndrome in cancer survivors will contribute to a more causal treatment than the current symptomatic treatment strategies. Meanwhile, the prevalence and the effects of the metabolic syndrome in survivors of cancer warrant its incorporation in future care plans.

Modifiable risk factors such as smoking, diet, age, and environmental pollution can be easily taken care of and can help to reduce the CVD and cancer disease burden. It has been documented that survivors of childhood- and adult-onset cancers have higher rates of cardiovascular risk factors and cardiovascular disease compared with matched controls. Targeting and managing known cardiac risk factors should be a main target in CVD prevention in cancer survivors. Keeping healthy body weight, regular physical activity, and balanced diet are the cornerstones of modifiable risk factors' control; the same applies to avoiding cigarette smoking and excessive alcohol consumption. As shown earlier, these lifestyle modifications are relevant not only in CVD prevention but also in decreasing cancer risk itself. Maintenance of adequate blood pressure control, glucose control, and low serum lipid levels is another main target in CVD prevention; when relevant and needed, pharmacological therapies should be considered and initiated. Further research should be directed toward defining better strategies of prevention, detection, and treatment of cardiovascular risk factors and conditions in cancer survivors.

5 Conclusion

To an extent, it might be right to say or visualize that both cancer and CVD are lifestyle-induced diseases. Your body will always manifest problems that are experienced internally. When you don't listen to the call of the body, then eventually and one by one, the different parts of the body will shut down and function incorrectly. Talking in terms of genetics, then it is very well documented that there is a very close association called gene-environment interaction. The exogenous insult to the body by feeding it with poor diet, sedentary lifestyle, and poor social habits (tobacco, smoking, alcohol) will cause endogenous damage, which at times is irreversible and irreparable. Epigenetic modifications induced due to environmental interactions will lead to the altered expression and regulation of several genes in our body which may lead to silencing of certain genes (e.g., tumor suppressor) and expression of other genes such as oncogenes. Proper diet, regular exercise, and controlled blood sugar, blood pressure, and cholesterol will help us to reduce the burden of several diseases.

As, both cancer and CVD, shared common risk factors therefore, once diagnosed with the disease and during the treatment for the same, it is important to keep a check on the occurence of metabolic syndromes, which will cumulatively, reduce the prognosis and simultaneously hamper the quality of life.

References

[1] J. Ferlay, M. Colombet, I. Soerjomataram, T. Dyba, G. Randi, M. Bettio, A. Gavin, O. Visser, F. Bray, Cancer incidence and mortality patterns in Europe: estimates for 40 countries and 25 major cancers in 2018, Eur. J. Cancer 103 (2018) 356–387.

[2] K.D. Miller, L. Nogueira, A.B. Mariotto, J.H. Rowland, K.R. Yabroff, C.M. Alfano, A. Jemal, J.L. Kramer, R.L. Siegel, Cancer treatment and survivorship statistics, 2019, CA Cancer J. Clin. 69 (2019) 363–385.

[3] K.M. Sturgeon, et al., A population-based study of cardiovascular disease mortality in US cancer patients, Eur. Heart J. 40 (2019) 3889–3897.

[4] P.T. Bradshaw, J. Stevens, N. Khankari, S.L. Teitelbaum, A.I. Neugut, M.D. Gammon, Cardiovascular disease mortality among breast cancer survivors, Epidemiology 27 (2016) 6–13.

[5] T. Mohammed, T. Parekh, A. Desai, Cardiovascular risk management in cancer survivors: are we doing it right? World J. Clin. Oncol. 12 (3) (2021) 144–149.

[6] A.D. Lopez, C.D. Mathers, M. Ezzati, D.T. Jamison, C.J. Murray, Global and regional burden of disease and risk factors, 2001: systematic analysis of population health data, Lancet 367 (9524) (2006) 1747–1757.

[7] R. Atun, T. de Jongh, F. Secci, K. Ohiri, O. Adeyi, Integration of targeted health interventions into health systems: a conceptual framework for analysis, Health Policy Plan. 25 (2009), czp055.

[8] A. Alwan, D.R. MacLean, L.M. Riley, E.T. d'Espaignet, C.D. Mathers, G.A. Stevens, D. Bettcher, Monitoring and surveillance of chronic non-communicable diseases: progress and capacity in high burden countries, Lancet 376 (9755) (2010) 1861–1868.

[9] J.L. Patnaik, T. Byers, C. DiGuiseppi, et al., Cardiovascular disease competes with breast cancer as the leading cause of death for older females diagnosed with breast cancer: a retrospective cohort study, Breast Cancer Res. 13 (2011) R64.

[10] J. Huang, L. Li, J. Lian, S. Schauer, P.W. Vesely, D. Kratky, G. Hoefler, R. Lehner, Tumor-induced hyperlipidemia contributes to tumor growth, Cell Rep. 15 (2) (2016) 336 318.

[11] J. Steinberger, C. Moorehead, V. Katch, A.P. Rocchini, Relationship between insulin resistance and abnormal lipid profile in obese adolescents, J. Pediatr. 126 (5 Pt 1) (1995) 690–695.

[12] A.P. Ferreira, C.E. Oliveira, N.M. Franca, Metabolic syndrome and risk factors for cardiovascular disease in obese children: the relationship with insulin resistance (HOMA-IR), J. Pediatr. 83 (1) (2007) 21–26.

[13] J.B. Cohen, A.S. Geara, J.J. Hogan, R.R. Townsend, Hypertension in cancer patients and survivors: epidemiology, diagnosis, and management, JACC CardioOncol. 1 (2019) 238–251.

[14] I.N. Daher, T.R. Daigle, N. Bhatia, J.-B. Durand, The prevention of cardiovascular disease in cancer survivors, Tex. Heart Inst. J. 39 (2) (2021) 190–198.
[15] World Health Organisation, Global Health Estimates 2015: Deaths by Cause, Age, Sex, by Country and by Region, 2000–2015, 2016.
[16] B.J. Petek, C. Greenman, J. Herrmann, M.S. Ewer, R.L. Jones, Cardio-oncology: an ongoing evolution, Future Oncol. 11 (14) (2015) 2059–2066.
[17] T. Gredner, G. Behrens, C. Stock, et al., Cancers due to infection and selected environmental factors, Dtsch. Arztebl. Int. 115 (35–36) (2018) 586–593.
[18] M. Ryberg, D. Nielsen, T. Skovsgaard, et al., Epirubicin cardiotoxicity: an analysis of 469 patients with metastatic breast cancer, J. Clin. Oncol. 16 (11) (1998) 3502–3508.
[19] J.M. Lubieniecka, J. Graham, D. Heffner, et al., A discovery study of daunorubicin induced cardiotoxicity in a sample of acute myeloid leukemia patients prioritizes P450 oxidoreductase polymorphisms as a potential risk factor, Front. Genet. 4 (2013) 231.
[20] K. Chatterjee, J. Zhang, N. Honbo, J.S. Karliner, Doxorubicin cardiomyopathy, Cardiology 115 (2) (2010) 155–162.
[21] P. Anderlini, R.S. Benjamin, F.C. Wong, et al., Idarubicin cardiotoxicity: a retrospective study in acute myeloid leukemia and myelodysplasia, J. Clin. Oncol. 13 (11) (1995) 2827–2834.
[22] N. Mohan, J. Jiang, M. Dokmanovic, W.J. Wu, Trastuzumab-mediated cardiotoxicity: current understanding, challenges, and frontiers, Antib. Ther. 1 (1) (2018) 13–17.
[23] T.K. Choueiri, E.L. Mayer, Y. Je, et al., Congestive heart failure risk in patients with breast cancer treated with bevacizumab, J. Clin. Oncol. 29 (6) (2011) 632–638.
[24] N. Aghel, D.H. Delgado, J.H. Lipton, Cardiovascular toxicities of BCR-ABL tyrosine kinase inhibitors in chronic myeloid leukemia: preventive strategies and cardiovascular surveillance, Vasc. Health Risk Manag. 13 (2017) 293–303.
[25] W. Ma, M. Liu, F. Liang, et al., Cardiotoxicity of sorafenib is mediated through elevation of ROS level and CaMKII activity and dysregulation of calcium homoeostasis, Basic Clin. Pharmacol. Toxicol. 126 (2) (2020) 166–180.
[26] S.E. Lipshultz, R. Karnik, P. Sambatakos, et al., Anthracycline-related cardiotoxicity in childhood cancer survivors, Curr. Opin. Cardiol. 29 (1) (2014) 103–112.
[27] S.J. Park, R. Bejar, Clonal hematopoiesis in aging, Curr. Stem Cell Rep. 4 (3) (2018) 209–219.
[28] J.T. Thackeray, S. Pietzsch, B. Stapel, et al., Insulin supplementation attenuates cancer induced cardiomyopathy and slows tumor disease progression, JCI Insight 2 (10) (2017), e93098.
[29] A.H. Berg, P.E. Scherer, Adipose tissue, inflammation, and cardiovascular disease, Circ. Res. 96 (2005) 939–949.
[30] K. Frayn, S. Bernard, K. Spalding, P. Arner, Adipocyte triglyceride turnover is independently associated with atherogenic dyslipidemia, J. Am. Heart Assoc. 1 (2012), e003467.
[31] B. Rodríguez-Iturbe, H. Pons, Y. Quiroz, R.J. Johnson, The immunological basis of hypertension, Am. J. Hypertens. 27 (2014) 1327–1337.
[32] K. Esposito, F. Nappo, R. Marfella, G. Giugliano, F. Giugliano, M. Ciotola, L. Quagliaro, A. Ceriello, D. Giugliano, Inflammatory cytokine concentrations are acutely increased by hyperglycemia in humans: role of oxidative stress, Circulation 106 (2002) 2067–2207.
[33] S. Ather, W. Chan, B. Bozkurt, et al., Impact of noncardiac comorbidities on morbidity and mortality in a predominantly male population with heart failure and preserved versus reduced ejection fraction, J. Am. Coll. Cardiol. 59 (11) (2012) 998–1005. 21.
[34] T. Hasin, Y. Gerber, S.M. McNallan, et al., Patients with heart failure have an increased risk of incident cancer, J. Am. Coll. Cardiol. 62 (10) (2013) 881–886.
[35] A. Banke, M. Schou, L. Videbaek, et al., Incidence of cancer in patients with chronic heart failure: a long-term follow-up study, Eur. J. Heart Fail. 18 (3) (2016) 260–266.
[36] T. Hasin, Y. Gerber, S.A. Weston, et al., Heart failure after myocardial infarction is associated with increased risk of cancer, J. Am. Coll. Cardiol. 68 (3) (2016) 265–271.

[37] T.J. Pfeffer, S. Pietzsch, D. Hilfiker-Kleiner, Common genetic predisposition for heart failure and cancer, Herz 45 (2020) 632–636. Department of Cardiology and Angiology, Hannover Medical School, Hannover, Germany.
[38] W.C. Meijers, M. Maglione, S.J.L. Bakker, et al., Heart failure stimulates tumor growth by circulating factors, Circulation 138 (7) (2018) 678–691.
[39] T.J. Pfeffer, S. Schlothauer, S. Pietzsch, et al., Increased cancer prevalence in peripartum cardiomyopathy, JACC CardioOncol. 1 (2) (2019) 196–205.
[40] A. Torkamani, G. Verkhivker, N.J. Schork, Cancer driver mutations in protein kinase genes, Cancer Lett. 281 (2009) 117–127.
[41] C. Greenman, P. Stephens, R. Smith, et al., Patterns of somatic mutation in human cancer genomes, Nature 446 (2007) 153–158.
[42] M.R. Stratton, Exploring the genomes of cancer cells: progress and promise, Science 331 (2011) 1553–1558.
[43] K.M. Wong, T.J. Hudson, J.D. McPherson, Unraveling the genetics of cancer: genome sequencing and beyond, Annu. Rev. Genomics Hum. Genet. 12 (2011) 407–430.
[44] P. Libby, R. Sidlow, A.E. Lin, et al., Clonal hematopoiesis: crossroads of aging, cardiovascular disease, and cancer: JACC review topic of the week, J. Am. Coll. Cardiol. 74 (4) (2019) 567–577.
[45] B.T. MacDonald, K. Tamai, X. He, Signalling: components, mechanisms, and diseases, Dev. Cell 17 (2009) 9–26.
[46] J.-J. Liu, X.-J. Dai, Y. Xu, P.-Q. Liu, Y. Zhang, X.-D. Liu, Z.-G. Fang, D.-J. Lin, R.-Z. Xiao, R.-W. Huang, H.-Q. Huang, Inhibition of lymphoma cell proliferation by peroxisomal proliferator-activated receptor-γ ligands via Wnt signaling pathway, Cell Biochem. Biophys. 62 (2012) 19–27.
[47] A. Mani, J. Radhakrishnan, H. Wang, A. Mani, M.-A. Mani, C. Nelson-Williams, K.S. Carew, S. Mane, H. Najmabadi, D. Wu, R.P. Lifton, LRP6 mutation in a family with early coronary disease and metabolic risk factors, Science 315 (2007) 1278–1282.
[48] A.R. Keramati, M. Fathzadeh, G.-W. Go, R. Singh, M. Choi, S. Faramarzi, S. Mane, M. Kasaei, K. Sarajzadeh-Fard, J. Hwa, K.K. Kidd, M.A. Babaee Bigi, R. Malekzadeh, A. Hosseinian, M. Babaei, R.P. Lifton, A. Mani, A form of the metabolic syndrome associated with mutations in DYRK1B, N. Engl. J. Med. 37 (2014) 1909–1919.
[49] E. Friedman, Mirk/dyrk1B kinase in ovarian cancer, Int. J. Mol. Sci. 14 (2013) 5560–5575.
[50] K. Westbrook, V. Stearns, Pharmacogenomics of breast cancer therapy: an update, Pharmacol. Ther. 139 (2013) 1–11.
[51] Q.S. Wells, O.J. Veatch, J.P. Fessel, et al., Genome-wide association and pathway analysis of left ventricular function after anthracycline exposure in adults, Pharmacogenet. Genomics 27 (7) (2017) 247–254.
[52] T.M. Okwuosa, S. Anzevino, R. Rao, Cardiovascular disease in cancer survivors, Postgrad. Med. J. 93 (2017) 82–90.
[53] P. Garcia-Pavia, Y. Kim, M.A. Restrepo-Cordoba, I.G. Lunde, H. Wakimoto, A.M. Smith, et al., Genetic variants associated with cancer therapy-induced cardiomyopathy, Circulation 140 (1) (2019) 31–41.
[54] J.W. Watters, A. Kraja, M.A. Meucci, et al., Genome-wide discovery of loci influencing chemotherapy cytotoxicity, Proc. Natl. Acad. Sci. U. S. A. 101 (32) (2004) 11809–11814.
[55] J.S. Ware, J. Li, E. Mazaika, et al., Shared genetic predisposition in peripartum and dilated cardiomyopathies, N. Engl. J. Med. 374 (3) (2016) 233–241.
[56] C.G. Broustas, H.B. Lieberman, DNA damage response genes and the development of cancer metastasis, Radiat. Res. 181 (2) (2014) 111–130.
[57] M. Sajjad, M. Fradley, W. Sun, et al., An exploratory study to determine whether BRCA1 and BRCA2 mutation carriers have higher risk of cardiac toxicity, Genes (Basel) 8 (2) (2017) 59.
[58] T. Force, K.L. Kolaja, Cardiotoxicity of kinase inhibitors: the predictionand translation of preclinical models to clinical outcomes, Nat. Rev. Drug Discov. 10 (2011) 111–126.

[59] E. Weiderpass, Lifestyle and cancer risk, J. Prev. Med. Public Health 43 (6) (2010) 459–471.
[60] Y.C. Chen, D.J. Hunter, Molecular epidemiology of cancer, Onkol. Dypl. 2 (2) (2005) 36–45.
[61] M. Arico, et al., Clinical outcome of children with newly diagnosed Philadelphia chromosome-positive acute lymphoblastic leukemia treated between 1995 and 2005, J. Clin. Oncol. 28 (31) (2010) 4755–4761.
[62] J. Monsalve, J. Kapur, D. Malkin, P.S. Babyn, Imaging of cancer predisposition syndromes in children, Radiographics 31 (2011) 263–280.
[63] L.M. Coussens, Z. Werb, Inflammation and cancer, Nature 420 (2002) 860–867.
[64] P. Libby, Inflammation and cardiovascular disease mechanisms, Am. J. Clin. Nutr. 83 (2006) 456S–460S.
[65] R. Medzhitov, Origin and physiological roles of inflammation, Nature 454 (2008) 428–435.
[66] E.R. Nelson, S.E. Wardell, J.S. Jasper, S. Park, S. Suchindran, M.K. Howe, N.J. Carver, R.V. Pillai, P.M. Sullivan, V. Sondhi, M. Umetani, J. Geradts, D.P. McDonnell, 27-hydroxycholesterol links hypercholesterolemia and breast cancer pathophysiology, Science 342 (2013) 1094–1098.
[67] B.S. Berlett, E.R. Stadtman, Protein oxidation in aging, disease, and oxidative stress, J. Biol. Chem. 272 (33) (1997) 20313–20316.
[68] R. Thanan, S. Oikawa, Y. Hiraku, S. Ohnishi, N. Ma, S. Pinlaor, P. Yongvanit, S. Kawanishi, M. Murata, Oxidative stress and its significant roles in neurodegenerative diseases and cancer, Int. J. Mol. Sci. 16 (2015) 193–217.
[69] C.-H. Chang, F.-Y. Yu, T.-S. Wu, L.-T. Wang, B.-H. Liu, Mycotoxin citrinin induced cell cycle G2/M arrest and numerical chromosomal aberration associated with disruption of microtubule formation in human cells, Toxicol. Sci. 119 (1) (2011) 84–92.
[70] C. Wang, G.B. Schuller Levis, E.B. Lee, et al., Platycodin D and D3 isolated from the root of Platycodon grandiflorum modulate the production of nitric oxide and secretion of TNF-α in activated RAW 264.7 cells, Int. Immunopharmacol. 4 (8) (2004) 1039–1049.
[71] S.S. Lim, T. Vos, A.D. Flaxman, et al., A comparative risk assessment of burden of disease and injury attributable to 67 risk factors and risk factor clusters in 21 regions, 1990–2010: a systematic analysis for the Global Burden of Disease Study 2010, Lancet 380 (2012) 2224–2260.
[72] World Health Organization, Ambient (Outdoor) Air Quality and Health, 2014, online http://www.who.int/mediacentre/factsheets/fs313/en/.
[73] G. Hoek, et al., Long-term air pollution exposure and cardio-respiratory mortality: a review, Environ. Health 12 (2013) 43.
[74] P.B. Morris, B.A. Ference, E. Jahangir, D.N. Feldman, J.J. Ryan, et al., Cardiovascular effects of exposure to cigarette smoke and electronic cigarettes, J. Am. Coll. Cardiol. 66 (12) (2015) 1378–1391.
[75] U.S. Department of Health and Human Services, The Health Consequences of Smoking—50 Years of Progress. A Report of the Surgeon General: Executive Summary, U.S. Department of Health and Human Services, Washington, DC, 2014.
[76] A.P. Owens, N. Mackman, Tissue factor and thrombosis: the clot starts here, Thromb. Haemost. 104 (3) (2010) 432–439.
[77] M.A. Pech-Amsellem, I. Myara, M. Storogenko, K. Demuth, A. Proust, et al., Enhanced modifications of low-density lipoproteins (LDL) by endothelial cells from smokers: a possible mechanism of smoking-related atherosclerosis, Cardiovasc. Res. 31 (6) (1996) 975–983.
[78] A. Chiolero, D. Faeh, F. Paccaud, J. Cornuz, Consequences of smoking for body weight, body fat distribution, and insulin resistance, Am. J. Clin. Nutr. 87 (4) (2008) 801–809.
[79] E.A. Bermudez, N. Rifai, J.E. Buring, J.E. Manson, P.M. Ridker, Relation between markers of systemic vascular inflammation and smoking in women, Am. J. Cardiol. 89 (9) (2002) 1117–1119.
[80] C.J. Smith, T.H. Fischer, Particulate and vapor phase constituents of cigarette mainstream smoke and risk of myocardial infarction, Atherosclerosis 158 (2) (2001) 257–267.

[81] U.S. Department of Health and Human Services, How Tobacco Smoke Causes Disease: The Biology and Behavioral Basis for Smoking-Attributable Disease: A Report of the Surgeon General, 2010, p. 12.
[82] World Health Organization, Global Status Report on Alcohol and Health, World Health Organization, Geneva, Switzerland, 2011.
[83] P. Boffetta, M. Hashibe, Alcohol and cancer. Review, Lancet 7 (2) (2006) 149–156.
[84] S. Costanzo, A. Di Castelnuovo, M.B. Donati, L. Iacoviello, G. de Gaetano, Cardiovascular and overall mortality risk in relation to alcohol consumption in patients with cardiovascular disease, Circulation 121 (2010) 1951–1959.
[85] T.T. Laitinen, K. Pahkala, C.G. Magnussen, J.S. Viikari, M. Oikonen, L. Taittonen, V. Mikkilä, E. Jokinen, N. Hutri-Kähönen, T. Laitinen, M. Kähönen, T. Lehtimäki, O.T. Raitakari, M. Juonala, Ideal cardiovascular health in childhood and cardiometabolic outcomes in adulthood: the Cardiovascular Risk in Young Finns Study, Circulation 125 (2012) 1971–1978.
[86] G. Pöschl, H.K. Seitz, Alcohol and cancer, Alcohol Alcohol. 39 (3) (2004) 155–165.
[87] H.K. Seitz, F. Stickel, Molecular mechanisms of alcohol-mediated carcinogenesis, Nat. Rev. Cancer 7 (8) (2007) 599–612.
[88] H.S. Yu, T. Oyama, T. Isse, K. Kitagawa, T.T. Pham, M. Tanaka, Kawamoto T formation of acetaldehyde-derived DNA adducts due to alcohol exposure, Chem. Biol. Interact. 188 (3) (2010) 367–375.
[89] P. Boffetta, M. Hashibe, Alcohol and cancer, Lancet Oncol. 7 (2) (2006) 149–156.
[90] K.W. Singletary, S.M. Gapstur, Alcohol and breast cancer: review of epidemiologic and experimental evidence and potential mechanisms, JAMA 286 (17) (2001) 2143–2151.
[91] R.R. Watson, P. Nixon, H.K. Seitz, R. Maclennan, Alcohol and cancer, Alcohol Alcohol. Suppl. 2 (1994) 453–455.
[92] Dietary patterns, foods and nutrients, and health outcomes, Part D, Chapter 2. Scientific Report of the 2015 Dietary Guidelines Advisory Committee.
[93] D. Aune, D.S. Chan, R. Lau, R. Vieira, D.C. Greenwood, E. Kampman, T. Norat, Dietary fibre, whole grains, and risk of colorectal cancer: systematic review and dose-response meta-analysis of prospective studies, BMJ 343 (2011), d6617.
[94] B. Magalhães, B. Peleteiro, N. Lunet, Dietary patterns and colorectal cancer: systematic review and meta-analysis, Eur. J. Cancer Prev. 21 (2012) 15–23.
[95] K.Y. Wolin, K. Carson, G.A. Colditz, Obesity and cancer, Oncologist 15 (6) (2010) 556–565.
[96] K.Y. Wolin, G.A. Colditz, Can weight loss prevent cancer? Br. J. Cancer 99 (7) (2008) 995–999.
[97] F. Aguilar, S.P. Hussain, P. Cerutti, Aflatoxin B1 induces the transversion of G- ->T in codon 249 of the p53 tumor suppressor gene in human hepatocytes, Proc. Natl. Acad. Sci. U. S. A. 90 (18) (1993) 8586–8590.
[98] H.Y. Huang, B. Caballero, S. Chang, et al., The efficacy and safety of multivitamin and mineral supplement use to prevent cancer and chronic disease in adults: a systematic review for a National Institutes of Health state-of-the-science conference, Ann. Intern. Med. 145 (5) (2006) 372–385.
[99] M. Stampfer, J.L. Jahn, Partnerships for promoting prevention, Circulation 127 (2013) 1267–1269.
[100] T. Wang, H.G. Lee, Advances in research on cis-9, trans-11 conjugated linoleic acid: a major functional conjugated linoleic acid isomer, Crit. Rev. Food Sci. Nutr. 55 (2015) 720–731.
[101] D.R. English, R.J. MacInnis, A.M. Hodge, J.L. Hopper, A.M. Haydon, G.G. Giles, Red meat, chicken, and fish consumption and risk of colorectal cancer, Cancer Epidemiol. Biomarkers Prev. 13 (2004) 1509–1514.
[102] A.D. Wood, A.A. Strachan, F. Thies, L.S. Aucott, D.M. Reid, A.C. Hardcastle, A. Mavroeidi, W.G. Simpson, G.G. Duthie, H.M. Macdonald, Patterns of dietary intake and serum carotenoid and tocopherol status are associated with biomarkers of chronic low-grade systemic inflammation and cardiovascular risk, Br. J. Nutr. 112 (2014) 1341–1352.

[103] S.M. Zhang, N.R. Cook, C.M. Albert, et al., Effect of combined folic acid, vitamin B6, and vitamin B12 on cancer risk in women: a randomized trial, JAMA 300 (17) (2008) 2012–2021.
[104] R. Baena Ruiz, H.P. Salinas, Diet and cancer: risk factors and epidemiological evidence, Maturitas 77 (2014) 202–208.
[105] C.M. Friedenreich, M.R. Orenstein, Physical activity and cancer prevention: etiologic evidence and biological mechanisms, J. Nutr. 132 (11 suppl) (2002) 3456S–3464S.
[106] M. Soliman, May 16th, Cancer Causing Chemicals, Cancer Causing Substances, Faik Atroshi, IntechOpen, 2018, https://doi.org/10.5772/intechopen.71560. Available from https://www.intechopen.com/chapters/57481.
[107] F. Ba, C.H. Selene, J. Chou, et al., Arsenic, in: G.F. Nordberg, B.A. Fowler, M. Nordberg, L.T. Friberg (Eds.), Handbook on the Toxicology of Metals, third ed., Elsevier, 2007, pp. 367–406.
[108] K.A. Moon, E. Guallar, J.G. Umans, et al., Association between exposure to low to moderate arsenic levels and incident cardiovascular disease: a prospective cohort study, Ann. Intern. Med. 159 (10) (2013) 649–659.
[109] L.N. Abhyankar, M.R. Jones, E. Guallar, A. Navas-Acien, Arsenic Exposure and hypertension: a systematic review, Environ. Health Perspect. 120 (2012) 494–500.
[110] F. Wu, P. Molinaro, Y. Chen, Arsenic exposure and subclinical endpoints of cardiovascular diseases, Curr. Environ. Health Rep. 1 (2014) 148–162.
[111] C.H. Wang, et al., Arsenic-induced QT dispersion is associated with atherosclerotic diseases and predicts long-term cardiovascular mortality in subjects with previous exposure to arsenic: a 17-year follow-up study, Cardiovasc. Toxicol. 10 (2010) 17–26.
[112] I. Mordukhovich, et al., Association between low-level environmental arsenic exposure and QT interval duration in a general population study, Am. J. Epidemiol. 170 (2009) 739–746.
[113] E.A. Maull, et al., Evaluation of the association between arsenic and diabetes: a National Toxicology Program workshop review, Environ. Health Perspect. 120 (2012) 1658–1670.
[114] C.C. Kuo, K. Moon, K.A. Thayer, A. Navas-Acien, Environmental chemicals and type 2 diabetes: an updated systematic review of the epidemiologic evidence, Curr. Diab. Rep. 13 (2013) 831–849.
[115] L. Zheng, et al., Arsenic and chronic kidney disease: a systematic review, Curr. Environ. Health Rep. 1 (2014) 192–207.
[116] L.Y. Zheng, et al., Urine arsenic and prevalent albuminuria: evidence from a population-based study, Am. J. Kidney Dis. 61 (2013) 385–394.
[117] B. Messner, et al., Cadmium is a novel and independent risk factor for early atherosclerosis mechanisms and in vivo relevance, Arterioscler. Thromb. Vasc. Biol. 29 (2009) 1392–1398.
[118] G. Choong, Y. Liu, D.M. Templeton, Interplay of calcium and cadmium in mediating cadmium toxicity, Chem. Biol. Interact. 211 (2014) 54–65.
[119] J.M. Woods, et al., Direct antiangiogenic actions of cadmium on human vascular endothelial cells, Toxicol. In Vitro 22 (2008) 643–651.
[120] B. Messner, D. Bernhard, Cadmium and cardiovascular diseases: cell biology, pathophysiology, and epidemiological relevance, Biometals 23 (2010) 811–822.
[121] C.C. Almenara, et al., Chronic cadmium treatment promotes oxidative stress and endothelial damage in isolated rat aorta, PLoS One 8 (2013), e68418.
[122] A. Navas-Acien, E. Guallar, E.K. Silbergeld, S.J. Rothenberg, Lead exposure and cardiovascular disease—a systematic review, Environ. Health Perspect. 115 (2007) 472–482.
[123] N.D. Vaziri, Mechanisms of lead-induced hypertension and cardiovascular disease, Am. J. Physiol. Heart Circ. Physiol. 295 (2008) H454–H465.
[124] C.W. Hanna, et al., DNA methylation changes in whole blood is associated with exposure to the environmental contaminants, mercury, lead, cadmium and bisphenol A, in women undergoing ovarian stimulation for IVF, Hum. Reprod. 27 (2012) 1401–1410.

[125] S.M. Tajuddin, et al., Genetic and non-genetic predictors of LINE-1 methylation in leukocyte DNA, Environ. Health Perspect. 121 (2013) 650–656.
[126] R.O. Wright, et al., Biomarkers of lead exposure and DNA methylation within retrotransposons, Environ. Health Perspect. 118 (2010) 790–795.
[127] C. Li, X. Yang, M. Xu, J. Zhang, N. Sun, Epigenetic marker (LINE-1 promoter) methylation level was associated with occupational lead exposure, Clin. Toxicol. (Phila.) 51 (2013) 225–229.
[128] A.C. Mertens, Y. Yasui, J.P. Neglia, et al., Late mortality experience in five-year survivors of childhood and adolescent cancer: the Childhood Cancer Survivor Study, J. Clin. Oncol. 19 (13) (2001) 3163–3172.
[129] K.C. Oeffinger, G.R. Buchanan, D.A. Eshelman, et al., Cardiovascular risk factors in young adult survivors of childhood acute lymphoblastic leukemia, J. Pediatr. Hematol. Oncol. 23 (7) (2001) 424–430.
[130] K.C. Oeffinger, A.C. Mertens, C.A. Sklar, et al., Chronic health conditions in adult survivors of childhood cancer, N. Engl. J. Med. 355 (15) (2006) 1572–1582.
[131] J. Nuver, A.J. Smit, D.T. Sleijfer, et al., Microalbuminuria, decreased fibrinolysis, and inflammation as early signs of atherosclerosis in long-term survivors of disseminated testicular cancer, Eur. J. Cancer 40 (2004) 701–706.
[132] E.C. de Haas, S.F. Oosting, J.D. Lefrandt, B.H. Wolffenbuttel, D.T. Sleijfer, J.A. Gietema, The metabolic syndrome in cancer survivors, Lancet Oncol. 11 (2) (2010) 193–203.
[133] E.C. de Haas, R. Altena, H.M. Boezen, et al., Early development of the metabolic syndrome after chemotherapy for testicular cancer, Ann. Oncol. 24 (3) (2013) 749–755.
[134] L.F. Van Gaal, I.L. Mertens, C.E. De Block, Mechanisms linking obesity with cardiovascular disease, Nature 444 (7121) (2006) 875–880.
[135] S.M. Grundy, Metabolic syndrome pandemic, Arterioscler. Thromb. Vasc. Biol. 28 (4) (2008) 629–636.
[136] R.H. Eckel, K.G. Alberti, S.M. Grundy, P.Z. Zimmet, The metabolic syndrome, Lancet 375 (9710) (2010) 181–183.
[137] N. Sattar, A. McConnachie, A.G. Shaper, et al., Can metabolic syndrome usefully predict cardiovascular disease and diabetes? Outcome data from two prospective studies, Lancet 371 (9628) (2008) 1927–1935.
[138] M.L. Irwin, Physical activity interventions for cancer survivors, Br. J. Sports Med. 43 (1) (2009) 32–38.
[139] C.H. Kroenke, W.Y. Chen, B. Rosner, M.D. Holmes, Weight, weight gain, and survival after breast cancer diagnosis, J. Clin. Oncol. 23 (7) (2005) 1370–1378.
[140] M. Arnold, N. Pandeya, G. Byrnes, et al., Global burden of cancer attributable to high body-mass index in 2012: a population-based study, Lancet Oncol. 16 (1) (2015) 36–46.
[141] O. Minton, A. Berger, A. Barsevick, et al., Cancer-related fatigue and its impact on functioning, Cancer 119 (Suppl. 11) (2013) 2124–2130.
[142] A. Lucia, C. Earnest, M. Perez, Cancer-related fatigue: can exercise physiology assist oncologists? Lancet Oncol. 4 (10) (2003) 616–625.
[143] K.G. Alberti, P. Zimmet, J. Shaw, Metabolic syndrome—a new world-wide definition: a consensus statement from the International Diabetes Federation, Diabet. Med. 23 (2006) 469–480.
[144] J.P. Despres, I. Lemieux, Abdominal obesity and metabolic syndrome, Nature 444 (2006) 881–887.
[145] R.H. Eckel, S.M. Grundy, P.Z. Zimmet, The metabolic syndrome, Lancet 365 (2005) 1415–1428.
[146] E.E. Calle, R. Kaaks, Overweight, obesity and cancer: epidemiological evidence and proposed mechanisms, Nat. Rev. Cancer 4 (8) (2004) 579–591.
[147] A.A. Samani, S. Yakar, D. LeRoith, P. Brodt, The role of the IGF system in cancer growth and metastasis: overview and recent insights, Endocr. Rev. 28 (1) (2007) 20–47.

[148] Y. Jiang, Y. Pan, P.R. Rhea, et al., A sucrose-enriched diet promotes tumorigenesis in mammary gland in part through the 12-lipoxygenase pathway, Cancer Res. 76 (1) (2016) 24–29.
[149] A.W. Campbell, The diabetes pandemic, Altern. Ther. Health Med. 17 (2011) 8–9.
[150] T. Karagiannis, P. Paschos, K. Paletas, D.R. Matthews, A. Tsapas, Dipeptidyl peptidase-4 inhibitors for treatment of type 2 diabetes mellitus in the clinical setting: systematic review and meta-analysis, BMJ 344 (2012), e1369.
[151] M.R. Soeters, P.B. Soeters, The evolutionary benefit of insulin resistance, Clin. Nutr. 31 (6) (2012) 1002–1007.
[152] J.I. Odegaard, A. Chawla, Pleiotropic actions of insulin resistance and inflammation in metabolic homeostasis, Science 339 (6116) (2013) 172–177.
[153] E.F. Wagner, M. Petruzzelli, Cancer metabolism: a waste of insulin interference, Nature 521 (7553) (2015) 430–431.
[154] H. Wang, J. Ye, Regulation of energy balance by inflammation: common theme in physiology and pathology, Rev. Endocr. Metab. Disord. 16 (1) (2015) 47–54.
[155] G. Wilcox, Insulin and insulin resistance, Clin. Biochem. Rev. 26 (2) (2005) 19–39.
[156] K.B. Gast, N. Tjeerdema, T. Stijnen, J.W. Smit, O.M. Dekkers, Insulin resistance and risk of incident cardiovascular events in adults without diabetes: meta-analysis, PLoS One 7 (12) (2012), e52036.
[157] G. Reaven, Insulin resistance and coronary heart disease in nondiabetic individuals, Arterioscler. Thromb. Vasc. Biol. 32 (8) (2012) 1754–1759.
[158] K.E. Bornfeldt, I. Tabas, Insulin resistance, hyperglycemia, and atherosclerosis, Cell Metab. 14 (5) (2011) 575–585.
[159] J.A. Davidson, C.G. Parkin, Is hyperglycemia a causal factor in cardiovascular disease? Does proving this relationship really matter? Yes, Diabetes Care 32 (Suppl 2) (2009) S331–S333.
[160] M. Laakso, J. Kuusisto, Insulin resistance and hyperglycaemia in cardiovascular disease development, Nat. Rev. Endocrinol. 10 (5) (2014) 293–302.
[161] A.M. Fiorenza, A. Branchi, D. Sommariva, Serum lipoprotein pro-file in patients with cancer. A comparison with non-cancer subjects, Int. J. Clin. Lab. Res. 30 (2000) 141–145.
[162] P.P. Naik, M.S. Ghadge, A.S. Raste, Lipid profile in leukemia and Hodgkin's disease, Indian J. Clin. Biochem. 21 (2006) 100–102.
[163] F. Bianchini, R. Kaaks, H. Vainio, Overweight, obesity and cancer risk, Lancet Oncol. 3 (2000) 565–574.
[164] E. Currie, A. Schulze, R. Zechner, T.C. Walther, R.V. Farese Jr., Cellular fatty acid metabolism and cancer, Cell Metab. 18 (2013) 153–161.
[165] S.M. Louie, L.S. Roberts, M.M. Mulvihill, K. Luo, D.K. Nomura, Cancer cells incorporate and remodel exogenous palmitate into structural and oncogenic signaling lipids, Biochim. Biophys. Acta 1831 (2013) 1566–1572.
[166] J. Stamler, D. Wentworth, J.D. Neaton, Is relationship between serum cholesterol and risk of premature death from coronary heart disease continuous and graded? Findings in 356,222 primary screenees of the Multiple Risk Factor Intervention Trial (MRFIT), JAMA 256 (1986) 2823–2828.
[167] M. Warner, J.A. Gustafsson, On estrogen, cholesterol metabolism, and breast cancer, N. Engl. J. Med. 370 (2014) 572–573.
[168] G.T. Armstrong, K.C. Oeffinger, Y. Chen, et al., Modifiable risk factors and major cardiac events among adult survivors of childhood cancer, J. Clin. Oncol. 31 (2013) 3673–3680.
[169] H. Izzedine, S. Ederhy, F. Goldwasser, et al., Management of hypertension in angiogenesis inhibitor-treated patients, Ann. Oncol. 20 (2009) 807–815.
[170] K.H. Fraeman, B.L. Nordstrom, W. Luo, S.H. Landis, S. Shantakumar, Incidence of new-onset hyper-tension in cancer patients: a retrospective cohort study, Int. J. Hypertens. 2013 (2013), 379252.

CHAPTER

3

Recent trends in cancer immunotherapy: Pathways and inhibitors

G. Vidya

Department of Zoology, St. Joseph's College (Autonomous) Irinjalakuda, Irinjalakuda, Kerala, India

1 Introduction

Cancer immunotherapy is one of the rapidly evolving fields in cancer therapeutics. Immunotherapy can be considered as the fifth weapon of cancer therapy joining the ranks of surgery, cytotoxic chemotherapy, radiation, and targeted therapy [1]. It is one of the promising areas of cancer research and therapy as in this the patient's own immune system is enlisted in the fight against cancer. The role of immune system in preventing cancer was first predicted by Paul Ehrlich in 1909 and was first demonstrated by Gross in 1953 [2]. Studies have shown that both cellular and humoral immunity are involved in the detection and elimination of tumor cells. The neoantigens present in the tumor cells, which are formed as a result of somatic mutation, are recognized by dendritic cells (DCs). DCs then migrate to lymph nodes to present captured antigens on MHC-I and MHC-II molecules to T lymphocytes. T-helper (Th) I produces interleukin-2 (IL-2) for clonal expansion of activated T lymphocytes which then recognizes and binds to tumor cells via T-cell receptors and MHC-I. Besides, Th2 cells induce humoral response leading to the activation and proliferation of B cells [3]. However, tumor cells have the ability to evade such immune responses by downregulating the molecules of major histocompatibility complex I (MHC-I), thereby preventing its recognition by cytotoxic T lymphocytes, expressing nonclassic human leukocyte antigen-G (HLA-G) molecules, hence protecting themselves from natural

Cardiovascular Toxicity and Therapeutic Modalities Targeting Cardio-oncology
https://doi.org/10.1016/B978-0-323-90461-2.00001-8

killer cells, triggering angiogenesis, and recruiting T-regulatory cells having immunosuppressive properties [4–6]. Thus, it is clear that cancer cells create a tumor microenvironment that enables them to survive and multiply. Hence, developing an immunotherapeutic method that has minimum side effects and can overcome the immunosuppressive activity of tumor cells is of utmost importance.

Immunotherapeutic methods act either by specifically eliminating the tumor cells by targeting tumor antigens or by boosting body's own immune system. Available therapeutic options now include cytokine therapy, monoclonal antibodies, oncolytic viruses, prophylactic and therapeutic vaccines, adoptive cell transfer, immune checkpoint inhibitors, and chimeric antigen receptor (CAR) T-cell therapy [7,8]. This chapter focuses on the comprehensive review of the literature on major pathways and inhibitors involved in cancer immunotherapy.

2 CTLA-4 and PD-1 pathways

The discovery of immune checkpoint proteins cytotoxic T lymphocyte-associated protein-4 (CTLA-4) and programmed cell death-1 (PD-1) was the major breakthrough in cancer immunotherapy [9]. Both the proteins are expressed in T lymphocytes. T lymphocytes/T cells provide cell-mediated immunity and recognize and specifically bind to the antigen. The development and maturation of T cells occur in the thymus. During maturation, T cells develop T-cell receptors (TCRs) that recognize specific antigen displayed in MHC on antigen-presenting cells (APCs) [10]. T cells that show strong reactivity toward self-proteins and those that show insufficient binding toward MHC are selectively removed in a process called central tolerance [11].

In order to avoid immune overreaction, the immune system maintains a delicate balance between the expressions of co-stimulatory and co-inhibitory signals. The first stimulatory signal is the recognition and binding of TCR to an antigen displayed in MHC on the surface of APCs. However, further immune reaction depends on multiple other factors that decide whether the T cell undergoes activation or anergy. The second stimulatory signal is the binding of T-cell CD28 receptor with B7-1 (CD80) or B7-2 (CD86) of APCs for proliferation [12]. Activated T cells express several co-inhibitory receptors such as CTLA-4 and PD-1 [13]. CTLA-4 and PD-1 are immune checkpoint pathways that regulate T cells at multiple steps in an immune response to prevent autoimmunity in a process called peripheral tolerance [11,14]. They operate at different stages of immune response. CTLA-4 regulates T-cell proliferation early in an immune response, primarily in lymph nodes [14,15], while PD-1 suppresses T cells late in immune response, in peripheral tissues [14]. PD-1 interacts with its ligand PD-L1 and CTLA-4 binds to receptor B7 and regulates T-cell responses to self-proteins [16–18]. This property of T cells is exploited by cancer cells, where some of the tumor types express PD-L1 that binds to PD-1 and B7 present in the APCs pairs with CTLA-4, thus inhibiting further activity of T cells.

As mentioned previously, CD28 receptor binds to B7 molecules (B7-1 and B7-2) on APCs and signals T-cell proliferation, increased T-cell survival, and production of interleukin-2 (IL-2). CTLA-4 and CD28 share high structural homology, and studies have shown that CTLA-4 binds to B7 molecules with much higher affinity [16]. However, they perform opposite functions. CD28 leads to stimulatory signal, while CTLA-4 leads to inhibitory signal that counteracts CD28:B7 and TCR:MHC binding [19,20]. It inhibits T-cell activation and limits IL-2 production and survival of T cells. CTLA-4 is located in the intracellular compartment of T cell, which comes to the cell surface by exocytosis upon stimulation signals from TCR and CD28:B7 binding. Weak stimulus results in positive activating signal, IL-2 production, and proliferation and increased survival of T cells. On the other hand, strong stimulus elicits more CTLA-4 translocation to the surface of T cell, resulting in net negative inhibitory signals. T-regulatory cells also play an important role in peripheral tolerance by controlling the functions of effector T cells. CTLA-4 is constitutively expressed in T-regulatory cells [21,22].

PD-1 is the member of the B7/CD28 family. They regulate T-cell activation by interacting with PD-L1 and PD-L2. PD-1/PD-L1 binding has shown to inhibit the production of interferon-gamma and IL-2, reducing T-cell proliferation and survival [18]. When both TCR and PD-1 bindings occur at the same time, PD-1 prevents phosphorylation of key TCR signaling intermediates, thus inhibiting TCR signaling [20,23]. As mentioned previously, PD-1 functions at a later stage of the immune response. Unlike CTLA-4, PD-1 expression is not only confined to T cells, but is also broadly expressed in activated T cells, B cells, and myeloid cells. PD-L1 and PD-L2 are widely expressed on leukocytes, nonhematopoietic cells, and nonlymphoid tissues. Besides, PD-L1 expression is also reported in different tumor types and tumor-infiltrating lymphocytes, thereby evading the immune response by peripheral tolerance.

2.1 Immune therapy

Immune checkpoints, especially co-inhibitory receptors CTLA-4, PD-1, and its ligand PD-L1, have gained much attention in the past few years in the field of immune therapy. In a normal scenario, these checkpoint proteins act by suppressing the immune system to prevent autoimmunity. However, during this process, they sometimes fail to recognize the tumor antigens, allowing the cancer cells to escape immune recognition.

Several preclinical studies showed a decrease in tumor growth and an increased survival rate after blocking CTLA-4 and PD-1 pathway, which encouraged the scientists to develop inhibitors against these immune checkpoint proteins. Monoclonal antibodies developed against CTLA-4 and PD-1 are now approved for several cancers including melanoma [24], lung cancer [25], kidney cancer [26], prostate cancer, and head and neck cancer [27]. The first checkpoint inhibitor approved for clinical use was ipilimumab. Ipilimumab is a monoclonal antibody that blocks the interaction of CTLA-4 with B7 and is approved by FDA for the treatment of melanoma in 2011 [28]. Later, it was found to be effective in various other cancers such as renal cell carcinoma [29], B-cell lymphoma [30], refractory metastatic colorectal cancer [31], and hepatic cellular carcinoma [32]. A combination of ipilimumab with other chemotherapeutic drugs have been shown to have better clinical outcome in advanced melanoma [33] and lung cancer [34]. Also, combination of ipilimumab with NF-κB ligand denosumab has shown to have a positive result in metastatic melanoma [35]. Tremelimumab is another monoclonal antibody to CTLA-4 demonstrated to have durable clinical responses in early trials and is now under phase II trials against lung cancer [31]. The exact mechanism with which the anti-CTLA-4 antibody induces antitumor response is not fully understood. However, studies have shown that CTLA-4 blockade resulted in the activation and proliferation of effector T cells and also there is a reduction in T-regulator cell-mediated suppression of T-cell response. There are reports of an increase in the diversity of peripheral T-cell pool in melanoma patients following CTLA-4 blockade. Studies have shown significant upregulation and autoreactivity of the immune system in CTLA-4 knockout mice [36]. Extensive tumor necrosis and lymphocyte tumor infiltration were observed in patients administered with the CTLA-4 antibody MDX-010. However, few side effects have been reported with the anti-CTLA-4 antibody. The most frequent adverse effects are immune-related after treatment with ipilimumab. Diarrhea was the most common adverse immune event.

At the same time when ipilimumab was approved by FDA, the clinical efficiency of anti-PD-1 agents was getting tested. Several preclinical studies have shown a better effect by anti-PD-1 agents as compared to those that target CTLA-4 in terms of tumor response and toxic effects. As mentioned previously, PD-L1 expression is seen in various tumor types and particularly abundant in melanoma, non-small-cell lung carcinoma, and ovarian cancer. Patient response to anti-PD-L1 blockade was strongest when PD-L1 was expressed by tumor-infiltrating immune cells. Several monoclonal antibodies including nivolumab, pembrolizumab, cemiplimab, atezolizumab, and avelumab were developed to block the interaction of PD-1 with PD-L1. Specifically, blocking PD-1:PD-L1 interaction while preserving PD-1:PD-L2 interaction reduces toxicity. Studies showed better results using nivolumab in metastatic melanoma cells [24], non-small-cell lung carcinoma [25], Hodgkin's lymphoma, and hepatocellular carcinoma [37]. Pembrolizumab showed positive results with metastatic urothelial carcinoma (19), head and neck squamous cell carcinoma, and non-Hodgkin's lymphoma. Pembrolizumab and nivolumab are approved by FDA in 2014 for patients with advanced melanoma who were nonresponsive to anti-CTLA-4 therapy.

3 PI3K/AKT/mTOR signaling pathways

The PI3K-AKT-mTOR signaling pathway has been very well characterized over the past years. It consists of three parts: PI3K and its downstream molecules PKB/AKT and mTOR. They control most of the hallmarks of cancer which include cell cycle, survival, protein synthesis and growth metabolism, cell proliferation, motility, and genomic stability. The molecular mechanism by which they regulate these processes is also very well studied [38–40]. Mammalian target of rapamycin (mTOR) is a multicomponent complex that has functionally and structurally different mTOR complex 1 (mTORC1) and mTOR complex II (mTORC2). They play a major role in many physiological processes such as protein synthesis, cytoskeleton remodeling, angiogenesis, and metabolism [41]. mTORC1 is rapamycin sensitive and regulates several processes required for cell growth and metabolism, including ribosome biogenesis, protein translation, and autophagy. mTORC2 is rapamycin insensitive and promotes phosphorylation of protein kinase B (AKT), serum and glucocorticoid kinase, and protein kinase C (PKC). It is also involved in the regulation of proliferation, survival, cytoskeleton remodeling, and cell migration [42–44]. mTORC1 is mainly activated by PI3K/AKT signaling. Growth factors, cytokines, etc., stimulate the activation of receptor tyrosine kinases (RTKs) which activates many intermediate molecules that inturn activates PI3K. PI3K then acts as kinase on lipid second messenger phosphatidylinositol-4,5-bisphosphate (PIP2) to produce phosphatidylinositol-3,4,5-triphosphate (PIP3) [45]. These lipids further recruit AKT, mTORC2, and phosphoinositide-dependent kinase I (PDK1). Either PDK1 or mTORC2 interacts with AKT and triggers its activation [46,47]. Activated AKT then activates mTORC1, which in turn promotes cell growth and protein synthesis [6]. Phosphatase and tensin homolog deleted on chromosome 10 (PTEN), a tumor repressor, reverses the action of PI3K, by dephosphorylating PIP3 to PIP2 [48,49]. mTOR is involved in the regulation of tumor-related vascular formation, by promoting angiogenesis, especially during hypoxic condition, by the expression of hypoxic stress response genes [50]. Studies have shown that deregulation of the PI3K-AKT-mTOR pathway is tightly connected to cancer initiation and progression, due to its involvement in many of the important physiological processes of the cell. Studies have shown that the pathway is deregulated in approximately 30% of the solid tumors [38]. Oncogenic activation can occur by mutation/amplification of gene encoding RTKs, subunits of PI3K, AKT, etc. [51–53]. Mutation in tumor suppressor genes such as PTEN can negatively regulate the function of the PI3K pathway [54,55].

A recent study has shown that treatment with PI3K-AKT-mTOR signaling inhibitors affects not only tumor progression but also tumor surveillance within the tumor microenvironment (TME). The activity of most immune cell types present in the TME is affected by the PI3K-AKT-mTOR pathway. TME consists of different cell types and paracrine factors. The involvement of TME in cancer progression and drug resistance has been well studied. TME is characterized by the presence of different cell types including cancer-associated fibroblasts, myeloid-derived suppressor cells (MDSCs), tumor-associated macrophages (TAMs), and regulatory T cells (Tregs) [56,57]. All these contribute to tumor progression in one or the other ways during cancer progression [58]. MDSCs inhibit immune cells by releasing IL-10, while TAMs promote tumor angiogenesis. Tregs are known to limit the activity of antitumor responses by the secretion of IL-10 and sequestration of IL-2 [59]. They play an important role in tumor immunity by suppressing the activity of effector T cells [60]. PI3K/mTOR inhibitors seem to modulate the number and function of specific TME cells that are involved in tumor progression. Earlier studies have shown the role of mTOR signaling in macrophage activation and differentiation [61]. When PI3K inhibitors were used, there was a decrease in TAM recruitment in TME [62]. mTOR has been identified as the regulator of memory $CD8^+$ T-cell differentiation. $CD8^+$ T cells have the ability to target and kill the cancer cells. mTOR negatively modulates memory T-cell formation [63]. Furthermore, in vitro and in vivo studies display that Tregs have a high dependence on the PI3K-AKT signaling pathway as compared to other T cells [64]. AKT and PI3K blockade selectively inhibits Treg proliferation with a minimal effect on the other T-cell population [64].

3.1 Immune therapy

Most of the immune therapies are focused on relieving T-cell suppression, such as monoclonal antibodies against CTLA-4, PD-1 on T cells, and its ligand PD-L1 that is expressed on tumor and immune cells [65]. Although blocking CTLA-4 and PD-1 checkpoints and other T-cell-mediated immune therapies exhibited clinical success to some extent, some patients failed to respond to immune therapy [66]. Only 50% patients responded, and in some cases, it was associated with severe toxicity [67]. Hence, it is necessary to know the molecular mechanism of resistance. mTOR plays a major role in coordinating environment cues and cell metabolism responses.

It also functions in immune cell homeostasis and activation [68]. Thus, it represents a potential target for cancer therapy. Besides, many studies suggest that there is a dual benefit in targeting the PI3K-AKT-mTOR signaling pathway, as it can inhibit tumor progression by limiting the proliferation and survival and can enhance tumor immunosurveillance by preventing the activation of immunosuppressive pathways and improving antitumor immune-intrinsic properties [69].

To date, six classes of compounds targeting the PI3K-AKT-mTOR pathway at different levels are reported, which include mTOR inhibitors, AKT inhibitors, Pan class I-PI3K inhibitors, and Pan-PI3K-mTOR inhibitors [70]. Many are under clinical trials. Rapamycin and its analogue are the first-generation mTOR inhibitors that inhibit the activity of mTORC1 by binding to FRBP-12 [71]. A novel rapamycin derivative, RAD001/everolimus, is developed which can inhibit mTORC1 but does not show any effect on mTORC2 [72]. Other mTOR inhibitors include temsirolimus, deforolimus, and ridaforolimus. These inhibitors are currently under clinical trials against a variety of human malignancies. Inhibitors such as temsirolimus and everolimus are now FDA-approved for the treatment of advanced renal cell carcinoma and subependymal renal cell astrocytoma and for progressive neuroendocrine tumors of pancreatic origin [73,74]. Several groups have shown that rapamycin displays anti-angiogenic properties in terms of reduction of proliferation, migration, tubular structure formation, and increase in apoptosis [50,75,76]. LY294002 and wortmannin are the two well-studied first-generation inhibitors [77]. Studies have shown that LY294002 can sensitize cancer cells to ionizing radiation by inactivating AKT, leading to cell cycle arrest [78]. However, there are certain disadvantages associated with this inhibitor, which include short half-life, numerous nontarget effects, and unacceptable toxic effects in animal studies [78]. These lead to the generation of other PI3K inhibitors such as BKM120, XL147, PX-866, and GDC-0941. Preclinical studies showed that BKM120 can actively inhibit cell proliferation and induce apoptosis [79]. Inhibition of the PI3K pathway using PI3K isoform inhibitors and Pan-PI3K inhibitors in various clinical models has been shown to enhance $CD8^+$ T-cell infiltration within the tumor tissue, thus reducing the tumor burden. AKT is the key transmitter of the PI3K/AKT/mTOR signaling pathway. AKT inhibitors include Palomid 529 (P529), perifosine, and MK2206 [80,81]. P529 is a novel and effective AKT inhibitor that did not exhibit any toxicity in vivo [80]. When combined with radiotherapy, it improved the radiosensitivity of tumor cells and induced DNA double-strand breaks and apoptosis [80]. Perifosine inhibits the phosphorylation of AKT. It is also proved to be effective and enhances the radiosensitivity of cancer cells. Studies have shown that the AKT inhibitor MK2206 and the mTOR inhibitor rapamycin significantly decrease PD-L1 mRNA transcripts in PTEN-mutant triple-negative breast cancer cell lines [81]. Tregs are known to limit the activity of antitumor responses by the secretion of IL-10 and sequestration of IL-2. Recent studies showed that using Pan-AKT/PI3K inhibitors, Tregs are selectively downregulated, enhancing the number of CD8 + T cells within the tumor tissue, and hence improving the tumor control. Overall, few PI3K-AKT-mTOR pathway inhibitors have reached late phases of drug development and approval.

4 Concluding remarks

In the past few years, several studies on co-stimulatory and co-inhibitory molecules led to a better understanding of the involvement of immune cells in cancer progression as well as cancer evasion. Hence, immune therapeutics against cancer appeared to be a much better therapeutic strategy and many have already reported to achieve clinical success against few cancers. Also, current studies on checkpoint blockage showed the efficacy of using checkpoint inhibitors in many patients. In many of these pathways, the molecules interact with a vast number of receptors/ligands and cause downstream signaling. More molecules are discovered with unique functions. Therefore, designing a combination therapy targeting multiple pathways and TME can offer a better treatment strategy. Besides, combination of immunotherapy with other traditional therapies can give better results and can overcome the limitations of using any particular therapy.

References

[1] S.J. Oiseth, M.S. Aziz, Cancer immunotherapy: a brief review of the history, possibilities, and challenges ahead, J. Cancer Metastasis Treat. 3 (2017) 250–261, https://doi.org/10.20517/2394-4722.2017.41.

[2] D. Ribatti, The concept of immune surveillance against tumors. The first theories, Oncotarget 4 (2017) 7175–7180, https://doi.org/10.18632/oncotarget.12739.

[3] D.S. Chulpanova, K.V. Kitaeva, A.R. Green, A.A. Rizvanov, V.V. Solovyeva, Molecular aspects and future perspectives of cytokine-based anti-cancer immunotherapy, Front. Cell Dev. Biol. 8 (2020), https://doi.org/10.3389/fcell.2020.00402. Article 402.

[4] E. Vivier, S. Ugolini, D. Blaise, C. Chabannon, L. Brossay, Targeting natural killer cells and natural killer T cells in cancer, Nat. Rev. Immunol. 12 (2012) 239–252, https://doi.org/10.1038/nri3174.

[5] N. Jørgensen, G. Persson, T.V.F. Hviid, The tolerogenic function of regulatory T cells in pregnancy and cancer, Front. Immunol. 10 (2019), https://doi.org/10.3389/fimmu.2019.00911. Article 911.

[6] A.M. Martelli, et al., The emerging multiple roles of nuclear Akt, Biochim. Biophys. Acta, Mol. Cell Res. 1823 (2012) 2168–2178, https://doi.org/10.1016/j.bbamcr.2012.08.017.

[7] I.Y. Filin, V.V. Solovyeva, K.V. Kitaeva, C.S. Rutland, A.A. Rizvanov, Current trends in cancer immunotherapy, Biomedicines 12 (2020) 621, https://doi.org/10.3390/biomedicines8120621.

[8] S. Azoury, D. Straughan, V. Shukla, Immune checkpoint inhibitors for cancer therapy: clinical efficacy and safety, Curr. Cancer Drug Targets 6 (2015) 452–462, https://doi.org/10.2174/156800961506150805145120.

[9] D.M. Altmann, A Nobel prize-worthy pursuit: cancer immunology and harnessing immunity to tumour neoantigens, Immunology 3 (2018) 283–284, https://doi.org/10.1111/imm.13008.

[10] L.E. Cano, D.E. Lopera, Introduction to T and B lymphocytes, in: Autoimmunity From Bench to Bedside, El Rosario University Press, Bogota, Colombia, 2013.

[11] A.W. Goldrath, M.J. Bevan, Selecting and maintaining a diverse T-cell repertoire, Nature 6759 (1999) 255–262, https://doi.org/10.1038/46218.

[12] E.I. Buchbinder, A. Desai, CTLA-4 and PD-1 pathways similarities, differences, and implications of their inhibition, Am. J. Clin. Oncol.: Cancer Clin. Trials 39 (2016) 98–106, https://doi.org/10.1097/COC.0000000000000239.

[13] R.J. Greenwald, G.J. Freeman, A.H. Sharpe, The B7 family revisited, Annu. Rev. Immunol. 23 (2005) 515–548, https://doi.org/10.1146/annurev.immunol.23.021704.115611.

[14] B.T. Fife, J.A. Bluestone, Control of peripheral T-cell tolerance and autoimmunity via the CTLA-4 and PD-1 pathways, Immunol. Rev. 22 (2008) 166–182, https://doi.org/10.1111/j.1600-065X.2008.00662.x.

[15] M.F. Krummel, J.P. Allison, CD28 and CTLA-4 have opposing effects on the response of T ceils to stimulation, J. Exp. Med. 2 (1995) 459–465, https://doi.org/10.1084/jem.182.2.459.

[16] C.A. Chambers, M.S. Kuhns, J.G. Egen, J.P. Allison, CTLA-4-mediated inhibition in regulation of T cell responses: mechanisms and manipulation in tumor immunotherapy, Annu. Rev. Immunol. 19 (2001) 565–594, https://doi.org/10.1146/annurev.immunol.19.1.565.

[17] A.V. Collins, et al., The interaction properties of costimulatory molecules revisited, Immunity 2 (2002) 201–210, https://doi.org/10.1016/S1074-7613(02)00362-X.

[18] M.E. Keir, M.J. Butte, G.J. Freeman, A.H. Sharpe, PD-1 and its ligands in tolerance and immunity, Annu. Rev. Immunol. 26 (2008) 677–704, https://doi.org/10.1146/annurev.immunol.26.021607.090331.

[19] J.G. Egen, M.S. Kuhns, J.P. Allison, CTLA-4: new insights into its biological function and use in tumor immunotherapy, Nat. Immunol. 7 (2002) 611–618, https://doi.org/10.1038/ni0702-611.

[20] R.V. Parry, et al., CTLA-4 and PD-1 receptors inhibit T-cell activation by distinct mechanisms, Mol. Cell. Biol. 21 (2005) 9543–9553, https://doi.org/10.1128/mcb.25.21.9543-9553.2005.

[21] P.S. Linsley, et al., Intracellular trafficking of CTLA-4 and focal localization towards sites of TCR engagement, Immunity 6 (1996) 535–543, https://doi.org/10.1016/S1074-7613(00)80480-X.

[22] M.F. Krummel, J.P. Allison, CTLA-4 engagement inhibits IL-2 accumulation and cell cycle progression upon activation of resting T cells, J. Exp. Med. 6 (1996) 2533–2540, https://doi.org/10.1084/jem.183.6.2533.

[23] F. Bennett, et al., Program Death-1 engagement upon TCR activation has distinct effects on costimulation and cytokine-driven proliferation: attenuation of ICOS, IL-4, and IL-21, but not CD28, IL-7, and IL-15 responses, J. Immunol. 2 (2003) 711–718, https://doi.org/10.4049/jimmunol.170.2.711.

[24] O. Hamid, et al., Safety and tumor responses with Lambrolizumab (anti–PD-1) in melanoma, N. Engl. J. Med. 2 (2013) 134 144, https://doi.org/10.1056/nejmoa1305133.

[25] J. Brahmer, et al., Nivolumab versus docetaxel in advanced squamous-cell non–small-cell lung cancer, N. Engl. J. Med. 373 (2015) 123–135, https://doi.org/10.1056/nejmoa1504627.

[26] D. Gallacher, et al., Pembrolizumab for previously treated advanced or metastatic urothelial cancer: an evidence review group perspective of a NICE single technology appraisal, PharmacoEconomics 1 (2019) 19–27, https://doi.org/10.1007/s40273-018-0689-3.

[27] M. Tahara, et al., Pembrolizumab in Asia-Pacific patients with advanced head and neck squamous cell carcinoma: analyses from KEYNOTE-012, Cancer Sci. 3 (2018) 771–776, https://doi.org/10.1111/cas.13480.

[28] E.J. Lipson, C.G. Drake, Ipilimumab: An anti-CTLA-4 antibody for metastatic melanoma, Clin. Cancer Res. 22 (2011) 6958–6962, https://doi.org/10.1158/1078-0432.CCR-11-1595.

[29] J.C. Yang, et al., Ipilimumab (anti-CTLA4 antibody) causes regression of metastatic renal cell cancer associated with enteritis and hypophysitis, J. Immunother. 8 (2007) 825–830, https://doi.org/10.1097/CJI.0b013e318156e47e.

[30] S.M. Ansell, et al., Phase I study of ipilimumab, an anti-CTLA-4 monoclonal antibody, in patients with relapsed and refractory B-cell non-Hodgkin lymphoma, Clin. Cancer Res. 20 (2009) 6446–6453, https://doi.org/10.1158/1078-0432.CCR-09-1339.

[31] K.Y. Chung, et al., Phase II study of the anti-cytotoxic T-lymphocyte-associated antigen 4 monoclonal antibody, tremelimumab, in patients with refractory metastatic colorectal cancer, J. Clin. Oncol. 21 (2010) 3485–3490, https://doi.org/10.1200/JCO.2010.28.3994.

[32] B. Sangro, et al., A clinical trial of CTLA-4 blockade with tremelimumab in patients with hepatocellular carcinoma and chronic hepatitis C, J. Hepatol. 1 (2013) 81–88, https://doi.org/10.1016/j.jhep.2013.02.022.

[33] J. Weber, et al., Randomized phase I pharmacokinetic study of ipilimumab with or without one of two different chemotherapy regimens in patients with untreated advanced melanoma, Cancer Immun. 13 (2013) 7.

[34] T.J. Lynch, et al., Ipilimumab in combination with paclitaxel and carboplatin as first-line treatment in stage IIIB/IV non-small-cell lung cancer: results from a randomized, double-blind, multicenter phase II study, J. Clin. Oncol. 17 (2012) 2046–2054, https://doi.org/10.1200/JCO.2011.38.4032.

[35] M.J. Smyth, H. Yagita, G.A. Mcarthur, Combination anti-CTLA-4 and anti-RANKL in metastatic melanoma, J. Clin. Oncol. 12 (2014) e104–e106, https://doi.org/10.1200/JCO.2013.51.3572.

[36] P. Waterhouse, et al., Lymphoproliferative disorders with early lethality in mice deficient in Ctla-4, Science 5238 (1995) 985–988, https://doi.org/10.1126/science.270.5238.985.

[37] I. Melero, et al., Nivolumab dose escalation and expansion in patients with advanced hepatocellular carcinoma (HCC): the CheckMate 040 study, J. Clin. Oncol. 35 (2017), https://doi.org/10.1200/jco.2017.35.4_suppl.226.

[38] D.A. Fruman, C. Rommel, PI3K and cancer: lessons, challenges and opportunities, Nat. Rev. Drug Discov. 2 (2014) 140–156.

[39] B. Vanhaesebroeck, L. Stephens, P. Hawkins, PI3K signalling: the path to discovery and understanding, Nat. Rev. Mol. Cell Biol. 13 (2012) 195–203, https://doi.org/10.1038/nrm3290.

[40] J.A. Engelman, J. Luo, L.C. Cantley, The evolution of phosphatidylinositol 3-kinases as regulators of growth and metabolism, Nat. Rev. Genet. 8 (2006) 606–619, https://doi.org/10.1038/nrg1879.

[41] F. Conciatori, et al., MTOR cross-talk in cancer and potential for combination therapy, Cancers 1 (2018) 23, https://doi.org/10.3390/cancers10010023.

[42] E. Jacinto, et al., Mammalian TOR complex 2 controls the actin cytoskeleton and is rapamycin insensitive, Nat. Cell Biol. 11 (2004) 1122–1128, https://doi.org/10.1038/ncb1183.

[43] V. Thomanetz, et al., Ablation of the mTORC2 component rictor in brain or Purkinje cells affects size and neuron morphology, J. Cell Biol. 2 (2013) 293–308, https://doi.org/10.1083/jcb.201205030.

[44] D.S. Dos, et al., Rictor, a novel binding partner of mTOR, defines a rapamycin-insensitive and raptor-independent pathway that regulates the cytoskeleton, Curr. Biol. 14 (2004) 1296–1302, https://doi.org/10.1016/j.cub.2004.06.054.

[45] N. Divecha, R.F. Irvine, Phospholipid signaling, Cell 2 (1995) 269–278, https://doi.org/10.1016/0092-8674(95)90409-3.

[46] B.M.T. Burgering, P.J. Coffer, Protein kinase B (c-Akt) in phosphatidylinositol-3-OH kinase signal transduction, Nature 6541 (1995) 599–602, https://doi.org/10.1038/376599a0.

[47] T.F. Franke, et al., The protein kinase encoded by the Akt proto-oncogene is a target of the PDGF-activated phosphatidylinositol 3-kinase, Cell 81 (1995) 727–736, https://doi.org/10.1016/0092-8674(95)90534-0.

[48] M. Milella, et al., PTEN: multiple functions in human malignant tumors, Front. Oncol. 5 (2015), https://doi.org/10.3389/fonc.2015.00024. Article 24.

[49] L. Ciuffreda, J. McCubrey, M. Milella, Signaling intermediates (PI3K/PTEN/AKT/mTOR and RAF/MEK/ERK pathways) as therapeutic targets for anti-cancer and anti-angiogenesis treatments, Curr. Signal Transduction Ther. 4 (2009) 130–143, https://doi.org/10.2174/157436209788167466.

[50] E.T. Shinohara, et al., Enhanced radiation damage of tumor vasculature by mTOR inhibitors, Oncogene 35 (2005) 5414–5422, https://doi.org/10.1038/sj.onc.1208715.

[51] X. Song, et al., ERBB3-independent activation of the PI3K pathway in EGFR-mutant lung adenocarcinomas, Cancer Res. 6 (2015) 1035–1045, https://doi.org/10.1158/0008-5472.CAN-13-1625.

[52] B. Karakas, K.E. Bachman, B.H. Park, Mutation of the PIK3CA oncogene in human cancers, Br. J. Cancer 4 (2006) 455–459, https://doi.org/10.1038/sj.bjc.6602970.

[53] K.H. Yi, J. Lauring, Recurrent AKT mutations in human cancers: functional consequences and effects on drug sensitivity, Oncotarget 4 (2016) 4241–4251, https://doi.org/10.18632/oncotarget.6648.

[54] P.A. Steck, et al., Identification of a candidate tumour suppressor gene, MMAC1, at chromosome 10q23.3 that is mutated in multiple advanced cancers, Nat. Genet. 4 (1997) 356–362, https://doi.org/10.1038/ng0497-356.

[55] J. Li, et al., PTEN, a putative protein tyrosine phosphatase gene mutated in human brain, breast, and prostate cancer, Science 5308 (1997) 1943–1947, https://doi.org/10.1126/science.275.5308.1943.

[56] D. Hanahan, L.M. Coussens, Accessories to the crime: functions of cells recruited to the tumor microenvironment, Cancer Cell 21 (2012) 309–322, https://doi.org/10.1016/j.ccr.2012.02.022.

[57] F. Chen, et al., New horizons in tumor microenvironment biology: challenges and opportunities, BMC Med. 13 (2015), https://doi.org/10.1186/s12916-015-0278-7. Article 45.

[58] F.R. Balkwill, M. Capasso, T. Hagemann, The tumor microenvironment at a glance, J. Cell Sci. 23 (2012), https://doi.org/10.1242/jcs.116392. Article 5591-6.

[59] B. Chaudhary, E. Elkord, Regulatory T cells in the tumor microenvironment and cancer progression: role and therapeutic targeting, Vaccine 3 (2016) 28, https://doi.org/10.3390/vaccines4030028.

[60] H. Zeng, mTOR signaling in immune cells and its implications for cancer immunotherapy, Cancer Lett. 408 (2017) 182–189, https://doi.org/10.1016/j.canlet.2017.08.038.

[61] V. Byles, et al., The TSC-mTOR pathway regulates macrophage polarization, Nat. Commun. 4 (2013) 2834, https://doi.org/10.1038/ncomms3834.

[62] M.C. Schmid, et al., Receptor tyrosine kinases and TLR/IL1Rs unexpectedly activate myeloid cell PI3Kγ, A single convergent point promoting tumor inflammation and progression, Cancer Cell 6 (2011) 715–727, https://doi.org/10.1016/j.ccr.2011.04.016.

[63] K. Araki, R. Ahmed, AMPK: a metabolic switch for CD8+T-cell memory, Eur. J. Immunol. 4 (2013) 878 881, https://doi.org/10.1002/eji.201343483.

[64] R. Abu-Eid, et al., Selective inhibition of regulatory T cells by targeting the PI3K-Akt pathway, Cancer Immunol. Res. 11 (2014) 1080–1089, https://doi.org/10.1158/2326-6066.CIR-14-0095.

[65] J.R. Brahmer, et al., Safety and activity of anti–PD-L1 antibody in patients with advanced cancer, N. Engl. J. Med. 26 (2012) 2455–2465, https://doi.org/10.1056/nejmoa1200694.

[66] M. Pancione, et al., Emerging insight into MAPK inhibitors and immunotherapy in colorectal cancer, Curr. Med. Chem. 14 (2017) 1383–1402, https://doi.org/10.2174/0929867324666170227114356.

[67] J.D. Wolchok, et al., Nivolumab plus Ipilimumab in advanced melanoma, N. Engl. J. Med. 377 (2013) 2503–2504, https://doi.org/10.1056/nejmoa1302369.
[68] J.S. O'Donnell, D. Massi, M.W.L. Teng, M. Mandala, PI3K-AKT-mTOR inhibition in cancer immunotherapy, redux, Semin. Cancer Biol. 48 (2018) 91–103, https://doi.org/10.1016/j.semcancer.2017.04.015.
[69] G. Xue, et al., Integrated Akt/PKB signaling in immunomodulation and its potential role in cancer immunotherapy, J. Natl. Cancer Inst. 107 (7) (2015), https://doi.org/10.1093/jnci/djv171.
[70] J.A. Engelman, Targeting PI3K signalling in cancer: opportunities, challenges and limitations, Nat. Rev. Cancer 8 (2009) 550–562, https://doi.org/10.1038/nrc2664.
[71] L.H. Meng, X.S. Zheng, Toward rapamycin analog (rapalog)-based precision cancer therapy, Acta Pharmacol. Sin. 10 (2015) 1163–1169, https://doi.org/10.1038/aps.2015.68.
[72] S. Brakemeier, et al., Everolimus in de novo kidney transplant recipients participating in the Eurotransplant senior program: results of a prospective randomized multicenter study (SENATOR), PLoS One 9 (2019), https://doi.org/10.1371/journal.pone.0222730, e0222730.
[73] R.J. Motzer, et al., Efficacy of everolimus in advanced renal cell carcinoma: a double-blind, randomised, placebo-controlled phase III trial, Lancet 9637 (2008) 449–456, https://doi.org/10.1016/S0140-6736(08)61039-9.
[74] T.E. Hutson, et al., Randomized phase III trial of temsirolimus versus sorafenib as second-line therapy after sunitinib in patients with metastatic renal cell carcinoma, J. Clin. Oncol. 8 (2014) 760–767, https://doi.org/10.1200/JCO.2013.50.3961.
[75] F. Viñals, J.C. Chambard, J. Pouysségur, p70 S6 kinase-mediated protein synthesis is a critical step for vascular endothelial cell proliferation, J. Biol. Chem. 274 (1999) 26776–26782, https://doi.org/10.1074/jbc.274.38.26776.
[76] O. Dormond, J.C. Madsen, D.M. Briscoe, The effects of mTOR-Akt interactions on anti-apoptotic signaling in vascular endothelial cells, J. Biol. Chem. 32 (2007) 23679–23686, https://doi.org/10.1074/jbc.M700563200.
[77] A. Duarte, G.G. Silveira, D.F. Soave, J.P.O. Costa, A.R. Silva, The role of the LY294002-a non-selective inhibitor of phosphatidylinositol 3-kinase (PI3K) pathway-in cell survival and proliferation in cell line SCC-25, Asian Pac. J. Cancer Prev. 11 (2019) 3377–3383, https://doi.org/10.31557/APJCP.2019.20.11.3377.
[78] A. Bavelloni, et al., Therapeutic potential of nvp-bkm120 in human osteosarcomas cells, J. Cell. Physiol. 7 (2019) 10907–10917, https://doi.org/10.1002/jcp.27911.
[79] J.D. Hainsworth, et al., Phase I/II study of bevacizumab with BKM120, an oral PI3K inhibitor, in patients with refractory solid tumors (phase I) and relapsed/refractory glioblastoma (phase II), J. Neuro-Oncol. 2 (2019) 303–311, https://doi.org/10.1007/s11060-019-03227-7.
[80] M.A. Weinberg, RES-529: A PI3K/AKT/mTOR pathway inhibitor that dissociates the mTORC1 and mTORC2 complexes, Anti-Cancer Drugs 6 (2016) 475–487, https://doi.org/10.1097/CAD.0000000000000354.
[81] E.A. Mittendorf, et al., PD-L1 expression in triple-negative breast cancer, Cancer Immunol. Res. 4 (2014) 361–370, https://doi.org/10.1158/2326-6066.CIR-13-0127.

CHAPTER

4

Stroke risk assessment and atrial fibrillation (AF) in cancer patients

Saima Wajid and Sana Kauser

Department of Biotechnology, School of Chemical and Life Sciences, Jamia Hamdard, New Delhi, India

1 Introduction

AF is the most common cardiac arrhythmia with the prevalence of 2%–4% in the general population and an expected 2.3-fold increment with age [1]. AF is associated with several cardiovascular conditions such as systemic embolic complications (including ischemic stroke), heart failure, hypertension, and noncardiovascular diseases, e.g., chronic pulmonary disease, thyroid disorder, diabetes, electrolyte abnormalities, sepsis, and obstructive sleep apnea [1]. AF tends to increase overall stroke risk by 5-fold, thromboembolism risk by 2-fold, and mortality risk by 3.5-fold [2,3]. As stroke risk is heterogeneous in nature that depends on specific risk modifiers, some prediction scores have been developed for the prediction of AF-associated ischemic stroke risk among AF patients. CHADS2 score is one of those prediction tools which is based on the history of other stroke risk factors including congestive heart failure [1 point], hypertension [1 point], age $\geq$75 [1 point], diabetes mellitus [1 point], and previous stroke/transient ischemic attack (TIA) [2 point] among AF patients [4]. Another more reliable score is the CHA2DS2-VASc (Congestive heart failure [1 point], Hypertension [1 point], Age [75 years] [2 point], Diabetes [1 point], Stroke [2 point], Vascular disease [1 point], Age [65–74 years] [1 point], Sex category [1 point]) score that is used for stratifying ischemic stroke risk among AF patients which allows selection of patients thought to be at high risk of stroke [5]. These scores are calculated on the basis of the presence of these risk factors in patients.

Cardiovascular Toxicity and Therapeutic Modalities Targeting Cardio-oncology
https://doi.org/10.1016/B978-0-323-90461-2.00002-X

Cancer is a well-known prothrombotic or hypercoagulable state due to the release of fibrinolytic agents and proinflammatory cytokines, and around 2%–5% cancer patients have AF at the time of diagnosis [6,7]. AF is one of the major risk factors causing ischemic stroke through venous thromboembolism (VTE) from the heart [1]. Recently, several studies have been carried out to investigate the incidence and risk of arterial thromboembolism (ATE) in cancer patients that increases the risk of myocardial infarction, ischemic stroke, and mortality all together. In a Danish population-based study and SEER (Surveillance, Epidemiology, and End Results)-based retrospective study, cancer patients had twice the risk of ATE, myocardial infarction, and ischemic stroke and thrice the risk of mortality within a period of 6–12 months. However, the risk of thromboembolic complications varied by age and different cancer type with lung, gastric, pancreatic, and bladder cancer patients pertaining to highest risk [8,9]. Aspberg et al. prompted higher occurrence of embolic events to be accountable for ischemic stroke in cancer patients accompanied by AF than nonembolic events, such that hypercoagulability or hemostatic disturbances in cancer may directly cause either cerebral arterial thrombus formation or left atrial/left atrial appendage thrombus formation in AF giving prominence to embolic stroke [10]. According to a study from Taiwan during January 1, 2000, to May 31, 2009, cancer patients with new-onset AF were at higher risk of thromboembolism and heart failure, and the study proposed higher CHADS2 score to be used as a predictive value for thromboembolism in cancer patients with baseline AF (AF that occurred before or at the same time of cancer diagnosis), but not in cancer patients with new-onset AF [6]. Similarly, higher value of CHADS2 score was found to be associated with high risk of stroke and mortality among cancer patients with preexisting AF [11,12]. A Danish population-based study reported lower CHA2DS2-VASc score to be associated with higher risk of thromboembolism in cancer and AF patients [13]. On the contrary, cancer patients with concomitant AF having CHA2DS2-VASc score > 5 had a 1.82-fold higher risk of ischemia stroke than non-AF cancer patients [14]. However, the prediction value of CHADS2 and CHA2DS2-VASc scores is under scrutiny to assess the risk of ischemia stroke among cancer patients with concomitant AF. Therefore, the efficacy of these clinical prediction tools for stroke risk assessment needs to be validated in a large population of oncologic patients with and without preexisting AF.

2 Association between cancer and atrial fibrillation (AF)

Several studies have elaborated the epidemiological evidence for the relationship between AF and cancer and suggested the possibility of development of either one after the diagnosis of the other. The bulk of studies revealed that surgical treatment of intrathoracic malignancies is associated with elevated risk of postoperative AF than nonthoracic malignancies with unclear etiology. The incidence of postoperative AF after pulmonary resection (thoracic surgery) in lung cancer patients has been estimated to be as high as 20–30% after a pneumonectomy [15]. Another primitive study reported the incidence of AF in patients with lung cancer and benign pulmonary diseases following thoracotomy to be increased by 19.6% and 3.1%, respectively ($P<.005$) [16]. In patients undergoing nonthoracic surgery for colorectal cancer and esophageal cancer, the incidence of postoperative AF was reported to be 4.4% after elective colectomy and 10.8% after esophagectomy [17,18]. An increased level of postoperative N-terminal pro-B-type natriuretic peptide (NT-proBNP) in patients undergoing thoracic surgery for lung cancer was found to be associated with higher incidence of AF (64% vs. 5% as compared with normal values; $P<.001$) and can be used as a predictor risk factor for the same [19]. Preoperative echocardiographic indexes such as mitral E/e′ ratio with a value greater than 8 were found to be highly sensitive and can be used as a predictor for postoperative AF in patients undergoing pulmonary resection for lung cancer [20]. Perioperative dosing of beta blockers such as metoprolol, landiolol, omega-3 fatty acids, and statins to the patients tends to reduce the risk of AF. More investigation is required to corroborate whether cancer is a contributing risk factor for the development of postoperative AF or not as a large portion of the studies had no direct comparison with noncancer patients undergoing the same surgical procedures.

Predisposition of AF with cancer being a risk factor is multifactorial, such as resultant adverse effects of chemotherapeutic drugs including alkylating agents (cisplatin, cyclophosphamide, ifosfamide, melphalan), anthracycline agents, topoisomerase II inhibitors, radiation, surgery, other cancer targeted therapies, and cancer-induced inflammation [21]. Chemotherapy causes cardiac dysfunction due to oxidative stress, myocyte damage, pericarditis, left ventricular dysfunction, heart failure, and diastolic dysfunction that finally contributes to the occurrence of AF [22]. With a synergistic effect, radiotherapy along with cardiotoxic chemotherapy increases the risk of AF through interstitial myocardial fibrosis and diastolic and systolic dysfunction [23]. Another potential pathophysiology behind the development of AF in cancer patients could be the prevalence of excess sympathetic activity or autonomic dysregulation due to pain, stress, metabolic irregularities, weight gain, sleep disturbances, and antineoplastic therapy in some subtypes of cancer [24,25]. This autonomic nervous system activation may induce significant changes in atrial electrophysiology and cause the initiation of AF by both reentry and triggered activity, probably through calcium-mediated mechanisms [26]. At the same time, alterations in atrial electrophysiology owing to AF development are often induced by disturbance in autonomic tone from both vagal and sympathetic influences [27]. AF may also be a consequence of the neoplastic behavior of primary or metastatic cardiac tumors or invasion of tumor to heart from adjacent tissues such as the lungs and esophagus [28].

AF in cancer can be manifested as a comorbidity due to additional circumstances such as hypoxia, advanced age, electrolyte abnormalities, and metabolic disorders [29]. Meanwhile, some recent studies have demonstrated associations between AF and cancer which were not surgery- or therapy-centric. An OPERA (Oulu Project Elucidating Risk of Atherosclerosis) study including middle-aged hypertensive subjects and age- and sex-matched controls ($N=1045$) was conducted to determine cancer as an independent risk factor for predisposition of AF. During a follow-up period of 16.3 years, cancer patients had higher probability of developing AF with a HR of 2.47 than noncancer patients [30]. A case–control study of 1247 patients admitted for surgical treatment of colorectal cancer without any previous cancer history or cancer therapy reported twice the risk of AF among cancer patients than among healthy controls with an adjusted odds ratio (OR) of 3.5 [31]. Another case-control study was conducted by the same group to evaluate the incidence of AF in patients with colorectal or breast cancer who never had cancer treatment ($N=1317$). On comparison of their electrocardiogram, cancer patients had a greater-than-twofold increased probability of having AF (3.6%) as compared with controls (1.6%) (OR = 3.3) [32]. A population-based case-control study including new-onset AF and freshly diagnosed colorectal cancer patients in northern Denmark during 1998–2006 reported higher risk of AF development among colorectal cancer patients than among controls (OR = 11.8) after 3 months of cancer diagnosis [33].

On the other hand, a large body of data suggested an inverse relationship such that AF could be a risk factor for the development of cancer as well as a marker for occult cancer. A cohort study based on the data from the Danish National Registry ($N=269{,}742$) demonstrated significantly increased probability of cancer diagnosis among newly diagnosed AF patients within 3 months of follow-up as 6656 patients developed cancer with an absolute risk of 2.5% (95% CI, 2.4%–2.5%) and standardized incidence ratios (SIRs) of 5.11. Also, within the 3 months of follow-up, relative risk of cancer diagnosis was clearly high for all types of cancer, but was even more prominent for lung, kidney, and colon cancers [34]. Another large prospective cohort study of 93,676 postmenopausal women by the Women's Health Initiative (WHI) from 1994 to 1998 has reported a 19% increased risk of invasive breast cancer among older women with AF. Though this study was unable to determine any significant association between AF and colorectal cancer incidence, it suggested the baseline use of cardiac glycosides to be associated by a 68% higher risk of invasive breast cancer with hazard ratio (HR) = 1.68 [35]. A large prospective cohort trial ($N=34{,}691$ women) of aspirin and vitamin E for the prevention of cardiovascular disease and cancer was done from 1993 to 2014 that illustrated the relationship between higher risk of subsequent cancer incidence and new onset of AF after multivariable adjustment (HR 1.48; 95% CI, $P < .001$). The risk was highest within the first 3 months of AF diagnosis (incidence, 3.8 per 100 person-years; adjusted HR: 3.54; 95% CI, $P < .001$) that remained significant during the past 1 year of AF diagnosis (incidence, 1.4 per 100 person-years; adjusted HR: 1.42; 95% CI, $P < .001$). The median follow-up period was 19.1 years, during which new-onset AF and malignant cancer were confirmed among 4.2% ($N = 1467$) and 14.8% ($N = 5130$) of individuals, respectively [36]. Similarly, another population-based cohort study including men ($N = 26{,}222$) and women ($N = 28{,}879$) registered in the Danish National Patient Registry and Danish Cancer Registry with a median follow-up period of 16–19 years reported a higher incidence rate of any type of cancer within the first 3 months after AF diagnosis (men: HR 2.89, 95% CI; women: HR 3.72, 95% CI). However, the risks of lung cancer (men: HR 7.70, 95% CI; women: HR 7.98, 95% CI) and colorectal cancer (men: HR 3.35, 95% CI; women: HR 5.91, 95% CI) were highest within the first 3 months after AF and remained higher during the initial 1 year for both sexes [37]. Two large prospective population-based case-control studies (Molecular Epidemiology of Colorectal Cancer (MECC), $N = 8383$; and Breast Cancer in Northern Israel Study (BCINIS), $N = 11{,}608$) reported a 1.85 times increased risk of cancer within the immediate period of 3 months after AF diagnosis in patients. Meanwhile, the risk of AF was 3.4-fold higher during the first 3 months of cancer diagnosis among patients [38]. Lastly, a meta-analysis suggests that available data still require strong evidence to justify the presence of occult cancer within 3 months of AF diagnosis [2].

Inflammation is an integral part of tumor microenvironment that increases the production of reactive oxygen species (ROS) and accelerates oxidative stress during cancer. Concurrently, the presence of these ROS and oxidative damage can bring about substantial atrial electrophysiology and structural remodeling leading to the initiation and progression of AF [39]. Oxidative stress through inflammatory cells is a common pathological process between AF and cancer. An inflammatory marker, C-reactive protein (CRP), whose circulatory level is elevated in many cancer, is also closely associated with the presence and future risk of developing AF [40,41]. Both inflammation and AF-related thrombogenesis are found to be associated with atrial fibrosis, left atrial thrombus formation, and dense spontaneous echo contrast, which, in turn, are associated with elevated CRP levels in AF patients. Moreover, atrial and ventricular biopsies of AF patients have displayed the presence of inflammatory infiltrates [42]. Therefore, inflammatory status of cancer patient can be crucial for pathogenesis and further progression of AF while preexisting AF before cancer diagnosis may result from chronic inflammation when cancer is

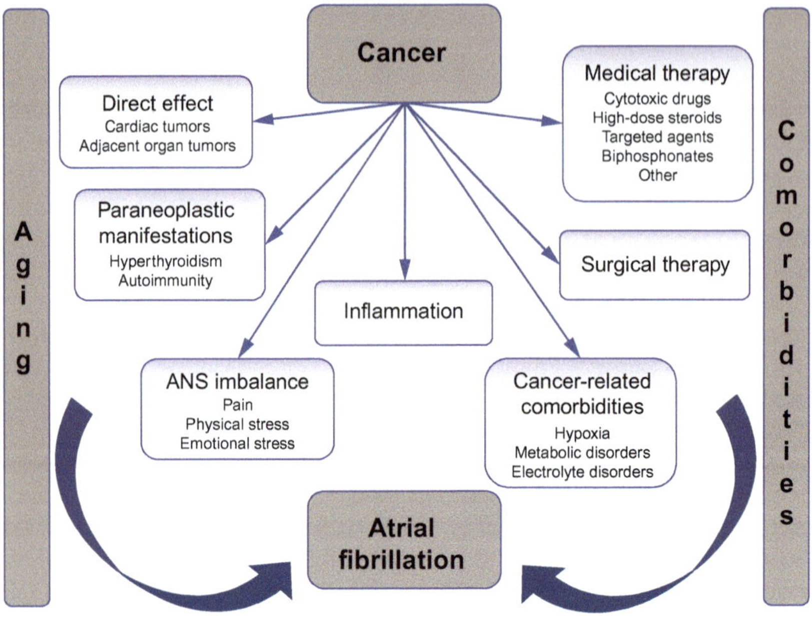

FIG. 1 Outline of different factors responsible for AF pathogenesis in cancer. *Adapted from D. Farmakis, J. Parissis, G. Filippatos, Insights into onco-cardiology: atrial fibrillation in cancer, J. Am. Coll. Cardiol. 63 (10) (2014) 945–953.*

in early subclinical developmental phase. AF in cancer patients can also be attributed to paraneoplastic disorders including hyperthyroidism and autoimmune reaction against atrial structures [43]. An outline of potential sources related to cancer-associated AF pathogenesis is depicted in Fig. 1.

3 Management and treatment modalities of AF

Management of AF includes an emphasis on two major perspectives, i.e., thromboembolic prophylaxis and effective stroke prevention with oral anticoagulation. On the basis of CHA2DS2-VASc score>_2 in men or>_3 in women and platelet count > 50,000/mm, AF patients in general population are recommended to be treated with oral anticoagulants (OACs) as the principal therapeutic approach for the stroke prevention [1]. But, treatment of AF with comorbid cancer poses a challenge for the selection of appropriate anticoagulants due to a plethora of risk factors associated with cancer such as VTE, nonembolic stroke, and bleeding due to local injury caused by tumor invasion, hematological malignancy-associated thrombocytopenia, chemotherapy aftermath, surgery or invasive procedures, and disseminated intravascular coagulation [44]. Few limited studies have demonstrated consistently higher bleeding risk (4% for the first 3 months, 4%–6% for the 6 months, and 10%–15% per year) in cancer patients undergoing anticoagulation that varies with varying cancer site, type, and stage (predominantly in gastrointestinal cancer) [45–47]. Prerequisition of balance between thromboembolic and bleeding risk complicates AF management in cancer patients. Therefore, decision on anticoagulants (OACs) for stroke prevention in cancer patients with comorbid AF should not only rely upon CHA2DS2-VASc and CHADS2 risk assessment scores that have not been validated for oncologic population. Rather, the European Society of Cardiology recommends complete individual assessment of the patient, that is evaluation of echocardiography, bleeding risk, and malignancy-related comorbidities are preferable while making decision on anticoagulation [23].

The choice of anticoagulant in the treatment of AF with comorbid cancer is a less explored topic in the literature. Three classes of drugs that are used in anticoagulation therapy include vitamin K antagonist (VKA; e.g., warfarin), non-VKA oral anticoagulant (NOAC), and low molecular weight heparin (LMWH). VKAs such as warfarin have been used as the prime therapeutic agent to reduce stroke or thromboembolic events in AF patients for many years. However, use of warfarin among cancer patients with concomitant AF tends to have certain limitations, e.g., drug-drug and food interaction via induction or inhibition of cytochrome P450 isozymes or disturbances in vitamin K status leads to difficulty in maintaining a stable international normalized ratio (INR) during chemotherapy which is already associated with improper food intake, nausea, or vomiting [48]. Moreover, cancer patients undergoing treatment with VKAs had a 3- to 6-fold increment in bleeding episodes [49]. Patients receiving chronic warfarin therapy that developed cancer had a significant reduction in percent time in therapeutic range in the first 6 months after cancer diagnosis [50]. Hence, their extended duration of action and delayed onset cause treatment cessation over invasive procedures and complicate thrombocytopenia management.

Non-vitamin K antagonist oral anticoagulants (NOACs) including factor Xa inhibitors (edoxaban, apixaban, and rivaroxaban) and direct thrombin inhibitor (dabigatran) have been recommended over VKAs by the European Society of Cardiology to be used in anticoagulation therapy among AF patients due to few advantages such as less drug-drug/food interaction and fast onset of action [1]. All NOACs are substrates for the excretory permeability glycoprotein (P-gp) system of the gastrointestinal epithelial cells, while apixaban and rivaroxaban are mainly metabolized via cytochrome P450 (CYP) 3A4 of the liver. Most of the chemotherapy drug interacts with P-gp system and cytochrome P450 (CYP) 3A4 as an inducer or inhibitor that influences the absorption and extent of anticoagulation by NOACs [48,51]. Utilization of NOACs along with drugs that are inhibitors or inducers of P-gp/CYP3A4 may result in unpredictable response toward anticoagulation. Thus, selection of the most suitable NOACs for cancer patient under treatment should be done with caution by keeping this limitation in mind. Several trials have revealed almost similar or better efficiency of NOACs as that of warfarin in the reduction of stroke in nonvalvular AF patients. Although clinical implication of NOACs in the management of cancer patients is still uncertain due to the scarcity and exclusion criteria of randomized clinical trials (RCTs) of NOACs as many patients with active cancer or lesser life expectancy were excluded, retrospective population-based cohort studies and post hoc analyses of RCTs have provided propitious evidences regarding the implication of NOACs among cancer patients for stroke prevention. ENGAGE AF-TIMI 48 (Effective Anticoagulation With Factor Xa Next Generation in Atrial Fibrillation-Thrombolysis in Myocardial Infarction Study 48) trial including patients with AF and malignancy ($N=1153$) demonstrated similar efficacy and safety of edoxaban as that of warfarin regardless of the development of any malignancy during the follow-up period [52]. The ARISTOTLE (Apixaban for Reduction in Stroke and Other Thromboembolic Events in Atrial Fibrillation) trial exploring efficacy, safety, and characteristic clinical outcome of apixaban vs. warfarin among patients with AF and concomitant cancer ($N=1236$) reported consistent apixaban and warfarin efficacy in patients with and without cancer. But, apixaban was found to have benefit over warfarin on ischemic composite endpoint in patients with active cancer [53]. Likewise, rivaroxaban was also found to be similarly effective and safe like warfarin on treating patients ($N=14{,}264$) with nonvalvular AF with or without cancer in the ROCKET AF (Rivaroxaban Once Daily Oral Direct Factor Xa Inhibition Compared with Vitamin K Antagonism for Prevention of Stroke and Embolism Trial in Atrial Fibrillation) randomized trial [54]. A retrospective population-based cohort study revealed better clinical outcomes (lesser ischemic stroke, major bleeding, and all-cause death) in the NOAC-treated group than in the warfarin group of nonvalvular AF patients with newly diagnosed cancer ($N=2568$) [55]. A comparative analysis comprising cancer patients with AF ($N=16{,}096$) identified in the MarketScan database presented similar rates of ischemic stroke and bleeding in NOAC and warfarin users, except for apixaban that reduced the risk of bleeding. Meanwhile, venous thromboembolism (VTE) complications were significantly reduced in NOAC (rivaroxaban, dabigatran, apixaban) users as compared to warfarin users [56]. Similarly, Sawant et al. reported a similar risk of ischemic stroke among NOAC (dabigatran, rivaroxaban, apixaban and edoxaban) and warfarin users. However, reductions in hemorrhagic stroke and mortality were significant among NOAC users [57]. Recently, Wu et al. demonstrated better effectiveness of NOACs (dabigatran, rivaroxaban, apixaban, or edoxaban) over warfarin such that cancer patients with AF ($N=336$) receiving NOACs had significantly lesser incidence of ischemic stroke, major bleeding, and intracranial hemorrhage [58]. The International Society on Thrombosis and Hemostasis guidelines also recommend that NOACs should be used over warfarin in cancer patients with new-onset AF, except if the patients have GI cancer or history of GI bleeding [59].

Randomized Comparison of Low-Molecular-Weight Heparin versus Oral Anticoagulant Therapy for the Prevention of Recurrent Venous Thromboembolism in Patients with Cancer (CLOT) investigators demonstrated better efficacy and performance of LMWH in reducing recurrent VTE in cancer patients, when compared with warfarin therapy [60]. After this CLOT trial, LMWH became the most preferred agent to treat cancer patients with VTE. However, LMWH-related few limitations complicate its further utilization, e.g., difficulty of self-injection and lack of reversal agent. Tafur et al. reported that cancer patients having warfarin interruption to undergo invasive procedure and subsequent bridging with LMWH increased risk of bleeding without decreasing VTE events in the patients [61]. Many retrospective and randomized trials have shown that LMWH increases the risk of major bleeding in AF patients when used as a bridging agent perioperatively while arterial thromboembolism risk remained unaffected [62,63]. Recently, meta-analysis of many randomized trials revealed that NOACs were associated lesser risk of VTE recurrence and similar risk of major bleeding in cancer patients, when compared with LMWH [64,65]. Given the lack of evidence of LMWH efficacy against AF in cancer patients, further investigation is required to be used in long-term anticoagulation therapy.

The American Society of Clinical Oncology proposed an algorithm (Fig. 2) for the management of AF in active cancer patients with the following recommendations:

- Estimation of thrombotic risk to aid patients with anticoagulation will solely depend on the prevalence of consistent stroke risk of 4%–5% per year without anticoagulation, and not on the risk of VTE, as several patients with a high risk of VTE tend to have high bleeding risks also.
- Patients with a CHA2DS2-VASc of 0 or 1 do not need to have anticoagulation. Meanwhile, patients with CHA2DS2-VASc of 2 or more need prior assessment of high bleeding-associated risk factors. Patients possessing high bleeding risk (any major or more than one minor risk factor) need to interrupt anticoagulation during chemotherapy, and restart later, once the reduction in the risk of thrombocytopenia as well as tumor burden is attained. At the same time, patients possessing a low bleeding risk, i.e., without any associated risk factor, will be suitable for anticoagulation. Patients with CHA2DS2-VASc of 6 or more (10% per year stroke risk) or possessing moderate risk, i.e., having only one minor risk factor, will be suitable to undergo anticoagulation.
- Cardiology discretion and follow-up are requisites for the management of new-onset AF in cancer patients. Emphasis should be given to resolve the triggering source instead of directly opting for anticoagulation.

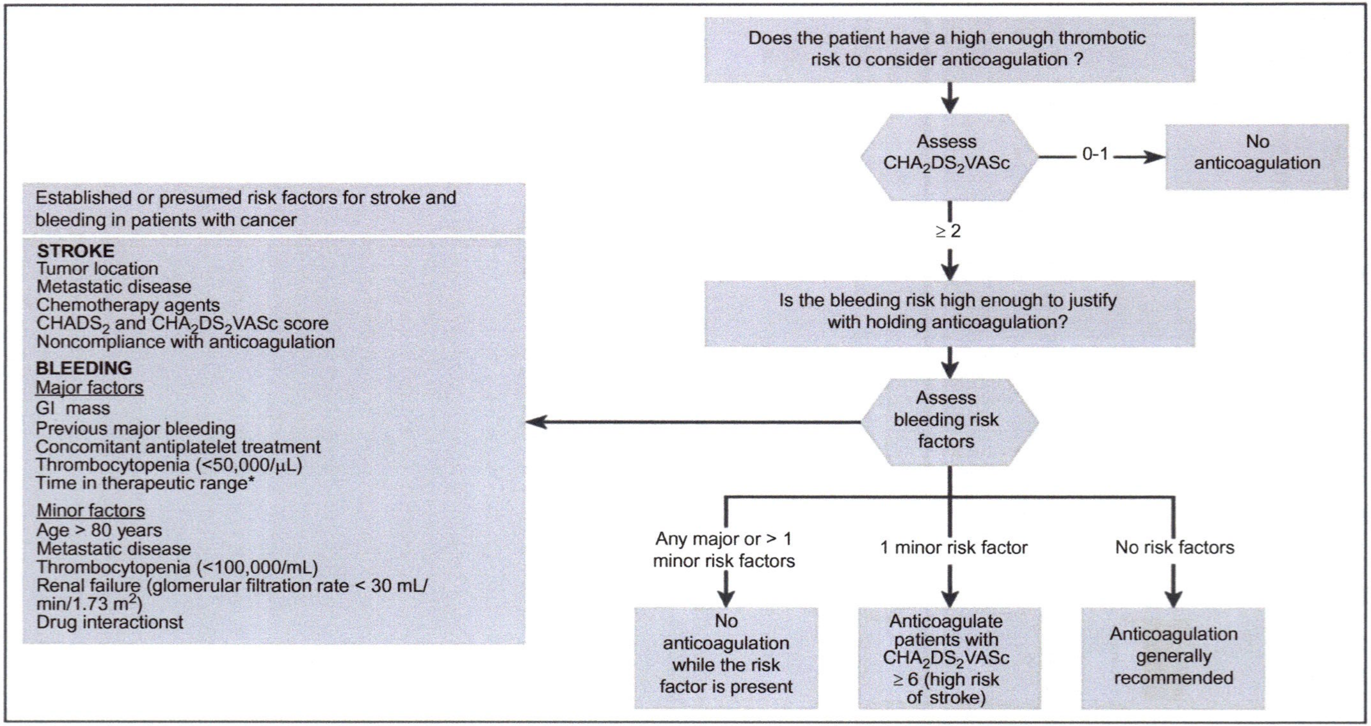

FIG. 2 Proposed algorithm for the management of anticoagulation among cancer patients with AF. *Adapted from M. Sorigue, M.D. Miljkovic, Atrial fibrillation and stroke risk in patients with cancer: a primer for oncologists, J. Oncol. Pract. 15 (12) (2019) 641–650.*

- Patients receiving anticoagulation could abstain from perioperative bridging. If suffering from major bleeding, anticoagulation may be terminated and underlying trigger (e.g., GI mass) should be resolved. It recommends prophylactic dose of LMWH to the cancer patient, once the bleeding has diminished, as the risk of VTE is high in the acute phase of the bleeding episode.
- AF patients may continue with the prescribed anticoagulant (either VKA or NOAC) even after cancer diagnosis, if anticoagulation is necessary. However, new-onset AF in cancer patients demands an individualized approach to choose appropriate anticoagulant, considering the risk/benefit range [7].

References

[1] G. Hindricks, T. Potpara, N. Dagres, E. Arbelo, J.J. Bax, C. Blomström-Lundqvist, et al., 2020 ESC guidelines for the diagnosis and management of atrial fibrillation developed in collaboration with the European Association for Cardio-Thoracic Surgery (EACTS) The Task Force for the diagnosis and management of atrial fibrillation of the European Society of Cardiology (ESC) Developed with the special contribution of the European Heart Rhythm Association (EHRA) of the ESC, Eur. Heart J. 42 (5) (2021) 373–498.

[2] N. Lateef, V. Kapoor, M.J. Ahsan, A. Latif, U. Ahmed, M. Mirza, et al., Atrial fibrillation and cancer; understanding the mysterious relationship through a systematic review, J. Community Hosp. Intern. Med. Perspect. 10 (2) (2020) 127–132.

[3] C. Magnussen, T.J. Niiranen, F.M. Ojeda, F. Gianfagna, S. Blankenberg, I. Njølstad, et al., Sex differences and similarities in atrial fibrillation epidemiology, risk factors, and mortality in community cohorts: results from the BiomarCaRE consortium (Biomarker for Cardiovascular Risk Assessment in Europe), Circulation 136 (17) (2017) 1588–1597.

[4] B.F. Gage, A.D. Waterman, W. Shannon, M. Boechler, M.W. Rich, M.J. Radford, Validation of clinical classification schemes for predicting stroke: results from the national registry of atrial fibrillation, JAMA 285 (22) (2001) 2864–2870.

[5] G.Y. Lip, R. Nieuwlaat, R. Pisters, D.A. Lane, H.J. Crijns, Refining clinical risk stratification for predicting stroke and thromboembolism in atrial fibrillation using a novel risk factor-based approach: the euro heart survey on atrial fibrillation, Chest 137 (2) (2010) 263–272.

[6] Y.F. Hu, C.J. Liu, P.M.H. Chang, H.M. Tsao, Y.J. Lin, S.L. Chang, et al., Incident thromboembolism and heart failure associated with new-onset atrial fibrillation in cancer patients, Int. J. Cardiol. 165 (2) (2013) 355–357.

[7] M. Sorigue, M.D. Miljkovic, Atrial fibrillation and stroke risk in patients with cancer: a primer for oncologists, J. Oncol. Pract. 15 (12) (2019) 641–650.

[8] B.B. Navi, A.S. Reiner, H. Kamel, C. Iadecola, P.M. Okin, M.S. Elkind, et al., Risk of arterial thromboembolism in patients with cancer, J. Am. Coll. Cardiol. 70 (8) (2017) 926–938.

[9] F.I. Mulder, E. Horváth-Puhó, N. van Es, L. Pedersen, H.R. Büller, H.E. Bøtker, H.T. Sørensen, Arterial thromboembolism in cancer patients: a Danish population-based cohort study, JACC: CardioOncol. 3 (2021) 205–218.

[10] S. Aspberg, L. Yu, B. Gigante, K.E. Smedby, D.E. Singer, Risk of ischemic stroke and major bleeding in patients with atrial fibrillation and cancer, J. Stroke Cerebrovasc. Dis. 29 (3) (2020) 104560.

[11] R. Patell, A. Gutierrez, L. Rybicki, A.A. Khorana, Usefulness of CHADS2 and CHA2DS2-VASc scores for stroke prediction in patients with cancer and atrial fibrillation, Am. J. Cardiol. 120 (12) (2017) 2182–2186.

[12] A.G. Bernal, R. Patell, L. Rybicki, A.A. Khorana, Predicting mortality in patients with cancer and atrial fibrillation, J. Am. Coll. Cardiol. 71 (11S) (2018) A471.
[13] M. D'Souza, N. Carlson, E. Fosbøl, M. Lamberts, L. Smedegaard, D. Nielsen, et al., CHA2DS2-VASc score and risk of thromboembolism and bleeding in patients with atrial fibrillation and recent cancer, Eur. J. Prev. Cardiol. 25 (6) (2018) 651–658.
[14] W.S. Hu, C.L. Lin, Impact of atrial fibrillation on the development of ischemic stroke among cancer patients classified by CHA2DS2-VASc score-a nationwide cohort study, Oncotarget 9 (7) (2018) 7623.
[15] D. Cardinale, M.T. Sandri, A. Colombo, M. Salvatici, I. Tedeschi, G. Bacchiani, et al., Prevention of atrial fibrillation in high-risk patients undergoing lung cancer surgery, Ann. Surg. 264 (2) (2016) 244–251.
[16] J. Beck-Nielsen, H.R. Sørensen, P. Alstrup, Atrial fibrillation following thoracotomy for non-cardiac diseases, in particular cancer of the lung, Acta Med. Scand. 193 (1–6) (1973) 425–429.
[17] C.W. Siu, H.M. Tung, K.W. Chu, M.H. Jim, C.P. Lau, H.F. Tse, Prevalence and predictors of new-onset atrial fibrillation after elective surgery for colorectal cancer, Pacing Clin. Electrophysiol. 28 (2005) S120–S123.
[18] J.H. Chin, Y.J. Moon, J.Y. Jo, Y.A. Han, H.R. Kim, E.H. Lee, I.C. Choi, Association between postoperatively developed atrial fibrillation and long-term mortality after esophagectomy in esophageal cancer patients: an observational study, PLoS ONE 11 (5) (2016), e0154931.
[19] D. Cardinale, A. Colombo, M.T. Sandri, G. Lamantia, N. Colombo, M. Civelli, et al., Increased perioperative N-terminal pro-B-type natriuretic peptide levels predict atrial fibrillation after thoracic surgery for lung cancer, Circulation 115 (11) (2007) 1339.
[20] T. Nojiri, H. Maeda, Y. Takeuchi, Y. Funakoshi, R. Maekura, K. Yamamoto, M. Okumura, Predictive value of preoperative tissue Doppler echocardiographic analysis for postoperative atrial fibrillation after pulmonary resection for lung cancer, J. Thorac. Cardiovasc. Surg. 140 (4) (2010) 764–768.
[21] W.L. Cheng, Y.H. Kao, S.A. Chen, Y.J. Chen, Pathophysiology of cancer therapy-provoked atrial fibrillation, Int. J. Cardiol. 219 (2016) 186–194.
[22] T.M. Suter, M.S. Ewer, Cancer drugs and the heart: importance and management, Eur. Heart J. 34 (15) (2013) 1102–1111.
[23] J.L. Zamorano, P. Lancellotti, D. Rodriguez Muñoz, V. Aboyans, R. Asteggiano, M. Galderisi, et al., 2016 ESC position paper on cancer treatments and cardiovascular toxicity developed under the auspices of the ESC committee for practice guidelines: the task force for cancer treatments and cardiovascular toxicity of the European Society of Cardiology (ESC), Eur. Heart J. 37 (36) (2016) 2768–2801.
[24] Y.J. Lu, Z.J. Geng, X.Y. Sun, Y.H. Li, X.B. Fu, X.Y. Zhao, B. Wei, Isoprenaline induces epithelial–mesenchymal transition in gastric cancer cells, Mol. Cell. Biochem. 408 (1) (2015) 1–13.
[25] S.G. Lakoski, L.W. Jones, R.J. Krone, P.K. Stein, J.M. Scott, Autonomic dysfunction in early breast cancer: incidence, clinical importance, and underlying mechanisms, Am. Heart J. 170 (2) (2015) 231–241.
[26] P.S. Chen, A.Y. Tan, Autonomic nerve activity and atrial fibrillation, Heart Rhythm. 4 (3) (2007) S61–S64.
[27] P. Coumel, Paroxysmal atrial fibrillation: a disorder of autonomic tone? Eur. Heart J. 15 (suppl_A) (1994) 9–16.
[28] D. Farmakis, J. Parissis, G. Filippatos, Insights into onco-cardiology: atrial fibrillation in cancer, J. Am. Coll. Cardiol. 63 (10) (2014) 945–953.
[29] P. Velagapudi, M.K. Turagam, A.G. Kocheril, Atrial fibrillation in cancer patients: an underrecognized condition, South. Med. J. 104 (9) (2011) 667–668.
[30] H. Kattelus, Y.A. Kesäniemi, H. Huikuri, O. Ukkola, Cancer increases the risk of atrial fibrillation during long-term follow-up (OPERA study), PLoS One 13 (10) (2018), e0205454.

[31] S. Guzzetti, G. Costantino, S. Sada, C. Fundarò, Colorectal cancer and atrial fibrillation: a case-control study, Am. J. Med. 112 (7) (2002) 587–588.
[32] S. Guzzetti, G. Costantino, A. Vernocchi, S. Sada, C. Fundarò, First diagnosis of colorectal or breast cancer and prevalence of atrial fibrillation, Intern. Emerg. Med. 3 (3) (2008) 227–231.
[33] R. Erichsen, C.F. Christiansen, F. Mehnert, N.S. Weiss, J.A. Baron, H.T. Sørensen, Colorectal cancer and risk of atrial fibrillation and flutter: a population-based case–control study, Intern. Emerg. Med. 7 (5) (2012) 431–438.
[34] E.B. Ostenfeld, R. Erichsen, L. Pedersen, D.K. Farkas, N.S. Weiss, H.T. Sørensen, Atrial fibrillation as a marker of occult cancer, PLoS One 9 (8) (2014), e102861.
[35] S. Wassertheil-Smoller, A.P. McGinn, L. Martin, B.L. Rodriguez, M.L. Stefanick, M. Perez, The associations of atrial fibrillation with the risks of incident invasive breast and colorectal cancer, Am. J. Epidemiol. 185 (5) (2017) 372–384.
[36] D. Conen, J.A. Wong, R.K. Sandhu, N.R. Cook, I.M. Lee, J.E. Buring, C.M. Albert, Risk of malignant cancer among women with new-onset atrial fibrillation, JAMA Cardiol. 1 (4) (2016) 389–396.
[37] N. Vinter, A.M. Christesen, M. Fenger-Grøn, A. Tjønneland, L. Frost, Atrial fibrillation and risk of cancer: a Danish population-based cohort study, J. Am. Heart Assoc. 7 (17) (2018), e009543.
[38] W. Saliba, H.S. Rennert, N. Gronich, S.B. Gruber, G. Rennert, Association of atrial fibrillation and cancer: analysis from two large population-based case-control studies, PLoS One 13 (1) (2018), e0190324.
[39] Y.F. Hu, Y.J. Chen, Y.J. Lin, S.A. Chen, Inflammation and the pathogenesis of atrial fibrillation, Nat. Rev. Cardiol. 12 (4) (2015) 230.
[40] R.J. Aviles, D.O. Martin, C. Apperson-Hansen, P.L. Houghtaling, P. Rautaharju, R.A. Kronmal, et al., Inflammation as a risk factor for atrial fibrillation, Circulation 108 (24) (2003) 3006–3010.
[41] K.H. Allin, B.G. Nordestgaard, Elevated C-reactive protein in the diagnosis, prognosis, and cause of cancer, Crit. Rev. Clin. Lab. Sci. 48 (4) (2011) 155–170.
[42] M. Harada, D.R. Van Wagoner, S. Nattel, Role of inflammation in atrial fibrillation pathophysiology and management, Circ. J. 79 (3) (2015) 495–502.
[43] R.B. Darnell, Onconeural antigens and the paraneoplastic neurologic disorders: at the intersection of cancer, immunity, and the brain, Proc. Natl. Acad. Sci. 93 (10) (1996) 4529–4536.
[44] P.W. Kamphuisen, J. Beyer-Westendorf, Bleeding complications during anticoagulant treatment in patients with cancer, Thromb. Res. 133 (2014) S49–S55.
[45] P. Prandoni, A.W. Lensing, A. Piccioli, E. Bernardi, P. Simioni, B. Girolami, et al., Recurrent venous thromboembolism and bleeding complications during anticoagulant treatment in patients with cancer and venous thrombosis, Blood J. Am. Soc. Hematol. 100 (10) (2002) 3484–3488.
[46] J. Trujillo-Santos, P. Prandoni, K. Rivron-Guillot, P. Román, R. Sánchez, G. Tiberio, et al., Clinical outcome in patients with venous thromboembolism and hidden cancer: findings from the RIETE registry, J. Thromb. Haemost. 6 (2) (2008) 251–255.
[47] M.C. Vedovati, M. Giustozzi, P. Verdecchia, L. Pierpaoli, S. Conti, M. Verso, et al., Patients with cancer and atrial fibrillation treated with doacs: a prospective cohort study, Int. J. Cardiol. 269 (2018) 152–157.
[48] N.J. Short, J.M. Connors, New oral anticoagulants and the cancer patient, Oncologist 19 (1) (2014) 82.
[49] B.A. Hutten, M.H. Prins, M. Gent, J. Ginsberg, J.G. Tijssen, H.R. Büller, Incidence of recurrent thromboembolic and bleeding complications among patients with venous thromboembolism in relation to both malignancy and achieved international normalized ratio: a retrospective analysis, J. Clin. Oncol. 18 (17) (2000) 3078–3083.
[50] D.B. Ambrus, J.I. Reisman, A.J. Rose, The impact of new-onset cancer among veterans who are receiving warfarin for atrial fibrillation and venous thromboembolism, Thromb. Res. 144 (2016) 21–26.

[51] J.L. Fitzgerald, L.G. Howes, Drug interactions of direct-acting oral anticoagulants, Drug Saf. 39 (9) (2016) 841–845.

[52] C.L. Fanola, C.T. Ruff, S.A. Murphy, J. Jin, A. Duggal, N.A. Babilonia, et al., Efficacy and safety of edoxaban in patients with active malignancy and atrial fibrillation: analysis of the ENGAGE AF-TIMI 48 trial, J. Am. Heart Assoc. 7 (16) (2018), e008987.

[53] C. Melloni, A. Dunning, C.B. Granger, L. Thomas, M.G. Khouri, D.A. Garcia, et al., Efficacy and safety of apixaban versus warfarin in patients with atrial fibrillation and a history of cancer: insights from the ARISTOTLE trial, Am. J. Med. 130 (12) (2017) 1440–1448.

[54] S.T. Chen, A.S. Hellkamp, R.C. Becker, S.D. Berkowitz, G. Breithardt, K.A. Fox, et al., Efficacy and safety of rivaroxaban vs. warfarin in patients with non-valvular atrial fibrillation and a history of cancer: observations from ROCKET AF, Eur. Heart J. Qual. Care Clin. Outcomes 5 (2) (2019) 145–152.

[55] K. Kim, Y.J. Lee, T.H. Kim, J.S. Uhm, H.N. Pak, M.H. Lee, B. Joung, Effect of non-vitamin K antagonist oral anticoagulants in atrial fibrillation patients with newly diagnosed cancer, Korean Circ. J. 48 (5) (2018) 406.

[56] S. Shah, F.L. Norby, Y.H. Datta, P.L. Lutsey, R.F. MacLehose, L.Y. Chen, A. Alonso, Comparative effectiveness of direct oral anticoagulants and warfarin in patients with cancer and atrial fibrillation, Blood Adv. 2 (3) (2018) 200–209.

[57] A.C. Sawant, A. Kumar, W. Mccray, S. Tetewsky, L. Parone, S. Sridhara, et al., Superior safety of direct oral anticoagulants compared to warfarin in patients with atrial fibrillation and underlying cancer: a national veterans affairs database study, J. Geriatr. Cardiol. 16 (9) (2019) 706.

[58] V.C.C. Wu, C.L. Wang, Y.T. Huang, W.C. Lan, M. Wu, C.F. Kuo, et al., Novel oral anticoagulant versus warfarin in cancer patients with atrial fibrillation: an 8-year population-based cohort study, J. Cancer 11 (1) (2020) 92.

[59] A. Delluc, T.F. Wang, E.S. Yap, C. Ay, J. Schaefer, M. Carrier, S. Noble, Anticoagulation of cancer patients with nonâ valvular atrial fibrillation receiving chemotherapy: guidance from the SSC of the ISTH, J. Thromb. Haemost. 17 (2019) 1247–1252.

[60] A.Y. Lee, M.N. Levine, R.I. Baker, C. Bowden, A.K. Kakkar, M. Prins, et al., Low-molecular-weight heparin versus a coumarin for the prevention of recurrent venous thromboembolism in patients with cancer, N. Engl. J. Med. 349 (2) (2003) 146–153.

[61] A.J. Tafur, W.E. Wysokinski, R.D. McBane, E. Wolny, E. Sutkowska, S.C. Litin, et al., Cancer effect on periprocedural thromboembolism and bleeding in anticoagulated patients, Ann. Oncol. 23 (8) (2012) 1998–2005.

[62] J.D. Douketis, A.C. Spyropoulos, S. Kaatz, R.C. Becker, J.A. Caprini, A.S. Dunn, et al., Perioperative bridging anticoagulation in patients with atrial fibrillation, N. Engl. J. Med. 373 (9) (2015) 823–833.

[63] J.D. Douketis, J.S. Healey, M. Brueckmann, J.W. Eikelboom, M.D. Ezekowitz, M. Fraessdorf, et al., Perioperative bridging anticoagulation during dabigatran or warfarin interruption among patients who had an elective surgery or procedure, Thromb. Haemost. 113 (03) (2015) 625–632.

[64] A. Elbadawi, M. Shnoda, K. Mahmoud, I.Y. Elgendy, Efficacy and safety of direct oral anticoagulants versus low molecular weight heparin for cancer related venous thromboembolism: a meta-analysis of randomized trials, Eur. Heart J. Cardiovasc. Pharmacother. 7 (2020) 380–388.

[65] M. Camilli, M. Lombardi, G.M. Vescovo, M.G. Del Buono, M. Galli, N. Aspromonte, et al., Efficacy and safety of novel oral anticoagulants versus low molecular weight heparin in cancer patients with venous thromboembolism: a systematic review and meta-analysis, Crit. Rev. Oncol. Hematol. 103074 (2020).

CHAPTER

5

Biochemistry of biomarkers in cardiotoxicity induced in chemotherapeutic drugs

Swaraj Mohanty and Yasmin Ahmad

Peptide and Proteomics Division, Defence Institute of Physiology & Allied Sciences (DIPAS), Defence Research & Development Organization (DRDO), Timarpur, Delhi, India

1 Introduction

Clinically approved biomolecules that help in the prediction and early diagnosis of a disease state are termed as biomarkers. These molecules can be any sort of molecules, ions, and chemical compounds that can be taken into consideration and estimated by the analysis of body fluids in any individuals. The characteristics of a potential biomarker is that its value shows a significance alteration during the normal biological process, onset, and progression of any disease and with therapeutic interventions so the biomarker science is an aid for the pharmaceutical and in the medical healthcare sectors [1]. In pharmaceuticals, it can reduce the investment cost during the discovery of a potent drug molecule, and in the case of healthcare science, it helps the physicians and scientists to diagnose and provide personalized treatment to individuals. In the last few years due to the discovery of biomarkers, early diagnosis, and personalized medicine, disease like cancer is showing promising results in recovery rate [2]. Biomarkers are discovered by targeting a particular biomolecule within a signaling pathway during a normal and disease condition and also by its alteration due to pharmaceutical drug intervention [3].

The development of carcinoma and hunt for its cure is always being in the zone of hot spot in medical research and development. The formulae of complete cure of the formation and growth of carcinoma are still

Cardiovascular Toxicity and Therapeutic Modalities Targeting Cardio-oncology
https://doi.org/10.1016/B978-0-323-90461-2.00013-4

a mystery for the scientific community. But with the extensive research in healthcare science, we can be able to manage the deadliest disease of cancer; simultaneously, the management strategies lead to various complications in patients like ischemia [4], acute coronary syndrome (ACS) [5], arrhythmias, thromboembolism, hypertension, and cardiotoxicity [6]. Among all the disorders mentioned just now, the heart valves failure and sudden cardiac arrest is the most common problem during the management of most of the cancer patients. Moreover, due to the surge of cancer patients worldwide, different strategies have been employed for its treatments. Basically, it has been dissipated that there are two main causes of cardiotoxicity, i.e., cancer radiation-induced cardiotoxicity (CRIC) and cancer drug-induced cardiotoxicity (CDIC) [7]. Although the main cause of cardiotoxicity is not clearly understood, the National Cancer Institute in collaboration with the National Heart, Lung and Blood Institute states that the cardiac injury is basically by death of myocyte (Type-I cardiotoxicity) or by reversible myocyte dysfunction (Type-II cardiotoxicity) [8].

2 Biomarker discovery pipeline

The most vital part of this biomarker discovery is a very time-consuming process and involves the following steps as generation of a hypothesis, collection of samples from a large number of individuals, generation of data, analysis of a large number of data, development, and validation of the assay and in the end regulatory approval before the clinical use. The strategies of biomarker development start by targeting the potential molecules from the cluster of biochemical pathways and tracking the level of the molecules and their fate during the signaling processes [9]. Different groups are being made from a large number of population based on their clinical history and few preliminary tests that help to categorize the individuals' body fluids that are being collected by well-trained personals with a consent of the individuals. There comes the development of the assay and validation of the potent molecules by means of different laboratory techniques with the use of high-thoughtful equipment for the data collection. In the last few decades, plethora of sophisticated instrumentation facilities and omics approaches are available, but based on the target and work plan, one can stick to the appropriate technique for the generation of data. Analysis of the data by means of various statistical tools, system biology, and mathematical modeling is also followed for finding the correlation with the disease state [10]. If few molecules show a promising clinical significance, then next comes the validation of the same among a large number of samples collected from different groups of individuals which is basically known as clinical utility validation. In the end, an approval from the regulatory agencies like Europe, Middle East, and Africa

(EMEA), and Food and Drug Administration (FDA) is an essence for the clinical and diagnostic use [11,12].

The whole process of biomarker discovery pipeline is graphically represented in the figure for better understanding (Fig. 1).

3 Prevalence of cancer and its management strategies

The statistical data collected by World Health Organization (WHO) describe the picture of the groups in the society affected by cancer and make the scientific society to put their effort toward control of this disease. The data represent that approximately 10 million death takes place annually and every sixth death in the world is due to cancer [13]. After years of extensive research, now we can able to manage this deadly disease and able to hold the mortality rate in this disease with the strategic managements. There are various methods available so far for the treatment of this disease like surgery [14], immunotherapy [15], stem cell transplantation [16], hormone therapy, chemotherapy, radiation therapy, and precision medicine [10,17,18]. Although the treatment depends upon different factors like age, health condition, localization, malignancy, and tolerance level of the individual, most widely used therapies are radiation therapy and chemotherapy.

3.1 Radiation therapy

A high beam of X-rays is allowed to penetrate where the tumor is detected which eventually damages the DNA of the cells which undergoing uncontrolled division. This high beam either kills the cells or slows down the growth of the cells of the areas and after which the tumor is removed from the site of occurrence by minor surgery if necessary. The radiation therapy is mainly subdivided into external and internal based on the beam of particles used [19]. In the external radiation therapy, the beam of radiation comes from the source that is placed outside the body and focused on the site of tumor, whereas in the internal radiation therapy, the radiation source is put near the tumor for its operation, if the sources of beam generation used from liquid form (systemic therapy) or solid form (brachytherapy) [20]. This radiation therapy depends upon various factors like type, size, location of the tumor, medical history, age, and the other treatment. The external radiation therapy is used to treat many local carcinomas developed on the surface of the body or as an extra growth, whereas the brachytherapy is effective in the cancer developed in head, eye, neck, cervix, breast, and prostate with the help of a solid source of beam [21]. And the systemic therapy uses radioactive isotopic molecules by injection and is effective to treat thyroid, neuroendocrine, and gastroenteropancreatic cancer [22].

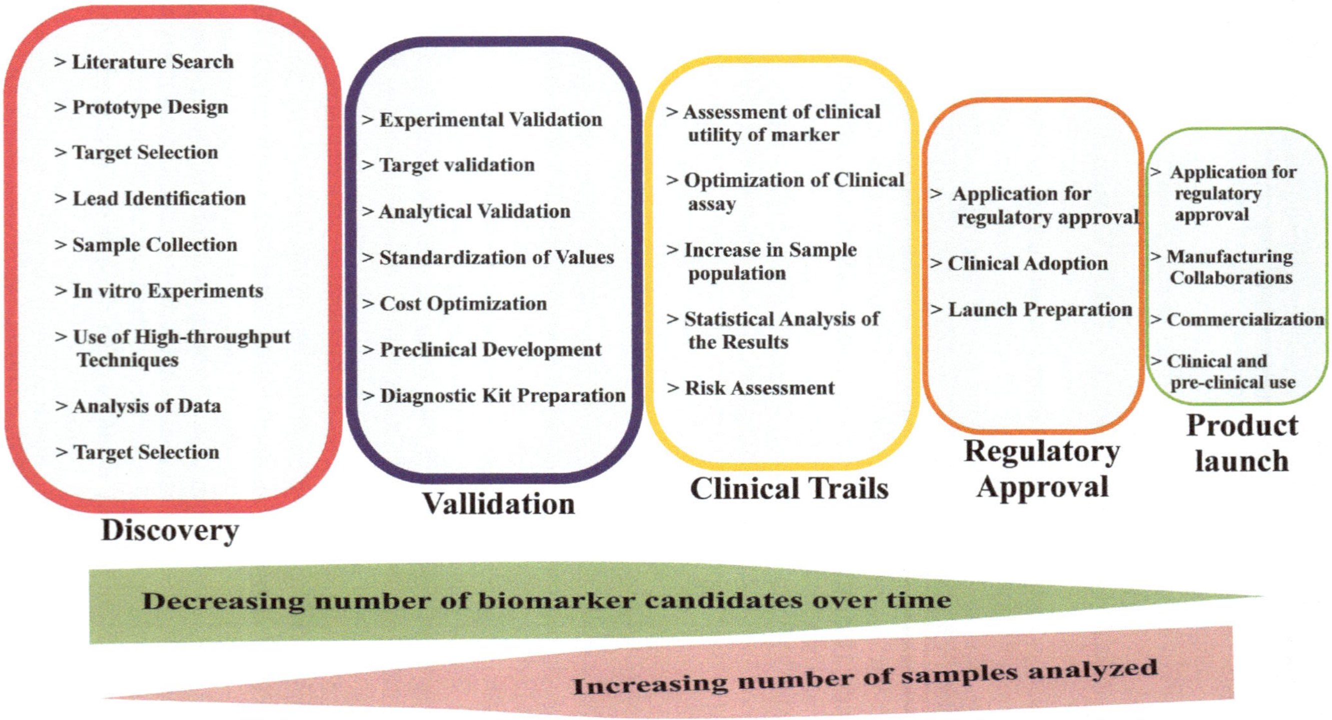

FIG. 1 Schematic view of biomarker discovery process pipeline.

The information of isotopes used as ion sources during different cancer therapy is given in Table 1.

3.2 Chemotherapy

A drug molecule is used in this therapeutic procedure which has the potential to kill the cancer cells by ceasing their growth or by shrinkage the tumor size when introduced into the body of patients. Based on the treatment time, it can be subdivided into neoadjuvant chemotherapy and adjuvant chemotherapy. If the tumor size is very large or rapid growth is observed, then the chemotherapeutic agent is administrated before the main treatment procedure to reduce the tumor size is known as neoadjuvant chemotherapy, whereas in the case of adjuvant chemotherapy, the chemotherapeutic agent is administrated after the surgical or radiation therapy for the complete elimination of rest of the cancerous cells from the body [33]. These chemotherapeutic agents can be in the form of cream, pills, capsules, and liquids, hence can be administrated topically, orally, or by injection (intravenously, intra-arterially, intrathecally, and intraperitonially). Few drawbacks like feeling exhausted, developing fatigueness, loss of hair, and loss of appetite persist for few hours as posttreatment symptoms after chemotherapy [34]. Hence, a professional oncologist clinically diagnoses different parameters of the patients and decides the interval and the dose one can accept along with regular monitoring and based on the tolerance level of the patient on a time basis. A number of chemotherapeutic drugs have already been approved from different pharmaceutical industries, and many are under clinical trial based on the type, localization, and progression of tumors.

The list of clinically approved and widely used drug molecules with the doses and medications used in different types of carcinomas is described in Table 2.

4 Cardiotoxicity due to anticancer therapy

In the last few decades, research and development in the anticancer therapies shows a very remarkable result in declining the mortality rate of patients suffering from different carcinoma. The use of these therapeutic interventions raises a new concern of cardiovascular death among the patients with history of anticancer therapy. The complications like thrombophilia, hypertension, arrhythmias, ventricular dysfunction, myocardial ischemia, and acute heart failure are the most challenging for an oncologist during the therapy [55]. Hence to closely monitor these side effects in the individuals, different monitoring techniques like radionuclide ventriculography, magnetic resonance imaging (MRI), and echocardiography (ECG)

TABLE 1 Isotopes used in different cancer therapies.

SI No.	Therapy	Ion sources	Half-life period	Type of cancer treated	References
1.	Brachytherapy	Iodine-125	60 days	Prostate cancer	[23]
		Iridium-192	74 days	Cervical cancer	[24]
		Gold-198	2.7 days	Skin, cervical, prostate, and breast cancer	[25]
		Palladium-103	17 days	Prostate cancer	[26]
		Cesium-137	30.17 years	Uterine, cervix, and endobronchial cancer	[27]
		Cobalt-60	5.3 years	Head, neck, and cervical cancer	[24]
2.	Systemic therapy	Bismuth-213	46 min	Ovarian cancer	[28]
		Lead-212	10.6 h	Breast and ovarian cancer	[29]
		Strontium-89	50.5 days	Bone and prostate cancer	[30,31]
		Yttrium-90	64.2 h	Liver cancer	[32]
		Rhenium-186	3.8 days	Bone cancer	[31]
		Samarium-153	1.9 days	Neuroectodermal and esophageal cancer	[31]

TABLE 2 Approved drugs for the treatment of carcinomas.

SI No.	Approved drugs	Doses	Uses	References
1.	Alemtuzumab	3–10 mg/day Intravenous (3 days in a week for 12 weeks)	Chronic lymphocytic leukemia	[35]
2.	Altretamine	260 mg/m^2/day Oral (14–21 days in a 28-day cycle)	Ovarian cancer	[36]
3.	Belinostat	1000 mg/m^2/day Intravenous (1–5 days in a 21-day cycle)	Peripheral T-cell lymphoma	[37]
4.	Bendamustine	100 mg/m^2/day Intravenous (Day 1 and 2 in a 28-day cycle)	B-cell lymphoma and chronic lymphocytic leukemia	[38]
5.	Bevacizumab	5–15 mg/kg Intravenous (every 3 weeks until disease progression)	Cervical, colorectal, breast, fallopian tube, ovarian cancer	[39]
6.	Blinatumomab	Days 1–7: 5 μg/m^2/day Days 8–28: 15 μg/m^2/day Days 8–28: 15 μg/m^2/day Intravenous	Refractory B-cell acute lymphoblastic leukemia	[40]
7.	Bortezomib	1.3 mg/m^2/day Intravenous (on days 1, 4, 8, and 11 every 3 weeks)	Multiple myeloma	[41]

Continued

TABLE 2 Approved drugs for the treatment of carcinomas—cont'd

SI No.	Approved drugs	Doses	Uses	References
8.	Capecitabine	1250 mg/m^2/day Oral (2 weeks on drug, 1 week of rest)	Gastrointestinal malignancies	[42]
9.	Carmustine	150–200 mg/m^2/day Intravenous (every 6–8 weeks)	Brain tumor and bone marrow transplant patients	[43]
10.	Cetuximab	400 mg/m^2/day over 120 min Intravenous (every 7 days)	Head, neck, and colorectal cancer	[44]
11.	Docetaxel	60–100 mg/m^2/day over 1 h Intravenous (every 21 days)	Pancreatic, esophageal, cervical, uterine, bladder, and ovarian cancer	[45]
12.	Estramustine	14 mg/kg/day Oral (in 3–4 equally divided doses)	Prostate and renal cell cancer	[46]
13.	Fludarabine	25 mg/m^2/day over 30 min for 5 days Intravenous (repeat every 28 days)	Acute myeloid leukemia and chronic lymphocytic leukemia	[47]
14.	Irinotecan	125 mg/m^2/day over 90 min Intravenous (once weekly for 4 weeks and then 2-week rest)	Lung, ovarian, and cervical cancers	[48]
15.	Lanreotide	120 mg every 4 weeks	Neuroendocrine and gastropancreatic cancer	[49]

16.	Mitomycin	20 mg/m^2/day Intravenous (every 6–8 weeks)	Head, gastric, neck, colorectal, and pancreatic adenocarcinomas	[50]
17.	Paclitaxel	80 mg/m^2/day Intravenous (every week)	Gastric, prostate, and bladder cancers	[51]
18.	Pembrolizumab	2 mg/kg Intravenously over 30 min (every 3 weeks)	Metastatic melanoma	[52]
19.	Temsirolimus	25 mg Intravenously over 30–60 min (weekly)	Renal cell carcinoma	[53]
20.	Vandetanib	300 mg/day Oral	Medullary thyroid cancer	[54]

are always advisable before and after the anticancer therapy. Clinically, cardiac toxicity has been classified into type-I and type-II myocardial injuries based on the severity level of the myocytes. The type-I injury is a pathophysiological condition in which the microstructures of the cardiac muscle cells get damaged and are irreversible in nature leading to necrosis and apoptosis. In contrast of type-II injury, the muscle cells of the heart stop functioning and can happen without any disturbance in structural integrity of the cells [56].

Sunitinib and trastuzumab are two chemotherapeutic agents used in the treatment of renal cell carcinoma and gastric cancer, respectively. Both these drugs are reported to initiate type-II injury in heart: sunitinib shows apoptosis by dysfunctioning mitochondria of myocytes, whereas trastuzumab blocks the human epidermal growth factor receptor 2 (HER2) and hence ceases the cell growth, differentiation, and proliferation of the myocytes. Daunorubicin and epirubicin are the drugs used in the treatment process of lungs, ovarian, uterine, and bladder cancer as it damages the DNA of the cells undergoing uncontrolled division, but the metabolites of these drugs show a side effect of free radical generation in the myocytes [57,58].

The major complications that have been reported in the individuals include the increase in the level of reactive oxygen species (ROS), thrombosis, and cardiac mitochondrial dysfunction leading to heart failure (Fig. 2).

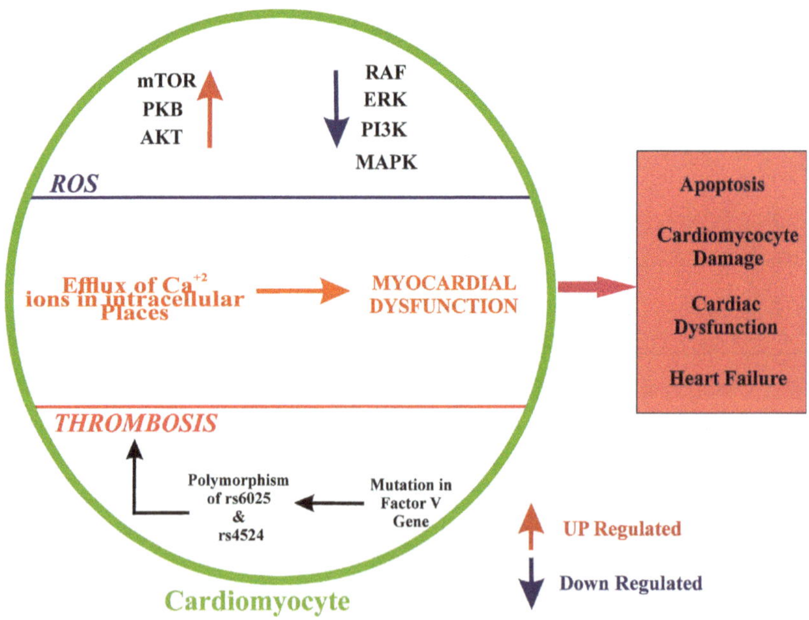

FIG. 2 Diagrammatic representation of molecular response in cardiomyocyte during myocardial dysfunction.

4.1 Effect of reactive oxygen species (ROS)

The mechanistic target of rapamycin (mTOR) and protein kinase B (PKB/AKT) are most crucial proteins whose level increases at the site of tumor formation and are responsible for various cellular processes like cell growth, proliferation, differentiation, and migration [59]. The downregulation of PI3K, rapidly accelerated fibrosarcoma (RAF), mitogen-activated protein kinase (MAPK), and extracellular-signal-regulated kinase (ERK) proteins is occurred by the oxidation of cysteine residue due to ROS generation [60]. Increasing level of ROS in myocytes causes membrane lipid peroxidation and activation of caspase 9 leading to apoptosis on the one side, and on the other side, Ca^{2+} concentration increases in a cell as the ATPase activity in sarcoplasmic reticulum is suppressed. Overdose of the drug like doxorubicin decreases the production of energy molecules (i.e., ATP) in mitochondria and stops the pumping function of the heart [61].

4.2 Thrombosis

The development of deep vein thrombosis (DVT) in cancer patients undergoing chemotherapy is very common among US and European population. Mutation in factor V gene results the polymorphism of rs6025 and rs4524 which increases the clot development in the leg and lungs [62]. At the very initial stage of therapy, the conditions like hypertension, thromboembolism, and vasospasm are very common which in the later stage develops into atherosclerosis causing death [63].

4.3 Cardiac mitochondrial dysfunction

As mitochondrion is the powerhouse of all kinds of cells due to the energy production from biochemical processes, its role in the cell is highly crucial for maintaining the homeostasis. Due course of anticancer therapy with trastuzumab and bevacizumab drugs, the mitochondrial homeostasis gets imbalanced giving rise to loss of membrane potential, ROS generation, and increase in the concentration of Ca^{2+} ion in the intracellular spaces leading to the mitochondrial swelling [64]. The alteration in intracellular Ca^{2+} ion dynamics activates pathological hypertrophic signaling pathways leading to apoptosis of myocytes due to the administration of imatinib [65].

5 Biomarkers and its molecular aspects in cardiotoxicity

The search for different biomarkers is always a challenging job for the oncologist and scientist of the healthcare research and development sector as it can pave the way as an indicator of pathophysiology and response of an individual to therapy. In the course of time, many biomarkers have already been discovered and found prognostic in the progression of cardiotoxicity. The list of biomarkers for cardiac toxicity such as cardiac troponins (cTn), natriuretic peptides (NPs), inflammatory markers, ischemic markers, and markers of endothelial dysfunction are clinically approved for the monitoring of the heart failure.

5.1 Cardiac troponins (cTn)

The most promising biomarkers in the case of the patients treated with anthracycline as its level increases with the cumulative dose administration. The early chronic cardiac toxicity and the chance of left ventricular dysfunction can be predicted by measuring of cTn level in the 24 h of therapy [66]. The prediction of toxicity can also be diagnosed within the first 90 days after therapy in the children suffering from acute lymphoblastic leukemia and may not be a sensitive biomarker for late onset of subclinical cardiotoxicity. The use of trastuzumab in breast cancer patients shows elevated cTn levels and acts as a biomarker for cardiotoxicity [67].

5.2 Natriuretic peptides (NPs)

During the myocardial wall stretch and pressure overload on the wall of heart, the level of NPs is released from myocytes. β-Type natriuretic peptide (BNP) level can be measured before the therapy and after 12 months of the therapy, and the lower the level of this BNP shows as a promising biomarker for early detection of cardiotoxicity in individuals. Myocytes release the precursor to BNP (pre-pro BNP) after cleavage produces BNP and N-terminal fragment (NT-pro BNP) act as a biomarker in the detection of left ventricular dysfunctioning and heart failure (HF) [68,69]. A recent study concluded that if the BNP level reaches 100 pg/mL, there will be cardiotoxicity due to HF.

5.3 Inflammatory biomarkers

Although the inflammatory molecules cannot be considered as a potential biomarker in the case of early detection of heart failure, interleukin-6 (IL-6) in association with Doppler imaging and antioxidant defense capacity can be acted as a biomarker of cardiotoxicity [70,71]. In the case of breast cancer patients treated with trastuzumab and doxorubicin, the leukocytes secrete myeloperoxidase (MPO) as an inflammatory marker for the prediction of heart failure due to toxicity and there is a 40% chance of HF [72].

5.4 Ischemic biomarkers

It has been reported that the level of fatty acid-binding protein (FABP) level elevates in the case of patients with possibility of cardiac dysfunction within 24h after treatment with an anthracycline dose of 300 $mg/m^2/day$ [73]. Glycogen phosphorylase isoenzyme BB (GPBB) level increases in patients with acute myeloid leukemia within 6 months after therapy is also a potent biomarker developing cardiotoxicity [74].

5.5 Biomarkers of endothelial dysfunction

Neuregulin-1 (NRG-1) level was measured in patients before and after chemotherapy with anthracycline and is found that the level of NRG-1 decreased in patients developing cardiotoxicity [75]. Higher level of plasminogen activator inhibitor-1 (PAI-1), high-sensitive c-reactive protein (hs-CRP), tissue plasminogen activator (tPA), fibrinogen, and von Willebrand factor (vWF) is found in the patients treated with cisplatin suffering from testicular cancer [76].

6 Future challenges and concluding remarks

The use of biomarkers for the disease detection is a promising tool for the clinicians to completely treat and recover the individuals, but the various aspects of few biomarkers are still a matter of study for finding the efficient biomarker. Plethora of the techniques available in bioscience research the use of high thoughtful omics techniques may be a reliable source to be followed in the biomarker discovery pipeline to minimize the time and effort for the biomarker discovery [77]. The clinical history of the groups of individuals needs to be explored before taking them as the subject in biomarker discovery with the help of genomics and transcriptomic techniques. Periodic proteomics analysis of the subjects should be checked to crosscheck the expression of different proteins within the signaling process. Emergence of superfast computing and the data science can be combined with the traditional biomarker valedictory methods for achieving a breakthrough in this biomarkers science. Although BNP is a satisfactory biomarker, its level fluctuation is not clearly understood yet which needs to be explored by increasing the number of individuals undergoing various chemotherapeutic agents. Many biomarkers like growth differentiation factor 15 (GDF-15) [78], galactin-3 [73], microRNA-208 (miR-208), miR-499 [79], peripheral blood mononuclear cells (PBMCs) [80], suppression of tumorigenicity-2 (ST-2) [81] are also in the clinical-stage pipeline for their approval and use.

Cardiovascular disease is the most serious disease and cancer is the most concerning disease among the society; hence, the therapy used should mainly focus on the consequences and side effects due to a therapy that varies from patient to patient. So the oncologist should regularly monitor and make substantial changes in the therapy based on the tolerance level and pathophysiological changes in the patients. As this field of novel biomarker discovery is open for research by scientists, the regular monitoring of the late onset of the symptoms leading to heart failure and the ease of survivability of the individual for a longer duration after treatment without any medication and finding for a permanent cure with a healthy lifestyle is the future insight.

Acknowledgment

Funding

SM is a recipient of DRDO-JRF fellowship, and the authors have not received any special funding for this work.

Conflict of interest

The authors state no conflict of interests.

Availability of data and materials

The figures and tables are not adopted from any sources.

Authors' contribution

YA conceptualized the manuscript. SM designed and wrote the entire manuscript and tables and figures for convenient representation. YA and SM critically evaluated the manuscript.

Ethics approval

Not applicable.

Consent for publication

All authors agreed.

References

[1] S. Naylor, Biomarkers: current perspectives and future prospects, Expert Rev. Mol. Diagn. 3 (5) (2003) 525–529.

[2] A. Ziegler, et al., Personalized medicine using DNA biomarkers: a review, J. Hum. Genet. 131 (10) (2012) 1627–1638.

[3] V.B. Kraus, Biomarkers as drug development tools: discovery, validation, qualification and use, Nat. Rev. Rheumatol. 14 (6) (2018) 354–362.

[4] R. Berardi, et al., State of the art for cardiotoxicity due to chemotherapy and to targeted therapies: a literature review, Crit. Rev. Oncol. Hematol. 88 (1) (2013) 75–86.

[5] P. Garg, et al., Cardiac biomarkers of acute coronary syndrome: from history to high-sensitivity cardiac troponin, Intern. Emerg. Med. 12 (2) (2017) 147–155.

[6] T.C. Tan, M. Scherrer-Crosbie, Cardiac complications of chemotherapy: role of imaging, Curr. Treat. Options Cardiovasc. Med. 16 (4) (2014) 1–19.

[7] L.-L. Tan, A.R. Lyon, Role of biomarkers in prediction of cardiotoxicity during cancer treatment, Curr. Treat. Options Cardiovasc. Med. 20 (7) (2018) 1–14.

[8] G. Riccio, et al., Trastuzumab and target-therapy side effects: is still valid to differentiate anthracycline type I from type II cardiomyopathies? Hum. Vaccin. Immunother. 12 (5) (2016) 1124–1131.

[9] C.H. Borchers, C.E. Parker, Improving the biomarker pipeline, Clin. Chem. 56 (12) (2010) 1786–1788.

[10] Y. Lin, et al., Computer-aided biomarker discovery for precision medicine: data resources, models and applications, Brief. Bioinform. 20 (3) (2019) 952–975.

[11] E. Drucker, K. Krapfenbauer, Pitfalls and limitations in translation from biomarker discovery to clinical utility in predictive and personalised medicine, EPMA J. 4 (1) (2013) 1–10.

[12] F. Goodsaid, M. Papaluca, Evolution of biomarker qualification at the health authorities, Nat. Biotechnol. 28 (5) (2010) 441–443.

[13] J. Ferlay, et al., Cancer statistics for the year 2020: an overview, Int. J. Cancer 149 (4) (2021) 778–789, https://doi.org/10.1002/ijc.33588.

[14] N.V. Christou, et al., Bariatric surgery reduces cancer risk in morbidly obese patients, Surg. Obes. Relat. Dis. 4 (6) (2008) 691–695.

[15] M. Rescigno, F. Avogadri, G. Curigliano, Challenges and prospects of immunotherapy as cancer treatment, Biochim. Biophys. Acta Rev. Cancer 1776 (1) (2007) 108–123.

[16] R.R. Jenq, M.R. Van den Brink, Allogeneic haematopoietic stem cell transplantation: individualized stem cell and immune therapy of cancer, Nat. Rev. Cancer 10 (3) (2010) 213–221.

[17] S. Mohanty, A. Dash, C.K. Pradhan, CRISPR-Cas9 technology: a magical tool for DNA editing, Int. J. Biosci. Bioeng. 1 (1:005) (2019) 1–11.

[18] C. Rubino, et al., Radiation dose, chemotherapy, hormonal treatment and risk of second cancer after breast cancer treatment, Br. J. Cancer 89 (5) (2003) 840–846.

[19] M. Sadeghi, M. Enferadi, A. Shirazi, External and internal radiation therapy: past and future directions, J. Cancer Res. Ther. 6 (3) (2010) 239.

[20] D. Vordermark, Radiotherapy of cervical cancer, Oncol. Res. Treat. 39 (9) (2016) 516–520.

[21] J. Skowronek, Current status of brachytherapy in cancer treatment—short overview, J. Contemp. Brachytherapy 9 (6) (2017) 581.

[22] N.H. Lester-Coll, et al., Cost-effectiveness of surgery, stereotactic body radiation therapy, and systemic therapy for pulmonary oligometastases, Int. J. Radiat. Oncol. Biol. Phys. 95 (2) (2016) 663–672.

[23] R.E. Peschel, et al., Iodine 125 versus palladium 103 implants for prostate cancer: clinical outcomes and complications, Cancer J. 10 (3) (2004) 170–174.

[24] T. Tantivatana, K. Rongsriyam, Treatment outcomes of high-dose-rate intracavitary brachytherapy for cervical cancer: a comparison of Ir-192 versus Co-60 sources, J. Gynecol. Oncol. 29 (5) (2018) e86, https://doi.org/10.3802/jgo.2018.29.e86.

[25] C.D. de Souza, et al., New gold-198 nanoparticle synthesis to be used in cancer treatment, Braz. J. Radiat. Sci. 9 (1(A)) (2019) 1–18, https://doi.org/10.15392/bjrs.v9i1A.1260.
[26] J.E. Sylvester, et al., Ten-year biochemical relapse-free survival after external beam radiation and brachytherapy for localized prostate cancer: the Seattle experience, Int. J. Radiat. Oncol. Biol. Phys. 57 (4) (2003) 944–952.
[27] N. Wachter-Gerstner, et al., Bladder and rectum dose defined from MRI based treatment planning for cervix cancer brachytherapy: comparison of dose–volume histograms for organ contours and organ wall, comparison with ICRU rectum and bladder reference point, Radiother. Oncol. 68 (3) (2003) 269–276.
[28] E.Y. Song, et al., Bismuth-213 radioimmunotherapy with C595 anti–MUC1 monoclonal antibody in an ovarian cancer ascites model, Cancer Biol. Ther. 7 (1) (2008) 76–80.
[29] R.F. Meredith, et al., Pharmacokinetics and imaging of 212Pb-TCMC-trastuzumab after intraperitoneal administration in ovarian cancer patients, Cancer Biother. Radiopharm. 29 (1) (2014) 12–17.
[30] S. Turner, et al., Less pain does equal better quality of life following strontium-89 therapy for metastatic prostate cancer, Br. J. Cancer 84 (3) (2001) 297–302.
[31] I.G. Finlay, M.D. Mason, M. Shelley, Radioisotopes for the palliation of metastatic bone cancer: a systematic review, Lancet Oncol. 6 (6) (2005) 392–400.
[32] R. Salem, et al., Yttrium-90 microspheres: radiation therapy for unresectable liver cancer, J. Vasc. Interv. Radiol. 13 (9) (2002) S223–S229.
[33] J.J. Meeks, et al., A systematic review of neoadjuvant and adjuvant chemotherapy for muscle-invasive bladder cancer, Eur. Urol. 62 (3) (2012) 523–533.
[34] T. Iwamoto, Clinical application of drug delivery systems in cancer chemotherapy: review of the efficacy and side effects of approved drugs, Biol. Pharm. Bull. 36 (5) (2013) 715–718.
[35] S.M. O'Brien, et al., Alemtuzumab as treatment for residual disease after chemotherapy in patients with chronic lymphocytic leukemia, Cancer Biother. Radiopharm. 98 (12) (2003) 2657–2663.
[36] J.K. Chan, et al., Oral altretamine used as salvage therapy in recurrent ovarian cancer, J. Gynecol. Oncol. 92 (1) (2004) 368–371.
[37] P. Reimer, S. Chawla, Long-term complete remission with belinostat in a patient with chemotherapy refractory peripheral T-cell lymphoma, J. Hematol. Oncol. 6 (1) (2013) 1–3.
[38] L.T. Hsiao, et al., Pharmacokinetic profile and first preliminary clinical evaluation of bendamustine in Taiwanese patients with heavily pretreated indolent B-cell non-Hodgkin lymphoma and mantle cell lymphoma, J. Hematol. Oncol. 33 (4) (2015) 136–144.
[39] S. Motl, Bevacizumab in combination chemotherapy for colorectal and other cancers, Am. J. Health Syst. Pharm. 62 (10) (2005) 1021–1032.
[40] H. Kantarjian, et al., Blinatumomab versus chemotherapy for advanced acute lymphoblastic leukemia, N. Engl. J. Med. 376 (9) (2017) 836–847.
[41] M. Furtado, et al., Addition of bortezomib to standard dose chop chemotherapy improves response and survival in relapsed mantle cell lymphoma, Br. J. Haematol. 168 (1) (2015) 55–62.
[42] Y.S. Heo, et al., Hand-foot syndrome in patients treated with capecitabine-containing combination chemotherapy, J. Clin. Pharmacol. 44 (10) (2004) 1166–1172.
[43] M.L. Affronti, et al., Overall survival of newly diagnosed glioblastoma patients receiving carmustine wafers followed by radiation and concurrent temozolomide plus rotational multiagent chemotherapy, Cancer 115 (15) (2009) 3501–3511.
[44] J.B. Vermorken, et al., Platinum-based chemotherapy plus cetuximab in head and neck cancer, N. Engl. J. Med. 359 (11) (2008) 1116–1127.
[45] A. Roth, J. Ajani, Docetaxel-based chemotherapy in the treatment of gastric cancer, Ann. Oncol. 14 (2003) ii41–ii44.
[46] G.M. Lubiniecki, et al., Thromboembolic events with estramustine phosphate-based chemotherapy in patients with hormone-refractory prostate carcinoma: results of a meta-analysis, Cancer 101 (12) (2004) 2755–2759.

[47] M. Hallek, B.F. Eichhorst, Chemotherapy combination treatment regimens with fludarabine in chronic lymphocytic leukemia, Hematol. J. 5 (2004) S20–S30.
[48] H. Okamoto, et al., A combination chemotherapy of carboplatin and irinotecan with granulocyte colony-stimulating factor (G-CSF) support in elderly patients with small cell lung cancer, Lung Cancer 53 (2) (2006) 197–203.
[49] K. Öberg, Chemotherapy and biotherapy in the treatment of neuroendocrine tumours, Ann. Oncol. 12 (2001) S111–S114.
[50] L. Smid, et al., Postoperative concomitant irradiation and chemotherapy with mitomycin C and bleomycin for advanced head-and-neck carcinoma, Int. J. Radiat. Oncol. Biol. Phys. 56 (4) (2003) 1055–1062.
[51] A. Bamias, et al., Adjuvant chemotherapy with paclitaxel and carboplatin in patients with advanced bladder cancer: a study by the Hellenic Cooperative Oncology Group, J. Urol. 171 (4) (2004) 1467–1470.
[52] R. Dummer, et al., A randomized controlled comparison of pembrolizumab and chemotherapy in patients with ipilimumab-refractory melanoma, J. Transl. Med. 13 (1) (2015) 1–2.
[53] J. Hadoux, S. Vignot, T. De La Motte Rouge, Renal cell carcinoma: focus on safety and efficacy of temsirolimus, Clin. Med. Insights Oncol. 4 (2010). CMO. S4482.
[54] C. Durante, et al., Vandetanib: opening a new treatment practice in advanced medullary thyroid carcinoma, Endocrine 44 (2) (2013) 334–342.
[55] X. Han, Y. Zhou, W. Liu, Precision cardio-oncology: understanding the cardiotoxicity of cancer therapy, NPJ Precis. Oncol. 1 (1) (2017) 1–11.
[56] M.S. Ewer, S.M. Ewer, Cardiotoxicity of anticancer treatments, Nat. Rev. Cardiol. 12 (9) (2015) 547.
[57] V. Mercurio, et al., Models of heart failure based on the cardiotoxicity of anticancer drugs, J. Card. Fail. 22 (6) (2016) 449–458.
[58] H. Joensuu, Cardiac toxicity of sunitinib, Lancet Oncol. 370 (9604) (2007) 1978–1980.
[59] A.S. Alzahrani, PI3K/Akt/mTOR inhibitors in cancer: at the bench and bedside, in: Seminars in Cancer Biology, Elsevier, 2019.
[60] X. Wang, et al., Ghrelin inhibits doxorubicin cardiotoxicity by inhibiting excessive autophagy through AMPK and p38-MAPK, Biochem. Pharmacol. 88 (3) (2014) 334–350.
[61] Y. Shi, et al., Mechanisms and management of doxorubicin cardiotoxicity, Herz 36 (4) (2011) 296–305.
[62] H. Haybar, et al., Mechanisms and biomarkers to detect chemotherapy-induced cardiotoxicity, Clin. Cancer Investig. J. 6 (5) (2017) 207.
[63] R. Moudgil, E.T. Yeh, Mechanisms of cardiotoxicity of cancer chemotherapeutic agents: cardiomyopathy and beyond, Can. J. Cardiol. 32 (7) (2016) 863–870.e5.
[64] D. Bovelli, G. Plataniotis, F. Roila, Cardiotoxicity of chemotherapeutic agents and radiotherapy-related heart disease: ESMO Clinical Practice Guidelines, Ann. Oncol. 21 (2010) v277–v282.
[65] T.P. Chambers, et al., Sab mediates mitochondrial dysfunction involved in imatinib mesylate-induced cardiotoxicity, Toxicology 382 (2017) 24–35.
[66] M. Adamcova, et al., Troponin as a marker of myocardiac damage in drug-induced cardiotoxicity, Expert Opin. Drug Saf. 4 (3) (2005) 457–472.
[67] K.S. Shah, et al., The role of biomarkers in detection of cardio-toxicity, Curr. Oncol. Rep. 19 (6) (2017) 1–8.
[68] L. Michel, et al., Cardiac biomarkers for the detection of cardiotoxicity in childhood cancer—a meta-analysis, ESC Heart Fail. 7 (2) (2020) 423–433.
[69] X. Lu, et al., BNP as a marker for early prediction of anthracycline-induced cardiotoxicity in patients with breast cancer, Oncol. Lett. 18 (5) (2019) 4992–5001.
[70] C. Pop, et al., Protective effects of a discontinuous treatment with alpha-lipoic acid in obesity-related heart failure with preserved ejection fraction, in rats, Antioxidants 9 (11) (2020) 1073.

[71] S. Steven, T. Münzel, A. Daiber, Exploiting the pleiotropic antioxidant effects of established drugs in cardiovascular disease, Int. J. Mol. Sci. 16 (8) (2015) 18185–18223.
[72] Z. Sun, et al., Vitexin attenuates acute doxorubicin cardiotoxicity in rats via the suppression of oxidative stress, inflammation and apoptosis and the activation of FOXO3a, Exp. Ther. Med. 12 (3) (2016) 1879–1884.
[73] D. Cardinale, et al., Using biomarkers to predict and to prevent cardiotoxicity of cancer therapy, Expert Rev. Mol. Diagn. 17 (3) (2017) 245–256.
[74] J. Horacek, et al., Glycogen phosphorylase BB could be a new circulating biomarker for detection of anthracycline cardiotoxicity, Ann. Oncol. 19 (9) (2008) 1656–1657.
[75] C.G. Lenneman, Neuregulin-1 signaling in the pathogenesis of chemotherapy-induced heart failure, Curr. Heart Fail. Rep. 11 (2) (2014) 134–138.
[76] C. Henri, T. Heinonen, J.-C. Tardif, The role of biomarkers in decreasing risk of cardiac toxicity after cancer therapy: supplementary issue: biomarkers and their essential role in the development of personalised therapies (A), Biomark. Cancer 8 (2016). BIC. S31798.
[77] J. Montaner, et al., Multilevel omics for the discovery of biomarkers and therapeutic targets for stroke, Nat. Rev. Neurol. 16 (5) (2020) 247–264.
[78] A. Vohra, A. Asnani, Biomarker discovery in cardio-oncology, Curr. Cardiol. Rep. 20 (7) (2018) 1–8.
[79] A. Bhagat, E.S. Kleinerman, Anthracycline-induced cardiotoxicity: causes, mechanisms, and prevention, in: Current Advances in Osteosarcoma, Springer, 2020, pp. 181–192.
[80] V.K. Todorova, et al., Biomarkers for presymptomatic doxorubicin-induced cardiotoxicity in breast cancer patients, PLoS One 11 (8) (2016), e0160224.
[81] N. Iqbal, et al., Novel biomarkers for heart failure, Expert Rev. Cardiovasc. Ther. 11 (9) (2013) 1155–1169.

CHAPTER

6

Cardiotoxic effects of tyrosine kinase inhibitors directed against VEGFR

Savneet Kaur

Department of Molecular and Cellular Medicine, Institute of Liver and Biliary Sciences (ILBS), New Delhi, India

1 Tumor angiogenesis

Tumor tissues have the ability to grow to a size of approximately 1–2 mm^3 beyond which they fail to increase in size due to the limitations of oxygen and nutrients. To develop further, tumor cells participate in the process of angiogenesis, whereby the neighboring endothelial cells give rise to new blood vessels to ensure the blood and nutrient requirements of the growing tumor. This is termed as an "angiogenic switch" of tumor cells and is the result of changes in the local equilibrium between positive and negative regulators of angiogenesis. Angiogenesis is the result of directed capillary in-growth from preexisting host endothelial or endothelial progenitor cells and forms the basis for the formation of cancerous or malignant tumors [1,2]. Angiogenesis is a hypoxia-driven and growth factor-dependent process. Angiogenesis occurs via several distinct mechanisms such as vasculogenesis, angiogenesis, arteriogenesis, and collateral vessel growth [2,3]. In adults, physiological angiogenesis occurs during the female reproductive cycle and in wound healing, whereas pathological angiogenesis can be seen in rheumatoid arthritis, liver cirrhosis, diabetic retinopathy, and tumor growth [2–4]. Insufficient vascular proliferation is implicated in the pathophysiology of myocardial and limb ischemia, where stimulation of angiogenesis is therapeutic [5].

The tumor cells produce an array of instructive proangiogenic factors, including vascular endothelial growth factor (VEGF), angiopoietin

Cardiovascular Toxicity and Therapeutic Modalities Targeting Cardio-oncology
https://doi.org/10.1016/B978-0-323-90461-2.00008-0

(ANGPT), basic fibroblast growth factor (bFGF), and platelet-derived growth factor (PDGF), that bind to their cognate receptors on neighboring endothelial cells [6]. Among them, the factors of VEGF family are reported to be the most pertinent. Factors of VEGF family participate in several key processes of angiogenesis including altering vascular permeability, endothelial cell proliferation, endothelial cell migration or invasion into the surrounding tissue, and finally capillary-like tube formation. The VEGF-related gene family comprises six secreted proteins, namely VEGF-A, VEGF-B, VEGF-C, VEGF-D, VEGF-E, and placenta growth factor (PIGF), and the VEGF receptor family consists of three related receptors, VEGFR-1, VEGFR-2, and VEGFR-3. While VEGFR-1 and VEGFR-2 act as regulators of endothelial cell functions, VEGF-C/VEGFR-3 signaling mainly controls the growth of the lymphatic vasculature [7,8]. Receptors of VEGFR family are type V receptor tyrosine kinases (RTKs) that comprise a family of transmembrane receptors with an extracellular part of seven immunoglobulin (Ig) homology domains. They may exist as monomers or dimers. RTKs function through the protein kinase domain located in the intracellular region of each RTK monomer. The binding of VEGF to the extracellular region of VEGFR leads to its dimerization and activation of the receptor's tyrosine kinase activity and in selective transautophosphorylation of tyrosine residues. Some of these phosphorylation sites maintain the active conformation of the kinase itself, while others become docking sites for various adaptor/effector scaffold proteins and enzymes which participate in a subsequent cascade of signaling events (Fig. 1).

2 VEGF signaling pathway in angiogenesis

Among the VEGF family members, VEGF-A, the major regulator of physiological and pathological angiogenesis, specifically binds to endothelial cells via either VEGFR-1 (flt-1) or VEGFR-2 (flk-1). VEGFR-1 binds VEGF-A with a 10-fold higher affinity than VEGFR-2. However, VEGFR-2 is the main receptor transducing the effects of VEGF-A into endothelial cells. The effects of VEGF-A in endothelial cells, such as cell survival and proliferation, cannot be seen in VEGFR-1 overexpressing cells lacking VEGFR-2. In several tumor models, angiogenesis is prevented by expression of a dominant-negative mutant VEGFR-2 or by use of VEGFR-2 blocking antibodies. Targeted inactivation of VEGFR-2 leads to embryonic lethality due to defects in the differentiation of endothelial and primitive hematopoietic cells and also the formation of mature blood vessels [9,10].

The binding of VEGF to its corresponding receptors initiates phosphorylation of the RTKs and leads to the activation of downstream signaling pathways, such as the phosphatidylinositol 3-kinase/protein kinase B (PI3K/Akt) and mitogen activated protein kinase or Ras/Raf/MEK/MAPK, which are involved in the proliferation, migration, and apoptosis of endothelial cells

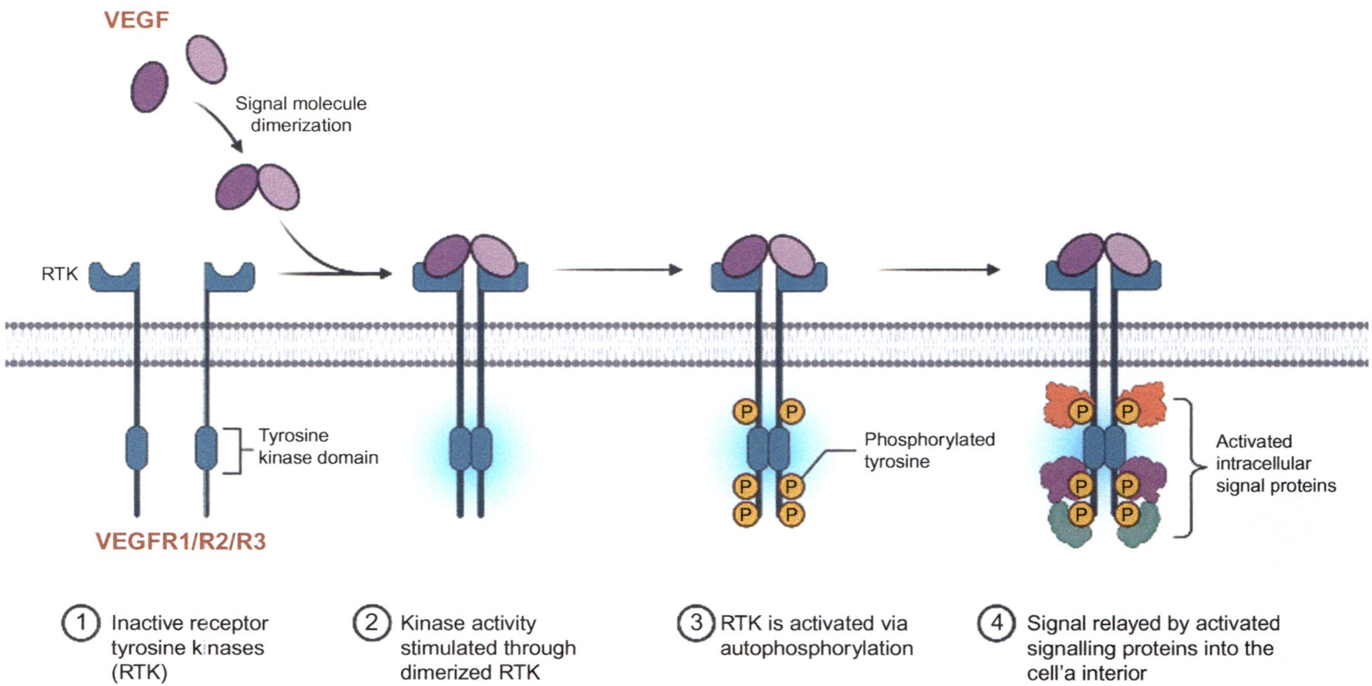

FIG. 1 Activation of receptor tyrosine kinases (RTK). When signaling molecules like VEGF bind to RTKs, they cause neighboring RTKs to associate with each other, forming dimers. Dimerization activates the tyrosine kinase domains in these RTKs, causing autophosphorylation of certain tyrosine residues. Tyrosine phosphorylation leads to subsequent activation of proteins inside the cell and the ensuing response.

[11]. The activation of the MAP kinase pathway by VEGF and subsequent cell proliferation has been documented for many types of endothelial cells. However, many studies have shown that the growth signal by phosphorylated receptor is mostly relayed through phospholipase C (PLC)γ-protein kinase C (PKC) and not via Ras pathway. Phosphorylated VEGFR-2 rapidly recruits and activates PLCγ which then releases sn-1,2-diacylglycerol (DAG) and inositol 1,4,5-trisphosphate (IP3) via the phosphatidylinositol 4,5-bisphosphate (PIP2) pathway [12]. These two secondary messengers are responsible for the direct activation of certain PKC isoforms and for the release of Ca^{2+} from intracellular stores, respectively. In addition to the stimulation of endothelial cell migration and proliferation, VEGF also provides a cell survival signal. VEGF protects endothelial cells against tumor necrosis factor α or apoptosis by inhibiting the cell caspases [13]. VEGFR-2-mediated survival effect is also transduced via the phosphatidylinositol 3-kinase (PI3-K), which further activates Akt or protein kinase B, a serine/threonine-specific protein kinase, and finally eNOS, which has vasoprotective effects [14] (Fig. 2).

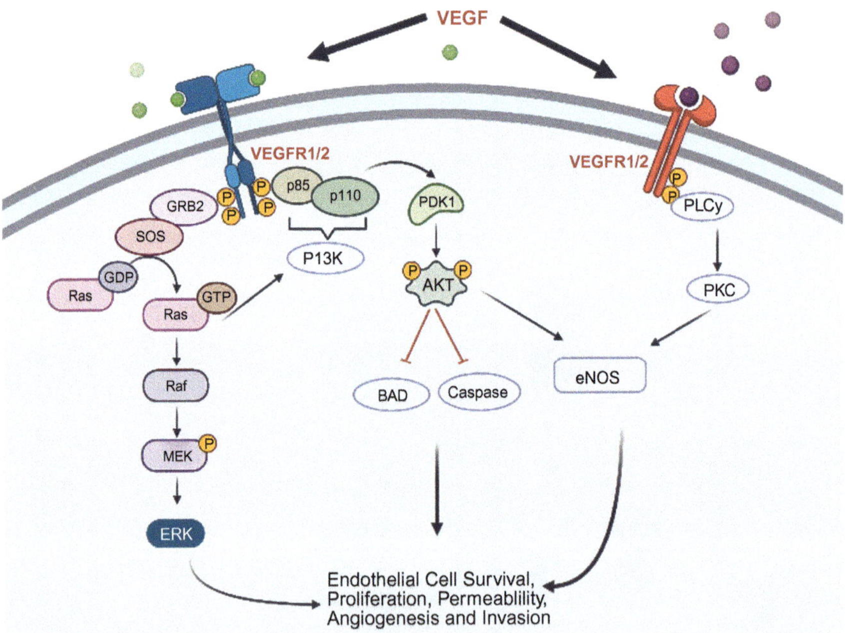

FIG. 2 VEGF/VEGFR signaling pathways. The binding of VEGF to its corresponding receptors initiates phosphorylation of the receptors. Activation of downstream signaling pathways, such as the PI3K/Akt and Ras/Raf/MEK/MAPK, is involved in the maintenance of endothelial cell survival (inhibition of apoptosis), proliferation, and migration. The VEGF growth signal by the phosphorylated receptor is also relayed through PLCγ-PKC pathway. This pathway is involved in the upregulation of eNOS, endothelial cell migration, and angiogenesis.

3 VEGF pathway inhibitors for cancer treatment

The VEGF pathway being the major perpetuator of tumor endothelial cell survival and angiogenesis represents the main target of antiangiogenic drugs in cancer therapy, which are applied as VEGF-A and/or VEGFR-2 inhibitors [15,16]. The first clinical trial in 2003 that demonstrated the benefits and prolonged survival of metastatic colorectal cancer patients was with humanized neutralizing antibodies targeting VEGF along with chemotherapy laid the foundation of antiangiogenic therapies to successfully treat cancers [17]. Subsequently, several antibodies and tyrosine kinase inhibitors (TKIs) that have been designed to target pro-angiogenic signaling have been approved as cancer therapies. TKIs have been developed to target VEGFRs and their downstream targets that inhibit tumor angiogenesis at various levels. These include sorafenib, sunitinib, axitinib, regorafenib, pazopanib, vandetanib, and cabozantinib (Table 1). These

TABLE 1 FDA-approved RTK inhibitors and antibodies.

Tyrosine kinase inhibitor	Target	Class	Clinical indication
Bevacizumab	VEGF	Monoclonal antibody	CRC, GBM, and advanced RCC
Ramucirumab	VEGF	Monoclonal antibody	Gastric cancer
Sunitinib	VEGFR-1/VEGFR-2 and PDGFR-α/β/KIT/FLT3/CSF1R	Small molecule	RCC and imatinib-resistant GIST
Sorafenib	RAF/VEGFR-2/PDGFR- β	Small molecule	Advanced HCC
Pazopanib	VEGFR-1/VEGFR-2/VEGFR-3, PDGFR-α/β, and KIT	Small molecule	Advanced RCC
Vandetanib	VEGFR-2, EGFR/RET kinase	Small molecule	MTC
Cabozantinib	VEGFR, RET, and MET	Small molecule	MTC, RCC, and HCC
Apatinib	VEGFR-2	Small molecule	Advanced gastric cancers
Axitinib	VEGFR	Small molecule	Advanced RCC
Regorafenib	VEGFR and EGFR	Small molecule	Advanced CRC, GIST, and HCC

CRC, colorectal cancer; *CSIFR*, receptor of macrophage-colony stimulating factor; *EGFR*, epidermal growth factor receptor; *FLT3*, FMS-like tyrosine kinase-3; *GBM*, glioblastoma; *GIST*, gastrointestinal stromal tumor; *HCC*, hepatocellular carcinoma; *KIT*, stem cell factor receptor; *MET*, hepatocyte growth factor receptor; *MTC*, medullary thyroid cancer; *PDGFR-α/β*, platelet-derived growth factor receptors (PDGFR-alpha and PDGFR-beta); *RCC*, renal cell carcinoma; *RET*: glial cell-line-derived neurotrophic factor receptor; *VEGF (R)*, vascular endothelial growth factor (receptor).

molecules can inhibit one or multiple angiogenic pathways. Anti-VEGF/VEGFR single-target drugs often lead to transient responses; however, tumor progression happens because other pathways, such as the PDGF/PDGFR, FGF/FGFR, and ANGPT/Tie-2, provide potential escape mechanisms [18–20]. Hence, antiangiogenic agents and TKIs inhibiting multiple signaling pathways seem more promising.

4 Cardiotoxic effects of VEGF pathway inhibitors

Although effective antiangiogenic treatment strategies have undoubtedly increased the number of cancer survivors, however, most of the angiogenesis inhibiting drugs and TKIs have been associated with adverse cardiovascular outcomes and cardiotoxicity in the long term. Cardiotoxicity is defined as one of the followings: (1) cardiomyopathy characterized by a decrease in cardiac left ventricular ejection fraction (LVEF), (2) symptoms of congestive heart failure (CHF), (3) tachycardia, (4) hypertension, (5) decline in LVEF of at least 5 to less than 55% with accompanying signs or symptoms of CHF, (6) prolonged QT interval (due to changes in action potential of ventricular myocytes and malfunction of ion channels), and (7) effects on blood vessels such as platelet aggregation, thrombosis, and atherosclerosis. The TKIs can have multiple effects on the physiology and functions of cardiomyocytes or endothelial cells [21] (Fig. 3). The cardiotoxicity of established agents

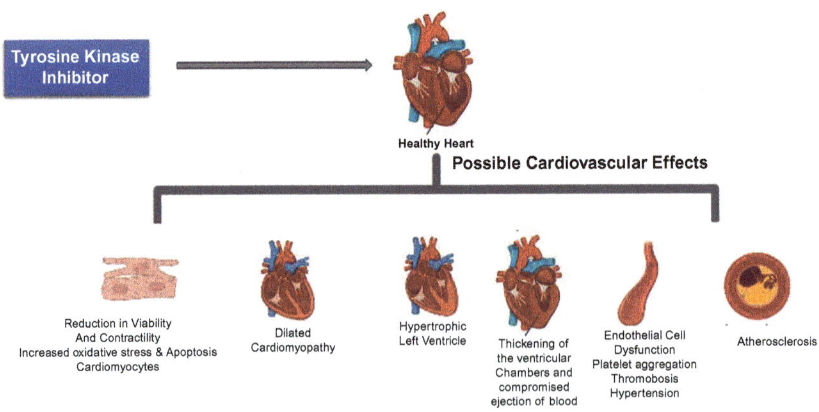

FIG. 3 Cardiotoxic effects of tyrosine kinase inhibitors (TKIs): Cardiotoxicity by TKIs can be cellular or physiological. They can lead to aberrant changes in the functions of cardiomyocytes or endothelial cells. Changes in cardiomyocytes may affect their viability and contractile functions. They may cause endothelial dysfunction, atherosclerosis, and thrombotic events. Physiologically, TKI may cause cardiomyopathy, hypertrophy of the left ventricle, thickening of the ventricular chambers, and a congested heart with a decrease in cardiac left ventricular ejection fraction.

such as anthracyclines, antimetabolites, and alkylating agents has been well established in the past years, and it is only in recent years, the adverse cardiovascular effects of TKIs, particularly VEGF pathway inhibitors have gained attention.

Inhibition of VEGF signaling not only blocks tumor angiogenesis and growth but also induces multiple deleterious effects in the cardiovascular system. VEGF/R inhibitors have been associated with an increased risk of adverse cardiovascular events including hypertension, thromboembolic events, myocardial ischemia, left ventricular dysfunction, heart failure, and QT interval prolongation (with possible associated arrhythmias). Cardiotoxicity in fact represents one of the most challenging side effects of VEGF/R inhibitors. Each TKI may act on the cardiovascular system in many specific ways. In the following paragraphs, we discuss the cardiotoxicity associated with the commonly used FDA-approved VEGF/R inhibitor drugs and TKIs providing both clinical evidence and mechanistic insights for each drug.

4.1 Bevacizumab

4.1.1 Clinical studies

Bevacizumab is a humanized monoclonal antibody that inhibits all VEGF-A isoforms to prevent tumor angiogenic processes. It is the first antiangiogenic drug approved for clinical treatments in cancer, and its intravenous administration has a therapeutic role in a variety of metastatic malignancies in colorectal cancer and glioblastoma [22,23]. Along with its antiangiogenic effects, bevacizumab has also been associated with a significant risk of cardiovascular complications, such as hypertension, cardiac ischemia, and CHF. Hypertension is a common adverse event occurring in patients treated with bevacizumab, with an overall incidence of 4%–35% reported in clinical trials. In a recent meta-analysis of randomized controlled trials, bevacizumab (5 mg/kg weekly) and chemotherapy together were associated with a statistically significant increase in hypertension in patients with renal cell carcinoma (RCC) and breast cancer [24].

Hypertension in patients may develop at any time during bevacizumab treatment, and the data suggest that there is a dose relationship. Hypertension risk is aggravated by three times with low doses and more than seven times with high doses of bevacizumab [25,26]. It should be, however, noted that hypertension is also considered as a predictive marker for clinical outcomes after bevacizumab therapy. Clinical trials with bevacizumab in various malignancies, such as metastatic breast, metastatic colorectal, nonsmall cell lung, and ovarian cancers, as well as malignant glioblastoma, have clearly shown beneficial effects and tumor progression-free survival and/or overall survival in patients who have developed bevacizumab-induced hypertension [27–29]. Discontinuation of bevacizumab has, however, been recommended if systolic BP is > 200 mmHg or if diastolic BP is > 100 mmHg. A previous history of hypertension, age > 65 years, smoking, and hypercholesterolemia have been identified as key risk factors for bevacizumab-induced hypertension [30]. Approximately 2%–4% of patients treated with bevacizumab are also known to develop CHF. The main mechanism responsible for bevacizumab-associated CHF is suggested to be uncontrolled hypertension, leading to LV hypertrophy [31]. Data, however, also indicate that the main predisposing factor for the development of CHF in bevacizumab-treated patients is due to previous therapies with cardiotoxic drugs, such as anthracyclines [32]. Recently, a study reported a case of a 62-year-old female with colorectal carcinoma who was given bevacizumab and achieved a stable disease during the entire course of treatment. Thereafter, she developed cardiotoxicity including grade 3 hypertension, tricuspid regurgitation, pulmonary hypertension, left ventricular diastolic dysfunction, and pericardial effusion after receiving bevacizumab-containing chemotherapy for more than 20 months. She was thereafter discontinued from the regimen with bevacizumab [33].

4.1.2 Mechanistic studies

The mechanisms underlying the cardiotoxic effects of bevacizumab are far from being understood. It has been proposed that VEGF inhibition by bevacizumab decreases nitric oxide production in the endothelium, leading to vasoconstriction and, therefore, to increased peripheral vascular resistance and blood pressure [34]. Reduced levels of nitric oxide also upregulate the expression of plasminogen-activator inhibitor 1, resulting in hypertension. Bevacizumab treatment in rats causes a high overexpression of inflammatory cytokines and monocyte chemoattractant protein, both in plasma and in myocardium [35], leading to increased inflammation and atherosclerotic instability, subsequent plaque rupture, and thrombus formation. Given the fact that VEGF also participates in endothelial cell survival, homeostasis, and regeneration, anti-VEGF therapy also severely compromises the regenerative capacity of endothelial cells, leading to endothelial cell dysfunction, increasing the risk of thrombosis and organ damage [36]. Using a murine model of bevacizumab-mediated cardiotoxicity, a study demonstrated loss of cellular integrity, increased oxidative stress, and increased cardiac apoptosis at day 14. The study also identified early LV systolic dysfunction by tissue velocity imaging (TVI) [37]. A miRNA array analysis reported two miRNAs, miRNA1254 and miRNA579, to be specifically elevated in patients with bevacizumab-induced cardiotoxicity, with miRNA1254 showing the strongest correlation to the clinical diagnosis of bevacizumab-induced cardiotoxicity [38].

4.2 Ramucirumab

Ramucirumab is a human immunoglobulin G1 monoclonal antibody that specifically blocks VEGFR-2 and is approved for the treatment of advanced gastric, nonsmall cell lung, and colorectal cancers. In a phase II study, patients receiving intravenous ramucirumab (10mg/kg) every 21days for 3cycles did not produce a statistically or clinically significant prolongation of QTcF (a measurement of ECG) [39]. There are yet no clinical or basic studies reporting the cardiotoxic effects of ramucirumab.

4.3 Sunitinib

4.3.1 Clinical studies

Sunitinib is a small-molecule inhibitor that blocks a repertoire of tyrosine kinase receptors such as VEGFR, PDGFR, stem cell factor receptor (KIT), colony-stimulating factor 1 receptor (CSF1R), receptor-type tyrosine-protein kinase (FLT3), and neurotrophic factor receptor (RET) [40]. Sunitinib is the gold standard of care in the treatment of several cancers. The concomitant inhibition of these receptors significantly attenuates tumor vascularization and also causes apoptosis of tumor cells. Owing to the multiple effects of sunitinib on cellular signal transduction, cell cycle regulation, and metabolism, its use has been associated with many unwanted effects on the cardiovascular system. Specifically, coronary microvessels have been implicated to be the main target of its toxicity [41]. Common cardiovascular side effects of sunitinib include hypertension, QT prolongation, left ventricular dysfunction (LVD), and heart failure [29]. Clinical studies illustrate that sunitinib treatment leads to LVEF reductions of 15% EF or more, in 20% of patients, hypertension in about 47% experienced, and congestive heart failure in 8% patients [42].

In a 1090-patient adjudicated database analysis, hypertension and symptomatic and asymptomatic LVEF decreases were more frequent in sunitinib group than in comparative interferon and placebo groups [41]. However, many studies have shown that cardiac dysfunction induced by sunitinib is reversible. After sunitinib withdrawal, hypertension and LVEF decreases were ameliorated. Many sunitinib-treated patients were able to resume sunitinib therapy following resolution of events [41]. In a recent multicenter prospective study of 90 patients with metastatic renal cell carcinoma (mRCC), echocardiography and biomarkers of cardiovascular injury and stress were quantified at baseline, 3.5, 15, and 33weeks following sunitinib treatment. Most patients receiving sunitinib exhibited only modest declines in LVEF and nonsignificant changes in high senstivity Troponin I (hsTnI) and brain natriuretic peptide (BNP). Approximately 9.7% to 18.9% of patients developed more substantive abnormalities but these changes were largely reversible with medical management [43].

In some studies, sunitinib-induced cardiac dysfunction was associated with improved survival rates [44]. Sunitinib-induced cardiac dysfunction was, however, also implicated as a positive prognostic biomarker and significantly associated with longer progression-free survival and overall survival [45,46]. It might be hypothesized that sorafenib being a multi-targeted inhibitor might be more cardiotoxic than the selective inhibitors like erlotinib. However, a contrasting study has suggested that inhibitor selectivity likely does not predict the molecular response in the heart. The study administered clinically relevant doses of sorafenib, sunitinib (cardio-toxic multitargeted kinase inhibitors), or erlotinib (a cardiosafe epidermal growth factor receptor inhibitor) to mice daily for 2 weeks and compared the effects of these three kinase inhibitors on the cardiac transcriptome using RNAseq and the cardiac kinome using multiplexed inhibitor beads coupled with mass spectrometry. The study chose erlotinib as it is a highly selective TKI without known cardiotoxicity while sunitinib and sorafenib are well known for their cardiotoxicity. Unexpectedly, broad changes were observed in the cardiac kinome activity of all 3 kinase inhibitors with erlotinib treatment causing changes in the kinome and transcriptome than either sunitinib or sorafenib. The cardiotoxic TKIs significantly affected transcriptional regulation and chromatin modification [47].

4.3.2 Mechanistic studies

Mechanistically, sunitinib in animal models upregulated the expression of hypoxia response genes, including cardiac prolyl hydroxylase domain-containing protein, an upstream regulator of HIF-1a [48–50]. It is well known that the chronic activation of HIF is sufficient to cause cardiomyopathies [50]. On isolated human heart muscles, however, oxidative stress brought by sunitinib exerted little or no effect on cardiomyocytes and contractile force [51,52]. Sunitinib-induced hypertension has been associated with activation of the endothelin-1 system, suppression of the renin-angiotensin system, and generalized microvascular dysfunction [53]. PDGFR inhibition by sunitinib damages the growth and survival of pericytes, a group of cells wrapped around capillary endothelial cells and dependent on intact PDGFR signaling. PDGFRs play diverse roles in several cellular functions of pericytes such as cell viability under pathological stress. In humans and mice, sunitinib treatment causes loss of pericytes and increased vascular permeability in the coronary vessels [41]. A recent study demonstrated the role of Sirt3 in sunitinib-induced pericyte loss and cardiotoxicity. Sunitinib-mediated pericyte death was accompanied by upregulation of Sirt3 in cardiac pericytes. The effects of sunitinib were greatly reduced in Sirt3-knockout mice, while increased in Sirt3-overexpressing mice. Sunitinib induced cardiac pericyte death through inhibition of GSTP1/JNK/autophagy pathway. Sirt3 interacted with and inhibited GSTP1, thus enhancing sunitinib-induced pericyte death [54].

Sunitinib treatment is also reported to cause changes in the energy metabolism of cells by alteration of the AMP-activated protein kinase (AMPK) activity. Since the heart requires high levels of energy, AMPK is crucial for the heart. Sunitinib has been reported to inhibit restoration of AMPK activity leading to energy-depletion-caused cell death in both in vivo and in vitro in mouse cardiomyocytes. In cultured rat neonatal cardiomyocytes, sunitinib decreased mitochondrial membrane potential and caused intracellular adenosine triphosphate (ATP) or energy depletion [55,56]. Sunitinib treatment in rats in vitro also led to a reduction of complex I- and complex II-dependent respiration in cardiomyocytes, and loss of mitochondrial membrane potential in cardiac microvascular endothelial cells [57]. Sunitinib caused inhibition of ribosomal S6 kinase (RSK), which transduces survival signals in the heart through inhibitory phosphorylation of the pro-apoptotic factor BCL2 associated agonist of cell death (BAD), and AMPK, at clinically relevant concentrations. An elegant study has proposed that mitochondrial damage with reactive oxygen species (ROS) accumulation is one of the key mechanisms of sunitinib-induced cardiotoxicity and apoptotic death of cardiomyocytes. Using H9c2 cardiac cells, the study showed that sunitinib reduced activities of mitochondrial enzyme complexes of the electron transport chain (ETC), increased mitochondrial ROS accumulation, and decreased the cellular GSH pool. Sunitinib also caused caspase 3 activation and DNA fragmentation in these cells. Sunitinib treatment in mice for 2 weeks led to cardiomyocyte damage and a decreased activity of enzyme complexes of the ETC and enhanced mitochondrial ROS, indicating cardiomyocyte apoptosis [57].

4.4 Sorafenib

4.4.1 Clinical studies

The multitargeted TKI, sorafenib inhibits VEGFR, PDGFR, and c-Kit receptors. It induces both tumor apoptosis and disruption of the tumor vasculature [58]. Sorafenib is approved for the treatment of RCC patients, as second-line treatment after failure of sunitinib, and for HCC [59,60]. In a pivotal trial, in advanced HCC, the incidence of cardiac events was similar between the sorafenib group and the placebo group, but in another observational, single-centered study, out of 25 sorafenib-treated patients, 14 were reported to have an effect on cardiac function (in terms of abnormal cardiac enzymes, cardiac arrhythmias, decreased LVEF, acute coronary syndrome, etc) [61]. The patients also showed enhanced levels of cardiac dysfunction markers, creatine kinase-MB, and cardiac troponin T.

4.4.2 Mechanistic studies

In isolated rat hearts, sorafenib caused ventricular arrhythmias and left ventricular dysfunction due to increased oxidative stress. The effects of sorafenib were attenuated by the antioxidant N-(2-mercaptopropionyl)-glycine (MPG). In isolated ventricular myocytes, sorafenib treatment led to an increase in diastolic intracellular Ca^{2+} levels, decreased Ca transients, and also the occurrence of Ca^{2+} waves. These changes were diminished by MPG, CaMKII inhibitor KN-93, and the mitochondrial permeability transition pore inhibitor, cyclosporin A. Sorafenib also reduced the activity of mitochondrial complex III and augmented mitochondrial ROS production. In vivo, rats treated with sorafenib displayed a decrease in the antioxidant defense. Histological changes such as hypertrophy, increased fibrosis, disordered myofibrils, and damaged mitochondria were also observed after treatment. All these effects were, however, protected by MPG [62]. Sorafenib markedly reduced the left ventricular pressure and heart rate and prolonged both the systolic and diastolic periods in treated rats [62].

Sorafenib cardiotoxicity seems to result from myocyte necrosis rather than from any direct effect on myocyte function. High doses and not low doses of sorafenib were reported to induce necrotic death of isolated myocytes in vitro. Histological analysis demonstrated an increased myocyte cross-sectional area despite smaller heart sizes after sorafenib treatment, a loss of the myocytes. Sorafenib also induced apoptotic cell death of cardiac- and bone-derived c-kit + stem cells in vitro and decreased the number of BrdU + (5-bromo-2′-deoxyuridine +) myocytes at the infarct border zone in formalin-fixed tissues. There was, however, no effect on infarct size, fibrosis, or post-MI neovascularization after sorafenib treatment [63].

4.5 Pazopanib

Pazopanib is a multikinase inhibitor acting on VEGFR-1/VEGFR-2/VEGFR-3, PDGFR-α/β, and c-Kit receptor [64]. Pazopanib in patients with advanced/metastatic renal cell carcinoma (mRCC) has been associated with hypertension, cardiomyopathy, and cardiac dysrhythmias. 57% of patients treated with pazopanib showed high BP [65]. It is also reported to produce a concentration-dependent decrease in heart rate and cause prolongation of the time-matched Fridericia-corrected QT (QTcF) interval. A study has described a case of fatal heart failure after treatment with gemcitabine and pazopanib in a 55-year-old female with sarcoma [66]. Another case report of a patient with spindle cell sarcoma described severe left ventricular systolic dysfunction after pazopanib therapy. Upon stopping therapy, however, there was a recovery of LVEF [67]. In animal studies, treatment with pazopanib for 2 weeks led to a significant increase in mean arterial pressure and a significant reduction in the cardiac output in mice as compared to control mice [68]. Mechanistically, however, a study has demonstrated negligible effects of pazopanib on the mitochondria of rat ventricular cardiomyocytes. There were no changes in mitochondrial membrane potential or intracellular ATP levels in these cells after pazopanib treatment [69].

4.6 Regorafenib

Regorafenib, a multikinase inhibitor targeting the RAS/RAF/MEK/ERK pathway, is used for the treatment of colorectal cancer, gastrointestinal stromal tumors, and hepatocellular carcinoma. The data on cardiotoxicity of regorafenib are limited. The recent network meta-analysis on a total of 20,027 patients from 45 randomized controlled trials associated with 9 FDA-approved VEGFR-TKIs (axitinib, cabozantinib, lenvatinib, nintedanib, pazopanib, regorafenib, sorafenib, sunitinib, and vandetanib) reported that regorafenib did not exhibit an increased risk of cardiac damage [70]. Mechanistically, regorafenib has been reported to cause loss of mitochondrial membrane potential, decreased mitochondrial DNA content, and structural abnormalities in mitochondria at high doses [71].

4.7 Vandetanib

Vandetanib, a multikinase inhibitor used for the treatment of various cancers, is known to induce several adverse cardiac effects. Two independent meta-analyses done with reported studies have shown that vandetanib had the highest risk for cardiotoxicity among all the VEGFR-TKIs [72]. Vandetanib had a relative risk of 7.71 with 95% CI 1.04–56.99 of causing cardiotoxicity [70]. A study has extensively investigated the mechanisms of vandetanib-induced cardiotoxicity using intracellular electrophysiological recordings on human-induced pluripotent stem cell-derived cardiomyocytes (hiPSC-CMs) and rabbit Purkinje fibers. Results show that in hiPSC-CMs, vandetanib markedly reduced the maximum rate of depolarization during the AP upstroke. Purkinje fiber assays and ion channel studies showed that vandetanib at concentrations of 1 and 3 μM prolonged the action potential duration [73].

4.8 Cabozantinib

Cabozantinib, targeting the MET, RET, and VEGFR-2, is used to treat medullary thyroid cancer, RCC, and HCC. A multicenter prospective study on metastatic RCC patients treated with cabozantinib between 2016 and November 2017 has revealed a modest risk of developing left ventricular systolic dysfunction related to cabozantinib [74].

4.9 Axitinib

Axitinib, a kinase inhibitor, inhibits VEGFR-1, VEGFR-2, VEGFR-3, PDGFR, and KIT. It has been approved for the treatment of metastatic renal cell carcinoma after failure of one prior systemic therapy. Axitinib has limited data on cardiotoxicity. A 70-year-old male patient with advanced

RCC treated first with interferon alpha-2b, then with sunitinib, and finally with axitinib (10 mg daily) presented with dyspnea and palpitations. The diagnostic workup showed a left ventricular dysfunction with an ejection fraction (EF) of 35% in echocardiography, indicating axitinib-associated cardiotoxicity [75].

5 Cardioprotective agents

The cardioprotective properties of several molecules against TKIs have been investigated. Statins or molecules that restore endothelial cell function, such as endothelial nitric oxide synthase/fibroblast growth factor-2 (eNOS-FGF-2) pathway, have the potential to mitigate the risk of anthracycline-induced cardiotoxicity [76]. The therapeutic effects of A_3 adenosine receptor agonist 2-chloro-N6-(3-iodobenzyl)-adenosine-5′-*N*-methyluronamide (IB-MECA) have been shown against sunitinib. A_3 adenosine receptor activation by IB-MECA attenuated sunitinib-induced cardiotoxicity through the involvement of PKCα [77]. In a recent report, oral administration of sunitinib (40 mg/kg/day) in mice progressively decreased the PI3K activity and cardiac function in 2 weeks with a significant decrease in the expression and activity of cardiac L-type Ca^{2+} channel (CaV1.2) and sarcoplasmic endoplasmic reticulum Ca^{2+}-ATPase (SERCA), both of which are key parameters of cardiomyocyte function and physiology. When cardiac-specific PI3K was overexpressed through adeno-associated virus 9-mediated gene delivery in mice, it markedly prevented the sunitinib-induced reduction in cardiac function, calcium transient, calcium current, and Cav1.2 expression. This study has implicated that increased PI3K activity is protective against sunitinib-induced calcium mishandling and contractile dysfunction [78]. Losartan, an angiotensin II receptor blocker that is used to treat hypertension, has the potential to exert a protective effect against sorafenib-induced cardiotoxicity [79]. Prophylactic administration of renin-angiotensin system inhibitors, including aliskiren, perindopril, or valsartan, prevented adverse cardiovascular remodeling and cardiac dysfunction in mice treated with bevacizumab or sunitinib [80]. Trimetazidine (TMZ) is an antiangina agent that has potential cardioprotective effects after sunitinib treatment. TMZ coadministration with sunitinib reversed cardiotoxicity in mice and attenuated cell viability loss in H9c2 cardiomyocytes. The AMPK/mTOR/autophagy pathway was also activated both in vivo and in vitro in presence of TMZ [81]. Rivaroxaban (RIVA), a factor Xa inhibitor, with cardioprotective action that inhibits atherosclerosis and numerous inflammatory cascades significantly mitigated sunitinib-induced cardiac injuries by restoring antioxidant enzyme levels and attenuating the proinflammatory cascades resulting from sunitinib-induced cardiac injuries [82].

6 Conclusion

Cardiotoxicity is one of the most challenging side effects of VEGFR-TKIs. Each TKI can affect the cardiovascular system via several overlapping and specific pathways. Regular monitoring via methods such as electroencephalogram and blood pressure should be undertaken in cancer patients undergoing treatment with TKIs. An early detection of cardiovascular toxicity using specific bloodstream biomarkers is also requisite. Also, before treatment with any TKI, the patients should be thoroughly evaluated for any past medical history and risk factors to clearly segregate those at heightened risk for TKI-induced cardiotoxicity. Some of the valuable strategies as a way forward to minimize the risk of TKI-induced cardiotoxicity include: (1) rigorous testing of TKI inhibitors in the in vitro and in vivo systems, (2) deciphering the molecular mechanisms underlying cardiotoxicity, (3) identification of predictive soluble biomarkers for early identification of cardiotoxicity in patients undergoing TKI treatment, (4) effective management of clinical symptoms of cardiotoxicity. An early identification of cardiotoxic effects and mechanism of each TKI in the in vitro studies would ensure its treatment dose optimization in cancer patients. Also, early prediction of cardiotoxicity using soluble biomarkers in patients would help to better manage these patients, thus preventing their undesirable effects on cardiovascular physiology. Lastly, the concomitant use of endothelial protective drugs alongside TKI treatment in cancer patients holds immense potential to attenuate the cardiotoxic effects of TKIs.

Acknowledgments

The figures have been created using biorender.com.

References

[1] S. Kaur, P. Bajwa, A 'tête-à tête' between cancer stem cells and endothelial progenitor cells in tumor angiogenesis, Clin. Transl. Oncol. 16 (2) (2014) 115–121.
[2] P. Carmeliet, Mechanisms of angiogenesis and arteriogenesis, Nat. Med. 6 (4) (2000) 389–395.
[3] J. Folkman, Angiogenesis, Annu. Rev. Med. 57 (2006) 1–18.
[4] S. Kaur, K. Anita, Angiogenesis in liver regeneration and fibrosis: "a double-edged sword", Hepatol. Int. 7 (4) (2013) 959–968.
[5] J.M. Isner, A. Takayuki, Therapeutic angiogenesis, Front. Biosci. 3 (1998) e49–e69.
[6] M. De Palma, D. Biziato, T.V. Petrova, Microenvironmental regulation of tumour angiogenesis, Nat. Rev. Cancer 17 (8) (2017) 457–474.
[7] P. Saharinen, L. Eklund, K. Pulkki, P. Bono, K. Alitalo, VEGF and angiopoietin signaling in tumor angiogenesis and metastasis, Trends Mol. Med. 17 (7) (2011) 347–362.
[8] S. Takahashi, Vascular endothelial growth factor (VEGF), VEGF receptors and their inhibitors for antiangiogenic tumor therapy, Biol. Pharm. Bull. 34 (12) (2011) 1785–1788.

[9] A. Álvarez-Aznar, L. Muhl, K. Gaengel, VEGF receptor tyrosine kinases: key regulators of vascular function, Curr. Top. Dev. Biol. 123 (2017) 433–482.
[10] N. Ferrara, H.P. Gerber, J. LeCouter, The biology of VEGF and its receptors, Nat. Med. 9 (6) (2003) 669–676.
[11] M. Simons, E. Gordon, L. Claesson-Welsh, Mechanisms and regulation of endothelial VEGF receptor signalling, Nat. Rev. Mol. Cell Biol. 17 (10) (2016) 611–625.
[12] X. Wang, A.M. Bove, G. Simone, B. Ma, Molecular bases of VEGFR-2-mediated physiological function and pathological role, Front. Cell Dev. Biol. 8 (2020), 599281.
[13] A.M. Byrne, D.J. Bouchier-Hayes, J.H. Harmey, Angiogenic and cell survival functions of vascular endothelial growth factor (VEGF), J. Cell. Mol. Med. 9 (4) (2005) 777–794.
[14] J. Karar, A. Maity, PI3K/AKT/mTOR pathway in angiogenesis, Front. Mol. Neurosci. 4 (2011) 51.
[15] P. Carmeliet, VEGF as a key mediator of angiogenesis in cancer, Oncology 69 (Suppl. 3) (2005) 4–10.
[16] M. Shibuya, Vascular endothelial growth factor (VEGF) and its receptor (VEGFR) signaling in angiogenesis: a crucial target for anti- and pro-angiogenic therapies, Genes Cancer 2 (12) (2011) 1097–1105.
[17] J.C. Yang, L. Haworth, R.M. Sherry, P. Hwu, D.J. Schwartzentruber, S.L. Topalian, S.M. Steinberg, H.X. Chen, S.A. Rosenberg, A randomized trial of bevacizumab, an anti-vascular endothelial growth factor antibody, for metastatic renal cancer, N. Engl. J. Med. 349 (5) (2003) 427–434.
[18] G.W. Prager, M. Poettler, M. Unseld, C.C. Zielinski, Angiogenesis in cancer: anti-VEGF escape mechanisms, Transl. Lung Cancer Res. 1 (1) (2012) 14–25.
[19] Y. Zhao, A.A. Adjei, Targeting angiogenesis in cancer therapy: moving beyond vascular endothelial growth factor, Oncologist 20 (6) (2015) 660–673.
[20] S. Qin, A. Li, M. Yi, S. Yu, M. Zhang, K. Wu, Recent advances on anti-angiogenesis receptor tyrosine kinase inhibitors in cancer therapy, J. Hematol. Oncol. 12 (1) (2019) 27.
[21] A. Seidman, C. Hudis, M.K. Pierri, S. Shak, V. Paton, M. Ashby, M. Murphy, S.J. Stewart, D. Keefe, Cardiac dysfunction in the trastuzumab clinical trials experience, J. Clin. Oncol. 20 (5) (2002) 1215–1221.
[22] N. Ferrara, K.J. Hillan, W. Novotny, Bevacizumab (Avastin), a humanized anti-VEGF monoclonal antibody for cancer therapy, Biochem. Biophys. Res. Commun. 333 (2) (2005) 328–335.
[23] N. Ferrara, K.J. Hillan, H.P. Gerber, W. Novotny, Discovery and development of bevacizumab, an anti-VEGF antibody for treating cancer, Nat. Rev. Drug Discov. 3 (5) (2004) 391–400.
[24] M.M. An, Z. Zou, H. Shen, P. Liu, M.L. Chen, Y.B. Cao, Y.Y. Jiang, Incidence and risk of significantly raised blood pressure in cancer patients treated with bevacizumab: an updated meta-analysis, Eur. J. Clin. Pharmacol. 66 (8) (2010) 813–821.
[25] X. Zhu, S. Wu, W.L. Dahut, C.R. Parikh, Risks of proteinuria and hypertension with bevacizumab, an antibody against vascular endothelial growth factor: systematic review and meta-analysis, Am. J. Kidney Dis. 49 (2) (2007) 186–193.
[26] F.F. Kabbinavar, J. Schulz, M. McCleod, T. Patel, J.T. Hamm, J.R. Hecht, R. Mass, B. Perrou, B. Nelson, W.F. Novotny, Addition of bevacizumab to bolus fluorouracil and leucovorin in first-line metastatic colorectal cancer: results of a randomized phase II trial, J. Clin. Oncol. 23 (16) (2005) 3697–3705.
[27] S.P. Gampenrieder, F. Romeder, C. Muß, M. Pircher, S. Ressler, G. Rinnerthaler, R. Bartsch, C. Sattlberger, B. Mlineritsch, R. Greil, Hypertension as a predictive marker for bevacizumab in metastatic breast cancer: results from a retrospective matched-pair analysis, Anticancer Res. 34 (1) (2014) 227–233.
[28] P. Österlund, L.M. Soveri, H. Isoniemi, T. Poussa, T. Alanko, P. Bono, Hypertension and overall survival in metastatic colorectal cancer patients treated with bevacizumab-containing chemotherapy, Br. J. Cancer 104 (4) (2011) 599–604.

[29] J. Zhong, A.N. Ali, A.D. Voloschin, Y. Liu, W.J. Curran Jr., I.R. Crocker, H.K. Shu, Bevacizumab-induced hypertension is a predictive marker for improved outcomes in patients with recurrent glioblastoma treated with bevacizumab, Cancer (2015) (Epub December 29, 2014).
[30] L. Horsley, K. Marti, G.C. Jayson, Is the toxicity of anti-angiogenic drugs predictive of outcome? A review of hypertension and proteinuria as biomarkers of response to anti-angiogenic therapy, Expert Opin. Drug Metab. Toxicol. 8 (3) (2012) 283–293.
[31] M.H. Chen, R. Kerkelä, T. Force, Mechanisms of cardiac dysfunction associated with tyrosine kinase inhibitor cancer therapeutics, Circulation 118 (1) (2008) 84–95.
[32] K.D. Miller, L.I. Chap, F.A. Holmes, et al., Randomized phase III trial of capecitabine compared with bevacizumab plus capecitabine in patients with previously treated metastatic breast cancer, J. Clin. Oncol. 23 (4) (2005) 792–799.
[33] J. Chen, F. Du, B. Hu, C. Chi, H. Chu, L. Jiang, P. Li, Z. Gong, Severe cardiotoxicity in a patient with colorectal cancer treated with bevacizumab, Anticancer Res. 37 (8) (2017) 4557–4561.
[34] T. Kamba, D.M. McDonald, Mechanisms of adverse effects of anti-VEGF therapy for cancer, Br. J. Cancer 96 (12) (2007) 1788–1795.
[35] J. Drímal, J. Zúrová-Nedelcevová, V. Knezl, R. Sotníková, J. Navarová, Cardiovascular toxicity of the first line cancer chemotherapeutic agents: doxorubicin, cyclophosphamide, streptozotocin and bevacizumab, Neuro Endocrinol. Lett. 27 (Suppl. 2) (2006) 176–179.
[36] S. Sundararajan, A. Kumar, M. Poongkunran, A. Kannan, N.J. Vogelzang, Cardiovascular adverse effects of targeted antiangiogenic drugs: mechanisms and management, Future Oncol. 12 (8) (2016) 1067–1080.
[37] K.A. Bordun, S. Premecz, M. daSilva, S. Mandal, V. Goyal, T. Glavinovic, M. Cheung, D. Cheung, C.W. White, R. Chaudhary, D.H. Freed, H.R. Villarraga, J. Herrmann, M. Kohli, A. Ravandi, J. Thliveris, M. Pitz, P.K. Singal, S. Mulvagh, D.S. Jassal, The utility of cardiac biomarkers and echocardiography for the early detection of bevacizumab- and sunitinib-mediated cardiotoxicity, Am. J. Physiol. Heart Circ. Physiol. 309 (4) (2015) H692–H701.
[38] Z. Zhao, J. He, J. Zhang, M. Liu, S. Yang, N. Li, X. Li, Dysregulated miR1254 and miR579 for cardiotoxicity in patients treated with bevacizumab in colorectal cancer, Tumour Biol. 35 (6) (2014) 5227–5235.
[39] A.J. Olszanski, D.C. Smith, L.H. Camacho, J. Thompson, S.S. Ramalingam, R.D. Harvey, L. Campos, D. Ferry, S. Tang, L. Gao, H. Safran, Electrocardiographic characterization of ramucirumab on the corrected QT interval in a phase II study of patients with advanced solid tumors, Oncologist 21 (4) (2016) 402–403.
[40] D.Y. Heng, C. Kollmannsberger, Sunitinib, Recent Results Cancer Res. 184 (2010) 71–82.
[41] T.F. Chu, M.A. Rupnick, R. Kerkela, S.M. Dallabrida, D. Zurakowski, L. Nguyen, K. Woulfe, E. Pravda, F. Cassiola, J. Desai, S. George, J.A. Morgan, D.M. Harris, N.S. Ismail, J.H. Chen, F.J. Schoen, A.D. Van den Abbeele, G.D. Demetri, T. Force, M.H. Chen, Cardiotoxicity associated with tyrosine kinase inhibitor sunitinib, Lancet 370 (9604) (2007) 2011–2019.
[42] V. Chintalgattu, M.L. Rees, J.C. Culver, A. Goel, T. Jiffar, J. Zhang, K. Dunner Jr., S. Pati, J.A. Bankson, R. Pasqualini, W. Arap, N.S. Bryan, H. Taegtmeyer, R.R. Langley, H. Yao, M.E. Kupferman, M.L. Entman, M.E. Dickinson, A.Y. Khakoo, Coronary microvascular pericytes are the cellular target of sunitinib malate-induced cardiotoxicity, Sci. Transl. Med. 5 (187) (2013) 187ra69.
[43] M.S. Ewer, T.M. Suter, D.J. Lenihan, L. Niculescu, A. Breazna, G.D. Demetri, R.J. Motzer, Cardiovascular events among 1090 cancer patients treated with sunitinib, interferon, or placebo: a comprehensive adjudicated database analysis demonstrating clinically meaningful reversibility of cardiac events, Eur. J. Cancer 50 (12) (2014) 2162–2170.

[44] V. Narayan, S. Keefe, N. Haas, L. Wang, I. Puzanov, M. Putt, A. Catino, J. Fang, N. Agarwal, D. Hyman, A.M. Smith, B.S. Finkelman, H.K. Narayan, S. Ewer, C. ElAmm, D. Lenihan, B. Ky, Prospective evaluation of sunitinib-induced cardiotoxicity in patients with metastatic renal cell carcinoma, Clin. Cancer Res. 23 (14) (2017) 3601–3609.
[45] J. Rautiola, F. Donskov, K. Peltola, H. Joensuu, P. Bono, Sunitinib-induced hypertension, neutropaenia and thrombocytopaenia as predictors of good prognosis in patients with metastatic renal cell carcinoma, BJU Int. 117 (1) (2016) 110–117.
[46] O.P. Hamnvik, T.K. Choueiri, A. Turchin, R.R. McKay, L. Goyal, M. Davis, M.D. Kaymakcalan, J.S. Williams, Clinical risk factors for the development of hypertension in patients treated with inhibitors of the VEGF signaling pathway, Cancer 121 (2) (2015) 311–319.
[47] T.J. Stuhlmiller, J.S. Zawistowski, X. Chen, N. Sciaky, S.P. Angus, S.T. Hicks, T.L. Parry, W. Huang, J.Y. Beak, M.S. Willis, G.L. Johnson, B.C. Jensen, Kinome and transcriptome profiling reveal broad and distinct activities of erlotinib, sunitinib, and sorafenib in the mouse heart and suggest cardiotoxicity from combined signal transducer and activator of transcription and epidermal growth factor receptor inhibition, J. Am. Heart Assoc. 6 (10) (2017) e006635.
[48] T. Eschenhagen, T. Force, M.S. Ewer, G.W. de Keulenaer, T.M. Suter, S.D. Anker, M. Avkiran, E. de Azambuja, J.L. Balligand, D.L. Brutsaert, G. Condorelli, A. Hansen, S. Heymans, J.A. Hill, E. Hirsch, D. Hilfiker-Kleiner, S. Janssens, S. de Jong, G. Neubauer, B. Pieske, P. Ponikowski, M. Pirmohamed, M. Rauchhaus, D. Sawyer, P.H. Sugden, J. Wojta, F. Zannad, A.M. Shah, Cardiovascular side effects of cancer therapies: a position statement from the Heart Failure Association of the European Society of cardiology, Eur. J. Heart Fail. 13 (1) (2011) 1–10.
[49] J. Moslehi, Y.A. Minamishima, J. Shi, D. Neuberg, D.M. Charytan, R.F. Padera, S. Signoretti, R. Liao, W.G. Kaelin Jr., Loss of hypoxia-inducible factor prolyl hydroxylase activity in cardiomyocytes phenocopies ischemic cardiomyopathy, Circulation 122 (10) (2010) 1004–1016.
[50] T. Force, D.S. Krause, R.A. Van Etten, Molecular mechanisms of cardiotoxicity of tyrosine kinase inhibition, Nat. Rev. Cancer 7 (5) (2007) 332–344.
[51] K.A. Henderson, R.B. Borders, J.B. Ross, T.B. Huwar, C.O. Travis, B.J. Wood, Z.J. Ma, S.P. Hong, T.M. Vinci, B.M. Roche, Effects of tyrosine kinase inhibitors on rat isolated heart function and protein biomarkers indicative of toxicity, J. Pharmacol. Toxicol. Methods 68 (1) (2013) 150–159.
[52] A.M. Thijs, S. El Messaoudi, J.C. Vos, A.C. Wouterse, V. Verweij, H. van Swieten, C.M. van Herpen, W.T. van der Graaf, L. Noyez, G.A. Rongen, Sunitinib does not attenuate contractile force following a period of ischemia in isolated human cardiac muscle, Target. Oncol. 10 (3) (2015) 439–443.
[53] M.H. Kappers, J.H. van Esch, W. Sluiter, S. Sleijfer, A.H. Danser, A.H. van den Meiracker, Hypertension induced by the tyrosine kinase inhibitor sunitinib is associated with increased circulating endothelin-1 levels, Hypertension 56 (4) (2010) 675–681.
[54] Y. Yang, N. Li, T. Chen, C. Zhang, J. Li, L. Liu, Y. Qi, X. Zheng, C. Zhang, P. Bu, Sirt3 promotes sensitivity to sunitinib-induced cardiotoxicity via inhibition of GTSP1/JNK/autophagy pathway in vivo and in vitro, Arch. Toxicol. 93 (11) (2019) 3249–3260.
[55] R. Kerkela, K.C. Woulfe, J.B. Durand, R. Vagnozzi, D. Kramer, T.F. Chu, C. Beahm, M.H. Chen, T. Force, Sunitinib-induced cardiotoxicity is mediated by off-target inhibition of AMP-activated protein kinase, Clin. Transl. Sci. 2 (1) (2009) 15–25.
[56] J.D. Cohen, J.E. Babiarz, R.M. Abrams, L. Guo, S. Kameoka, E. Chiao, J. Taunton, K.L. Kolaja, Use of human stem cell derived cardiomyocytes to examine sunitinib mediated cardiotoxicity and electrophysiological alterations, Toxicol. Appl. Pharmacol. 257 (1) (2011) 74–83.
[57] J. Bouitbir, A. Alshaikhali, M.V. Panajatovic, V.F. Abegg, F. Paech, S. Krähenbühl, Mitochondrial oxidative stress plays a critical role in the cardiotoxicity of sunitinib: running title: sunitinib and oxidative stress in hearts, Toxicology 426 (2019), 152281.

[58] S.M. Wilhelm, C. Carter, L. Tang, D. Wilkie, A. McNabola, H. Rong, C. Chen, X. Zhang, P. Vincent, M. McHugh, Y. Cao, J. Shujath, S. Gawlak, D. Eveleigh, B. Rowley, L. Liu, L. Adnane, M. Lynch, D. Auclair, I. Taylor, R. Gedrich, A. Voznesensky, B. Riedl, L.E. Post, G. Bollag, P.A. Trail, BAY 43–9006 exhibits broad spectrum oral antitumor activity and targets the RAF/MEK/ERK pathway and receptor tyrosine kinases involved in tumor progression and angiogenesis, Cancer Res. 64 (2004) 7099–7109.

[59] B. Escudier, T. Eisen, W.M. Stadler, C. Szczylik, S. Oudard, M. Siebels, S. Negrier, C. Chevreau, E. Solska, A.A. Desai, F. Rolland, T. Demkow, T.E. Hutson, M. Gore, S. Freeman, B. Schwartz, M. Shan, R. Simantov, R.M. Bukowski, TARGET study group. Sorafenib in advanced clear-cell renal-cell carcinoma, N. Engl. J. Med. 356 (2007) 125–134.

[60] J.M. Llovet, S. Ricci, V. Mazzaferro, P. Hilgard, E. Gane, J.F. Blanc, A.C. de Oliveira, A. Santoro, J.L. Raoul, A. Forner, M. Schwartz, C. Porta, S. Zeuzem, L. Bolondi, T.F. Greten, P.R. Galle, J.F. Seitz, I. Borbath, D. Häussinger, T. Giannaris, M. Shan, M. Moscovici, D. Voliotis, J. Bruix, SHARP Investigators Study Group, Sorafenib in advanced hepatocellular carcinoma, N. Engl. J. Med. 359 (378–390) (2008) 20.

[61] M. Schmidinger, C.C. Zielinski, U.M. Vogl, A. Bojic, M. Bojic, C. Schukro, M. Ruhsam, M. Hejna, H. Schmidinger, Cardiac toxicity of sunitinib and sorafenib in patients with metastatic renal cell carcinoma, J. Clin. Oncol. 26 (2008) 5204–5212.

[62] W. Ma, M. Liu, F. Liang, L. Zhao, C. Gao, X. Jiang, X. Zhang, H. Zhan, H. Hu, Z. Zhao, Cardiotoxicity of sorafenib is mediated through elevation of ROS level and CaMKII activity and dysregulation of calcium homoeostasis, Basic Clin. Pharmacol. Toxicol. 126 (2) (2020) 166–180.

[63] J.M. Duran, C.A. Makarewich, D. Trappanese, P. Gross, S. Husain, J. Dunn, H. Lal, T.E. Sharp, T. Starosta, R.J. Vagnozzi, R.M. Berretta, M. Barbe, D. Yu, E. Gao, H. Kubo, T. Force, S.R. Houser, Sorafenib cardiotoxicity increases mortality after myocardial infarction, Circ. Res. 114 (11) (2014) 1700–1712.

[64] E.I. Heath, J. Infante, L.D. Lewis, T. Luu, J. Stephenson, A.R. Tan, S. Kasubhai, P. LoRusso, B. Ma, A.B. Suttle, J.F. Kleha, H.A. Ball, M.M. Dar, A randomized, double-blind, placebo-controlled study to evaluate the effect of repeated oral doses of pazopanib on cardiac conduction in patients with solid tumors, Cancer Chemother. Pharmacol. 71 (3) (2013) 565–573.

[65] D. Pinkhas, T. Ho, S. Smith, Assessment of pazopanib-related hypertension, cardiac dysfunction and identification of clinical risk factors for their development, Cardiooncology 3 (2017) 5.

[66] D. Di Lisi, G. Manno, C. Filorizzo, T. Guarino, G. Santanelli, M. Lunetta, G. Badalamenti, A. Russo, G. Novo, Fatal heart failure induced by pazopanib in a sarcoma patient previously treated with gemcitabine, J. Saudi Heart Assoc. 32 (2) (2020) 285–287.

[67] K. Gajjar, K. Balakumaran, A.S. Kim, Reversible left ventricular systolic dysfunction secondary to pazopanib, Cureus 10 (10) (2018), e3517.

[68] A. Kempton, C. Justice, A. Guo, M. Cefalu, M. Makara, P. Janssen, T.H. Ho, S.A. Smith, Pazopanib for renal cell carcinoma leads to elevated mean arterial pressures in a murine model, Clin. Exp. Hypertens. 40 (6) (2018) 524–533.

[69] K.J. French, R.W. Coatney, J.P. Renninger, C.X. Hu, T.L. Gales, S. Zhao, L.M. Storck, C.B. Davis, J. McSurdy-Freed, E. Chen, K.S. Frazier, Differences in effects on myocardium and mitochondria by angiogenic inhibitors suggest separate mechanisms of cardiotoxicity, Toxicol. Pathol. 38 (5) (2010) 691–702.

[70] W. Hou, M. Ding, X. Li, X. Zhou, Q. Zhu, A. Varela-Ramirez, C. Yi, Comparative evaluation of cardiovascular risks among nine FDA-approved VEGFR-TKIs in patients with solid tumors: a Bayesian network analysis of randomized controlled trials, J. Cancer Res. Clin. Oncol. 147 (8) (2021) 2407–2420.

[71] T. Boran, A.G. Akyildiz, A.T. Jannuzzi, B. Alpertunga, Extended regorafenib treatment can be linked with mitochondrial damage leading to cardiotoxicity, Toxicol. Lett. 336 (2021) 39–49.

[72] M. Santoni, F. Guerra, A. Conti, A. Lucarelli, S. Rinaldi, L. Belvederesi, A. Capucci, R. Berardi, Incidence and risk of cardiotoxicity in cancer patients treated with targeted therapies, Cancer Treat. Rev. 59 (2017) 123–131.

[73] H.A. Lee, S.A. Hyun, B. Byun, J.H. Chae, K.S. Kim, Electrophysiological mechanisms of vandetanib-induced cardiotoxicity: comparison of action potentials in rabbit Purkinje fibers and pluripotent stem cell-derived cardiomyocytes, PLoS ONE 13 (4) (2018), e0195577.

[74] R. Iacovelli, C. Ciccarese, G. Fornarini, F. Massari, D. Bimbatti, C. Mosillo, S.E. Rebuzzi, V. Di Nunno, M. Grassi, E. Fantinel, A. Ardizzoni, G. Tortora, Cabozantinib-related cardiotoxicity: a prospective analysis in a real-world cohort of metastatic renal cell carcinoma patients, Br. J. Clin. Pharmacol. 85 (6) (2019) 1283–1289.

[75] O. Tanriverdi, S. Ates, K.K. Sandal, S. Uylas, I.C. Bosna, A. Alkan, Left ventricular dysfunction associated with axitinib and nivolumab experience in an advanced renal cell carcinoma, J. Oncol. Pharm. Pract. 26 (7) (2020) 1765–1768.

[76] L. Morbidelli, S. Donnini, M. Ziche, Targeting endothelial cell metabolism for cardio-protection from the toxicity of antitumor agents, Cardiooncology 2 (1) (2016) 3.

[77] H. Sandhu, S. Cooper, A. Hussain, C. Mee, H. Maddock, Attenuation of sunitinib-induced cardiotoxicity through the A3 adenosine receptor activation, Eur. J. Pharmacol. 814 (2017) 95–105.

[78] C. Li, R. Zou, H. Zhang, Y. Wang, B. Qiu, S. Qiu, W. Wang, Y. Xu, Upregulation of phosphoinositide 3-kinase prevents sunitinib-induced cardiotoxicity in vitro and in vivo, Arch. Toxicol. 93 (6) (2019) 1697–1712.

[79] A.A. Abdelgalil, O.Y. Mohamed, S.R. Ahamad, F.I. Al-Jenoobi, The protective effect of losartan against sorafenib induced cardiotoxicity: ex-vivo isolated heart and metabolites profiling studies in rat, Eur. J. Pharmacol. 882 (2020), 173229.

[80] V. Mozolevska, A. Schwartz, D. Cheung, V. Goyal, B. Shaikh, B. Dingman, E. Kim, I. Mittal, C.Y. Asselin, A. Edel, A. Ravandi, J. Thliveris, P.K. Singal, P. Czaykowski, D.S. Jassal, Role of renin-angiotensin system antagonists in the prevention of bevacizumab- and sunitinib-mediated cardiac dysfunction, Am. J. Physiol. Heart Circ. Physiol. 316 (3) (2019) H446–H458.

[81] Y. Yang, N. Li, T. Chen, C. Zhang, L. Liu, Y. Qi, P. Bu, Trimetazidine ameliorates sunitinib-induced cardiotoxicity in mice via the AMPK/mTOR/autophagy pathway, Pharm. Biol. 57 (1) (2019) 625–631.

[82] F. Imam, N.O. Al-Harbi, M.R. Khan, W. Qamar, M. Alharbi, A.A. Alshamrani, H.N. Alhamami, N.B. Alsaleh, K.S. Alharbi, Protective effect of RIVA against sunitinib-induced cardiotoxicity by inhibiting oxidative stress-mediated inflammation: probable role of TGF-β and Smad signaling, Cardiovasc. Toxicol. 20 (3) (2020) 281–290.

CHAPTER

7

Critical insights into cardiotoxicity of anthracyclines

Hanumappa Ananda[a] *and Imteyaz Qamar*[b]

[a]Department of Clinical Embryology, Kasturba Medical College, Manipal, Manipal Academy of Higher Education, Manipal, Karnataka, India, [b]School of Biotechnology, Gautam Buddha University, Greater Noida, Uttar Pradesh, India

1 Introduction

Anthracyclines are cytostatic antibiotics, among the most generally used and highly effective drugs in cancer treatment. In the past several years, they have become an integral component of adjunctive and palliative therapy for various cancers. Anthracyclines are anticancer drugs that were originally derived from *Streptomyces* bacteria [1]. Anthracyclines are belonged to a group of antineoplastic antibiotics, which interfere with cell cycle replication by interfering with the DNA at various levels, thereby showing an adverse effect in every phase of the cell cycle. These drugs act mainly by interacting with DNA and interfering with DNA metabolism and RNA production. The primary structure of anthracyclines is that of a tetracyclic molecule with an anthraquinone strength connected to a sugar moiety by a glycosidic linkage. When taken up by a cell, the four-ring structure intercalates between DNA base pairs, while the sugar sits within the minor groove and interacts with adjacent base pairs. Doxorubicin, daunorubicin, and epirubicin are anthracyclines. Tan et al. [2] described that the pathognomonic appearance of anthracycline-induced cardiotoxicity is a hypokinetic cardiomyopathy increasingly important to heart failure. Preclinically, anthracycline-induced cardiomyopathy appears to negatively affect the outcome of patients, as chemotherapeutic options are limited [3]. Cardiovascular-related side effects are standard as possibly the major leading cause of treatment-related morbidity and mortality among cancer fighters [4]. The following cardiotoxic events are

Cardiovascular Toxicity and Therapeutic Modalities Targeting Cardio-oncology
https://doi.org/10.1016/B978-0-323-90461-2.00009-2

TABLE 1 Classification, risk factors, screening, and prevention for anthracycline-induced cardiotoxicity.

Classification	Risk factors	Screening	Prevention
Acute cardiotoxicity (1%) (1 day–2 weeks)	Age > 65 or < 18	Regular screening	Dexrazoxane
Early-onset chronic cardiotoxicity (98%) Within 1 year	Sex		Liposomal anthracyclines
Late-onset chronic cardiotoxicity (1%) ➢ 1 year	American African	History and physical examination	Statins
	Cumulative dose	Electrocardiogram	ACE inhibitors
	Left ventricular dysfunction	Biomarkers (troponin and natriuretic peptide)	RAS inhibitors
	Hypertension	Echocardiography	Beta blockers
	Renal failure	Magnetic resonance	Vincristine
	Personal health habits Comorbidities		PDE5 inhibitors
	Preexisting cardiac risk factors		Vincristine
	Concomitant therapy		
	Rate of anthracycline administration		

mainly categorized by their time of onset as acute (occurring during or immediately after treatment), early (occurring within 1 year of exposure), or late (occurring one or more years after initial exposure), risk factors, screening, and prevention for cardiac abnormalities in cancer survivors (Table 1).

2 Mechanism underlying anthracycline-induced cardiotoxicity

The cardiomyocyte has always been measured the significant cellular objective of anthracycline toxic effect in the heart, as their devastation results in the advanced growth of cardiac dysfunction. Anthracyclines lead to gather in the powerhouse of mitochondria, which relatively elucidates their tendency to formation of myocardial tissue, which is characterized by a high mitochondrial density due to its high ATP demand [5]. However, other various cell types, such as cardiac progenitor cells, cardiac fibroblasts, and endothelial cells, have been recognized as probable additional targets, in which paracrine effects mediated by microRNA (miRNA) and other cell indications could similarly be involved in cardiomyocyte injury, generating additional complex and interesting situation in the pathogenesis of anthracycline-induced cardiomyopathy [6]. The anthracyclines have been extensively studied for their communications with molecular components and their influence on cellular processes. The detailed mechanism of action of anthracycline cardiotoxicity remains very undistinguishable [7]. Anthracyclines inhibit cancer through multiple pathways. They promote creation of free radicals, which destroy nucleic acids and other biomolecules. The possible mechanism is the generation of reactive oxygen species (ROS) and changes in iron metabolism, DNA damage, senescence, and cell death. The following mechanisms that occur at clinically relevant drug concentrations are very important.

2.1 DNA intercalation

Anthracyclines are occupied by cells and localized to the nucleus. The chromophore moiety of anthracyclines has intercalating function and inserts in between the adjacent base pair of DNA. DNA intercalating drugs from the keystone of common anticancer regimes. Prominent between these are the anthracyclines, a group of compounds including daunorubicin and doxorubicin [8]. The DNA intercalators function by introducing between the base pairs of the DNA double helix, thereby altering its cellular topology. DNA-related proteins are inhibited and lead to a reduction of DNA replication, causing the death of actively dividing cancer cells. Unfortunately, anthracyclines can cause severe cardiotoxicity. The DNA intercalation of anthracyclines with DNA is always reversible, and no covalent binding is necessary. Hydrophobic interactions, hydrogen bonds to the phosphate groups of the DNA, and the insertion of the daunosamine sugar into the small groove of the DNA with an attraction to the CpG-complex and transcriptional energetic sites of the DNA lead to a static drug-DNA-complex with a very long half-life.

2.2 Topoisomerase II poison

The most extensively recognized for the achievement of anthracyclines is their interaction with topoisomerase II. The ternary complex, topoisomerase II, is an enzyme that creates temporary double-stranded DNA (dsDNA) breaks and reseals them after managing torsion of DNA supercoils. The anthracycline compounds intercalated in DNA form a rigid anthracycline-DNA-topoisomerase II ternary complex, thereby poisoning the enzyme and hindering the fusion of double-stranded DNA breaks [9]. It is very important and possible that intercalation of anthracyclines induces an alteration in three-dimensional conformation of DNA that arrests the cycle of TOPO-II action at the point of DNA cleavage. Numerous studies have shown that anthracyclines induce topoisomerase II-medicated DNA damage at drug absorptions that are clinically significant. With respect to DNA intercalation and inhibition of topoisomerase II, the anthracyclines act as chemically inactive compounds by their capacity to distort the three-dimensional geometry of the targets DNA and TOPO-II [10].

2.3 Reactive oxygen species

The important mechanism of chemotherapy-induced cardiotoxicity comprises the generation of ROS. High level of ROS that causes cellular damages and changes in responses, referred to as oxidative stress, to a delicate balance of the ROS-generating system and antioxidant defense system. The chemical compound quinone moiety of anthracyclines can undergo redox reactions [11]. To produce more quantity of reactive oxygen species (ROS) in the presence of oxidoreductive enzymes such as cytochrome P450 reductase, xanthine oxidase, and NADH dehydrogenase. Converting quinone to semiquinone produces free radicals that actively react with oxygen to generate superoxides, hydroxyl radicals, and peroxides [12]. In addition to direct action on DNA, RNA, proteins, and lipids, ROS can also act as secondary signaling molecules in innumerable pathways that are complicated in homeostasis, including cell proliferation and cell death [13]. In the heart, oxidative stress could lead to cellular hypertrophy [14], extracellular matrix [15], gene expression changes [16], activation of cell apoptosis [17], calcium transient perturbation [18], and ventricular remodeling [17], all that lead to the pathophysiological changes that result in cardiomyopathy and heart failure.

2.4 DNA adduct formation

Anthracyclines can also form adducts with DNA by a single covalent bond through an aminal linkage from the 3′-amino of daunosamine to the exocyclic amino of guanine (N-C-N) linkage to the exocyclic amino group of guanine residues [19].

3 Molecular mechanism of DOX-mediated cardiotoxicity

Doxorubicin-induced toxicity first reported was in 1967 with clinical presentation of heart failure in children treated with doxorubicin for leukemia. The scientific studies revealed persuasively that cardiac abnormality could be the right supplement with repeated doxorubicin administration [20]. The cytotoxic effect of doxorubicin (DOX) on tumor cells and its contaminated action on different organs are thought to be related to nucleotide base intercalation and cell membrane lipid binding activities of doxorubicin. Generally, the doxorubicin-induced cardiotoxicity is divided into three different methods: early-onset chronic (within days or weeks) acute, and chronic progressive cardiotoxicity (weeks to months after drug administration). Toxic effects of doxorubicin occur during administration of the drug and/or immediately after [21]. The increasing cumulative dose of doxorubicin that leads to thinning of the ventricular wall and then systolic performance will be decreased and reflected a progressive cardiomyocyte injury. The changes in histopathology are very unique and comprised of inflation of the sarcoplasmic reticulum of myocytes, swelling of mitochondria, cytoplasmic vacuolization inflammation of mitochondria, and myofibril disturbance and loss [22]. Doxorubicin's detailed mechanism of cardiotoxicity is multifactorial and involves free radical stress, calcium overload, and mitochondrial dysfunction. These are major causes of doxorubicin-induced cardiotoxicity. However, the change in gene activation and expression of ubiquitin-ligase proteasome system and cell death are all involved in its cardiotoxicity. Furthermore, the level of oxidative stress stimulated by doxorubicin is up to10 times more than in the level in the liver, spleen, and kidney [23].

As in the situation of cellular respiration and energy production, mitochondria play an important role in adaptable pathophysiological activities, such as proliferation, differentiation, cell growth, damage, repair, and cell death [24]. Recent data have shown that the primary mechanism of DOX-induced mitochondrial dysfunction strength be connected to mitochondrial permeability transition (MPT). DOX could impair NF-kB signaling accompanied by mitochondrial death protein such as Bcl2/19 kDa-related protein [25]. The latest study revealed that DOX could also increase mitochondrial fragmentation and accelerate mitochondrial degradation by the lysosome, ultimately leading to cardiotoxicity [26]. Although the action of DOX on mitochondrial injury in the heart has been broadly studied, the exact mechanism and pathways by which DOX induces mitochondrial permeability transition pore (MPTP) showing and mitochondrial fragmentation need to be further discovered.

Endoplasmic/sarcoplasmic reticulum stress and calcium dyshomeostasis in cardiac cell changes in the action of endoplasmic reticulum (ER) lead to the formation of stretched or misfolded proteins and lead to a cellular condition called ER stress [27]. After initiation of ER stress is sustained or aggravated, the c-JUN NH2- terminal kinase (JNK), C/EBP homologous protein (CHOP), and caspase-12 signaling pathways-mediated cell death arise [28]. Additional study displayed that DOX could proliferate transcriptional intermediary factor 1 (TIF1) communication but inhibit

GRP78 expression to motivate ATF6 and IRE1 and lead to ER stress [29]. The specialized form of the ER in cardiac cells is sarcoplasmic reticulum (SR), which is predominantly responsible for the regulation of Ca^{2+} fluxes and subsequently the regulation of excitation-contraction coupling [30]. Llach et al. revealed the cardiac regulation with a decrease in ejection fraction and minor expansion at 15 weeks post the end of the treatment, which could be linked with a shortage in the intracellular free Ca^{2+} concentration [31]. Less amount of GRP78 overexpression could safeguard cardiomyocytes from DOX-encouraged cell death by decreasing CaMKII stimulation. The above CaMKII initiation and p53 formation qws are further explained by the Ca^{2+}-dependent manner in which GRP78 exerts its positive effects in the normalization of Ca^{2+} management [32]. Finally, in DOX-induced cardiotoxicity ER stress and $Ca2^{+}$ dyshomeostasis are very essential. However, the useful relationship between the ER and SR in cardiomyocytes is yet unclear; therefore, the interface between ER stress in DOX-induced cardiotoxicity needs additional investigation.

Programmed cell death in multicellular organisms is named as apoptosis. It plays a very important role in the development of organisms, stability of the inner atmosphere, and common cell renewal [33]. A study described that DOX-induced cardiotoxicity is linked to nuclear translocation of p53 and p53-dependent apoptosis. The kinases such as extracellular signal-regulated kinases (ERKs) could be initiated by DOX to phosphorylate p53 at Ser15. The phosphorylation of p53 could lead to cardiomyocyte apoptosis via downregulation of antiapoptotic B-cell lymphoma 2 (Bcl-2); upregulation of pro-apoptotic Bcl-2-associated X protein (Bax); and activation of caspases-3 caspase-9, and ADP-ribose polymerase [34]. Besides, DOX-induced oxidative stress unlocks the voltage-dependent anion channel (VDAC) to outcome in MPT and consequently discharge cytochrome complex (Cyt C) from mitochondria [35]. Cyt C modulates the allosteric activation and hepta-oligomerization of the adaptor molecule apoptosis-protease-activating factor 1 to produce in apoptosome in the presence of ATP, which recruits the dimers of caspase-9 to enable the stimulation of caspase-3, ultimately commencing the apoptotic deprivation stage [36]. Even though the effects and mechanisms of DOX on cardiomyocyte apoptosis have been studied widely, the exact signaling pathways involved in the DOX-induced apoptosis are considered complicated.

4 Physiological functions of the redox system in CV toxicity

Anthracyclines generally used for treatments have improved overall survival and progression-free survival in the treatment of an increasing number of cardiotoxicity deaths. However, various types of anthracycline drugs can cause a broad range of cardiovascular toxicity, particularly in long-term cancer survivors [37]. Cardiotoxicity comprises vasospastic and thromboembolic ischemia, hypertension, dysrhythmia, myocarditis, and left ventricular dysfunction, leading to heart failure [38]. More newly, immune myocarditis has entered as a very grand challenge in the cardio-oncologic arena, due to an accumulative number of patients cured with immune checkpoint inhibitors that unlash immune responses [39]. Anthracyclines (doxorubicin, epirubicin, and daunorubicin), broadly used to treat various types of cancers, are recognized as prototype of type 1 cardiotoxicity since the 1960s [2]. ANT can damage CTX via a series of numerous cellular and molecular mechanisms. Administration of anthracyclines can modify redox homeostasis in cardiomyocytes and tissue residents (e.g., fibroblasts, endothelial cells, mast cells, and macrophages) and mixing inflammatory cells (e.g., neutrophils and eosinophils) in the heart by manufacturing ROS [40].

5 Strategies to limit doxorubicin cardiotoxicity

The current trend of a number of cancer survivors continues to be high as the success of cancer treatment modalities increases. Anthracyclines are still the important chemotherapeutic drug class of choice for considering many cancers [41]. Cardio-oncology is a widespread, dynamic, and new arena of medicine and physiology. As identified and considered in the most prominent published studies, such as prevention and treatment of anthracycline-induced cardiotoxicity. Here, we have summarized the strengths, weaknesses, prospects, and fears associated with the field as identified in the most relevant published results with special emphasis on the inhibition of anthracycline-induced cardiotoxicity. Cancer and the resultant cardiotoxicity from both current and conventional treatment considerably affect a cumulative number of survivors. The prime strategy for preventing and managing chemotherapy-induced cardiotoxicity remains unknown. We would resist that the routine use of neurohormonal antagonists for primary cardio protection in this population is not presently reasonable, given only peripheral profits and associated contrary events, mainly with long-term use. On the contrary, dexrazoxane provides effective primary cardioprotection against anthracycline-induced cardiotoxicity, and its use beyond the existing FDA-permitted indications should be investigated further in detail. In the longitudinal direction, studies are very much needed to elucidate the prognostic value of subclinical markers of management connected cardiovascular injury on the long-term risk of cardiovascular disease.

6 Summary

In summary, it looks that the numerous mechanisms of action that have been ascribed to the anthracyclines may be associated with the consumption of various drug concentrations under a varied investigational situation. The difference in mechanisms of action at the level of the tumor cell and the heart at a clinically more appropriate dose delivers a possibility for the development of real approaches for limiting cardiotoxicity while maintaining antitumor effectiveness. The use of the anthracycline class of therapeutics continues to be hindered due to dose-dependent cardiac toxicity, the true mechanism of action of which remains uncertain. The latest research and development discerns this mechanism, and it has become more evident that anthracycline cardiotoxicity is due to more than uniquely cell death, but comprises effects on both organization and purpose of cardiac myocytes. Different studies have defined anthracyclines to conquer transcription and stimulate degradation of very important sarcomeric proteins, prominent to myocyte sarcopenia that can finally shrinkage contractility. The risk advantage ratio of several approaches and/or treatments to abrogate cardiotoxicity has to be critically analyzed to ensure that endorsements follow a sound empirical in pediatric population as well as in adults who are cumulative in number from cancer survivorship as battling cancer should not expose them to improved cardiovascular risk and/or events. In the end, more advanced clinical research is warranted to address this issue.

References

[1] A. Fujiwara, T. Hoshino, J.W. Westley, Anthracycline antibiotics, Crit. Rev. Biotechnol. 3 (1985) 133–157.
[2] C. Tan, H. Tasaka, K.P. Yu, M.L. Murphy, D.A. Karnofsky, Daunomycin, an antitumor antibiotic, in the treatment of neoplastic disease. Clinical evaluation with special reference to childhood leukemia, Cancer 20 (1967) 333–353.
[3] G. Curigliano, D. Cardinale, S. Dent, C. Criscitiello, O. Aseyev, D. Lenihan, C.M. Cipolla, Cardiotoxicity of anticancer treatments: epidemiology, detection, and management, CA Cancer J. Clin. 66 (2016) 309–325.
[4] H. Abdel-Qadir, P.C. Austin, D.S. Lee, E. Amir, J.V. Tu, P. Thavendiranathan, K. Fung, G.M. Anderson, A population-based study of cardiovascular mortality following early-stage breast cancer, JAMA Cardiol. 2 (2017) 88–93.
[5] A.B. Anderson, G. Xiong, E.A. Arriaga, Doxorubicin accumulation in individually electrophoresed organelles, J. Am. Chem. Soc. 126 (2004) 9168–9169.
[6] D. Cappetta, F. Rossi, E. Piegari, F. Quaini, L. Berrino, K. Urbanek, A. De Angelis, Doxorubicin targets multiple players: a new view of an old problem, Pharmacol. Res. 127 (2018) 4–14.
[7] D.A. Gewirtz, A critical evaluation of the mechanisms of action proposed for the antitumor effects of the anthracycline antibiotics adriamycin and daunorubicin, Biochem. Pharmacol. 57 (1999) 727–741.

[8] K.R. Hande, Clinical applications of anticancer drugs targeted to topoisomerase II, Biochim. Biophys. Acta 1400 (1998) 173–184.

[9] M. Binaschi, M. Bigioni, A. Cipollone, C. Rossi, C. Goso, C.A. Maggi, G. Capranico, F. Animati, Anthracyclines: selected new developments, Curr. Med. Chem.—Anti-Cancer Agents 1 (2001) 113–130.

[10] B.R. Abdella, J. Fisher, A chemical perspective on the anthracycline antitumor antibiotics, Environ. Health Perspect. 64 (1985) 4–18.

[11] C. Panis, A.C.S.A. Herrera, V.J. Victorino, F.C. Campos, L.F. Freitas, T. De Rossi, A.N. Colado Simão, A.L. Cecchini, R. Cecchini, Oxidative stress and hematological profiles of advanced breast cancer patients subjected to paclitaxel or doxorubicin chemotherapy, Breast Cancer Res. Treat. 133 (2012) 89–97.

[12] P. Angsutararux, S. Luanpitpong, S. Issaragrisil, Chemotherapy-induced cardiotoxicity: overview of the roles of oxidative stress, Oxidative Med. Cell. Longev. 2015 (2015), 795602.

[13] D.G. Deavall, E.A. Martin, J.M. Horner, R. Roberts, Drug-induced oxidative stress and toxicity, J. Toxicol. 2012 (2012), 645460.

[14] A. Sabri, H.H. Hughie, P.A. Lucchesi, Regulation of hypertrophic and apoptotic signaling pathways by reactive oxygen species in cardiac myocytes, Antioxid. Redox Signal. 5 (2003) 731–740.

[15] D.A. Siwik, P.J. Pagano, W.S. Colucci, Oxidative stress regulates collagen synthesis and matrix metalloproteinase activity in cardiac fibroblasts, Am. J. Phys. Cell Phys. 280 (2001) C53–C60.

[16] A. Siwik Deborah, D. Tzortzis John, R. Pimental David, L.F. Chang Donny, J. Pagano Patrick, K. Singh, B. Sawyer Douglas, S. Colucci Wilson, Inhibition of copper-zinc superoxide dismutase induces cell growth, hypertrophic phenotype, and apoptosis in neonatal rat cardiac myocytes in vitro, Circ. Res. 85 (1999) 147–153.

[17] S.H. Kwon, D.R. Pimentel, A. Remondino, D.B. Sawyer, W.S. Colucci, H2O2 regulates cardiac myocyte phenotype via concentration-dependent activation of distinct kinase pathways, J. Mol. Cell. Cardiol. 35 (2003) 615–621.

[18] N.S. Dhalla, R.M. Temsah, T. Netticadan, Role of oxidative stress in cardiovascular diseases, J. Hypertens. 18 (2000).

[19] S.M. Cutts, A. Nudelman, A. Rephaeli, D.R. Phillips, The power and potential of doxorubicin-DNA adducts, IUBMB Life 57 (2005) 73–81.

[20] H. Al-kuraishy, A. Al-Gareeb, Cardio-protective effects of cyclosporine in doxorubicin induced cardiotoxicity and assessment of Interleukin-17 as biomarker of cardiac injury: an animal model study, Adv. Biomed. Pharma. 2 (3) (2015) 138–145.

[21] D.D. Von Hoff, M.W. Layard, P. Basa, H.L. Davis, A.L. Von Hoff, M. Rozencweig, F.M. Muggia, Risk factors for doxorubicin-induced congestive heart failure, Ann. Intern. Med. 91 (1979) 710–717.

[22] M.J. Adams, S.E. Lipshultz, Pathophysiology of anthracycline- and radiation-associated cardiomyopathies: implications for screening and prevention, Pediatr. Blood Cancer 44 (2005) 600–606.

[23] J. Diamanti, B. Mezzetti, F. Giampieri, J.M. Alvarez-Suarez, J.L. Quiles, A. Gonzalez-Alonso, C. Ramirez-Tortosa Mdel, S. Granados-Principal, A.M. Gonzáles-Paramás, C. Santos-Buelga, M. Battino, Doxorubicin induced oxidative stress in rats is efficiently counteracted by dietary anthocyanin differently enriched strawberry (*Fragaria × ananassa* Duch.), J. Agric. Food Chem. 62 (2014) 3935–3943.

[24] J. Nunnari, A. Suomalainen, Mitochondria: in sickness and in health, Cell 148 (2012) 1145–1159.

[25] R. Dhingra, M. Guberman, I. Rabinovich-Nikitin, J. Gerstein, V. Margulets, H. Gang, N. Madden, J. Thliveris, L.A. Kirshenbaum, Impaired NF-κB signalling underlies cyclophilin D-mediated mitochondrial permeability transition pore opening in doxorubicin cardiomyopathy, Cardiovasc. Res. 116 (2020) 1161–1174.

[26] M.P. Catanzaro, A. Weiner, A. Kaminaris, C. Li, F. Cai, F. Zhao, S. Kobayashi, T. Kobayashi, Y. Huang, H. Sesaki, Doxorubicin-induced cardiomyocyte death is mediated by unchecked mitochondrial fission and mitophagy, FASEB J. 33 (2019) 11096–11108.
[27] C. Hetz, The unfolded protein response: controlling cell fate decisions under ER stress and beyond, Nat. Rev. Mol. Cell Biol. 13 (2012) 89–102.
[28] I. Tabas, D. Ron, Integrating the mechanisms of apoptosis induced by endoplasmic reticulum stress, Nat. Cell Biol. 13 (2011) 184–190.
[29] Z.-M. Xu, C.-B. Li, Q.-L. Liu, P. Li, H. Yang, Ginsenoside Rg1 prevents doxorubicin-induced cardiotoxicity through the inhibition of autophagy and endoplasmic reticulum stress in mice, Int. J. Mol. Sci. 19 (2018) 3658.
[30] M. Michalak, M. Opas, Endoplasmic and sarcoplasmic reticulum in the heart, Trends Cell Biol. 19 (2009) 253–259.
[31] A. Llach, M. Mazevet, P. Mateo, O. Villejouvert, A. Ridoux, C. Rucker-Martin, M. Ribeiro, R. Fischmeister, B. Crozatier, J.-P. Benitah, Progression of excitation-contraction coupling defects in doxorubicin cardiotoxicity, J. Mol. Cell. Cardiol. 126 (2019) 129–139.
[32] H. Tscheschner, E. Meinhardt, P. Schlegel, A. Jungmann, L.H. Lehmann, O.J. Müller, P. Most, H.A. Katus, P.W. Raake, CaMKII activation participates in doxorubicin cardiotoxicity and is attenuated by moderate GRP78 overexpression, PLoS ONE 14 (2019), e0215992.
[33] G. Takemura, H. Fujiwara, Doxorubicin-induced cardiomyopathy: from the cardiotoxic mechanisms to management, Prog. Cardiovasc. Dis. 49 (2007) 330–352.
[34] J. Liu, W. Mao, B. Ding, C.-S. Liang, ERKs/p53 signal transduction pathway is involved in doxorubicin-induced apoptosis in H9c2 cells and cardiomyocytes, Am. J. Phys. Heart Circ. Phys. 295 (2008) H1956–H1965.
[35] A.C. Childs, S.L. Phaneuf, A.J. Dirks, T. Phillips, C. Leeuwenburgh, Doxorubicin treatment in vivo causes cytochrome C release and cardiomyocyte apoptosis, as well as increased mitochondrial efficiency, superoxide dismutase activity, and Bcl-2: Bax ratio, Cancer Res. 62 (2002) 4592–4598.
[36] C. Garrido, L. Galluzzi, M. Brunet, P. Puig, C. Didelot, G. Kroemer, Mechanisms of cytochrome c release from mitochondria, Cell Death Differ. 13 (2006) 1423–1433.
[37] S.H. Armenian, C. Lacchetti, A. Barac, J. Carver, L.S. Constine, N. Denduluri, S. Dent, P.S. Douglas, J.-B. Durand, M. Ewer, C. Fabian, M. Hudson, M. Jessup, L.W. Jones, B. Ky, E.L. Mayer, J. Moslehi, K. Oeffinger, K. Ray, K. Ruddy, D. Lenihan, Prevention and monitoring of cardiac dysfunction in survivors of adult cancers: American Society of Clinical Oncology clinical practice guideline, J. Clin. Oncol. 35 (2016) 893–911.
[38] T.M. Suter, M.S. Ewer, Cancer drugs and the heart: importance and management, Eur. Heart J. 34 (2013) 1102–1111.
[39] D.B. Johnson, J.M. Balko, M.L. Compton, S. Chalkias, J. Gorham, Y. Xu, M. Hicks, I. Puzanov, M.R. Alexander, T.L. Bloomer, Fulminant myocarditis with combination immune checkpoint blockade, N. Engl. J. Med. 375 (2016) 1749–1755.
[40] C.G. Tocchetti, C. Cadeddu, D. Di Lisi, S. Femmino, R. Madonna, D. Mele, I. Monte, G. Novo, C. Penna, A. Pepe, From molecular mechanisms to clinical management of antineoplastic drug-induced cardiovascular toxicity: a translational overview, Antioxid. Redox Signal. 30 (2019) 2110–2153.
[41] M. Valcovici, F. Andrica, C. Serban, S. Dragan, Cardiotoxicity of anthracycline therapy: current perspectives, Arch. Med. Sci. 12 (2016) 428–435.

CHAPTER

8

Noninvasive cardiovascular imaging techniques

Bipul Kumar Sinha[a], *Uditi*[b], *and Aman Rathore*[b]

[a]**Fortis Escorts Heart Institute & Research Centre, New Delhi, Delhi, India,**
[b]**School of Biotechnology, Gautam Buddha University, Greater Noida, Uttar Pradesh, India**

1 Introduction

A major boon offered by these noninvasive imagine techniques is the evaluation of therapeutics effects of these approaches anatomically and physiologically, evaluation of treatment response, and progression of disease along with the apt diagnosis and by providing the accurate qualitative and quantitative measurements while working in in vivo systems [1].

This noninvasive cardiac imaging refers to a combination of various techniques serving the purpose of obtaining images giving inferences regarding structure and function of heart in order to avoid the invasive measurement methods requiring the insertion of catheters [2]. Having the advantage of being noninvasive, these techniques are easier to perform and can be used to detect various heart conditions [3]. These techniques help in identifying or ruling out various forms of heart diseases in accordance with the person's symptoms, in establishing the risk of future heart disease developments, for example, heart attack possibility, and also help in deciding the need for further medical therapies and treatment options for the patient, for example, deciding with the treatment option of placing stents in the arteries of heart or opting for a bypass surgery as an aggressive treatment with medications associated with a patent with an indication of coronary artery disease [4,5].

Although all the symptoms like chest discomfort, jaw or left shoulder-arm pain, shortness of breath, or light-headedness or dizziness or fatigue correspond to cardiac testing, these imaging techniques help the

Cardiovascular Toxicity and Therapeutic Modalities Targeting Cardio-oncology
https://doi.org/10.1016/B978-0-323-90461-2.00012-2

physician to develop the correlation between the symptoms with a particular disease and come up with the best treatment plan [6]. The choice of these techniques depends upon the type of evaluation required by the patient for heart disease being evaluated in accordance with the medical history, for example, the patients with the history of coronary heart disease require testing for abnormalities in blood flow, whereas the lower risk patients with no cardiac history cardiac computed tomography are considered for investigatory purposes [7,8].

1.1 Pretest considerations: Patient history and physical examination

Before attempting to implement any of these imaging techniques, the most important step from a clinical perspective is taking the patient history into account and physical examination that lights the path forward for the diagnosis using any of the techniques appearing suitable. Also, patient history and physical examination can guide a great deal about the possible suspicious disease or early onset of any disease coinciding with the symptoms and most importantly can warn about any precautions or restrictions the physician might have to entertain for any particular patient such as patient with metal implants like pacemakers and defibrillators. Patient with kidney function abnormalities could not take a certain test, and the physician might have to find an alternative diagnostic path for a particular patient.

This chapter aims to provide the comprehensive understanding of different types of noninvasive cardiovascular imaging techniques along with the advantages, limitations, and threats posed by them along with the clinical basis of their applicability under indicated conditions.

2 Echocardiology

Echocardiology is a widely used diagnostic procedure for the investigation of a cardiovascular disorder where cardiac echo or echocardiograms are created using two-dimensional, three-dimensional, and Doppler ultrasound in order to get the comprehensive and diagnostic images of heart [9]. Cardiac echo or ultrasound evaluates the overall heart function, such as a calculation of the cardiac output, ejection fraction, and diastolic function, assessment of heart size and function, valvular function, and pericardial diseases [10]. The principle behind the ultrasound-based imaging technique is that a piezoelectric transducer converts electrical signals to acoustic waves that oscillate at frequencies from 2 to 30 MHz and when these waves transmit into the body, they are reflected as they pass through the layers of different tissues [11]. The vibrations created by the transducer are converted into the electrical signal and are sent to the scanner for the image processing [12].

The echocardiogram generated is processed on the basis of echo duration, frequency, and signal intensity of test; most importantly the magnitude of excitation frequency is directly proportional to the penetration depth of the acoustic wave and to the spatial resolution of the image. Enhancement of blood-myocardial interfaces, improvement of measurement of tissue perfusion, and blood flow are achieved by contrast-enhanced ultrasound by using gas-filled microbubbles conjugated with targeted ligands that bind the receptors and enabling targeted delivery of contrast agents to the sites of disease under the relative measurement of Doppler ultrasound flow and velocity [13].

Recent developments in the field have shown promise of the technique's applicability in delivering and tracking cells using the Doppler ultrasound along with determining the presence of many types of indicated heart defects, such as valve disease, myocardial disease, pericardial disease, infective endocarditic, cardiac masses, and congenital heart disease, as can also be seen in the figure illustrating different types and applications of technique as well (Fig. 1) [9]. M-mode echocardiography displays a one-dimensional slice of cardiac structure varying over time. It is used mostly for the measurement of cardiac dimensions (wall thickness and chamber size) and cardiac function (fractional shortening and valve thickening). M-mode echocardiography is also used for assessing the motion of intracardiac structures (opening and closing of valves and movement of free valves and septa) and the anatomy of valves (Fig. 1).

There are five different types of echocardiography techniques.

2.1 Transthoracic echocardiography (TTE).
2.2 Transesophageal echocardiography (TEE).
2.3 Stress echocardiography
  a. Stress echocardiography
  b. Dobutamine stress echocardiography (DSE)
2.4 Three-dimensional echocardiography (3D Echo).
2.5 Contrast echocardiography.

Principle: Echocardiology uses ultrasound waves of frequency of 1-10MHZ which are emitted from piezoelectric crystals that are kept in a transducer microcomputer. From the transducer, ultrasound waves traverse the internal body structure, interact with the tissues, get reflected back to the transducer, and here with the help of microcomputers, images are generated. These images suggest the cardiac structure and function. Early ultrasound machines are used to send a single beam which showed single scan line that was noted with the help of moving paper or a screen. This method was known as **M-mode echocardiography**. Now, 2-D imaging method is used [14].

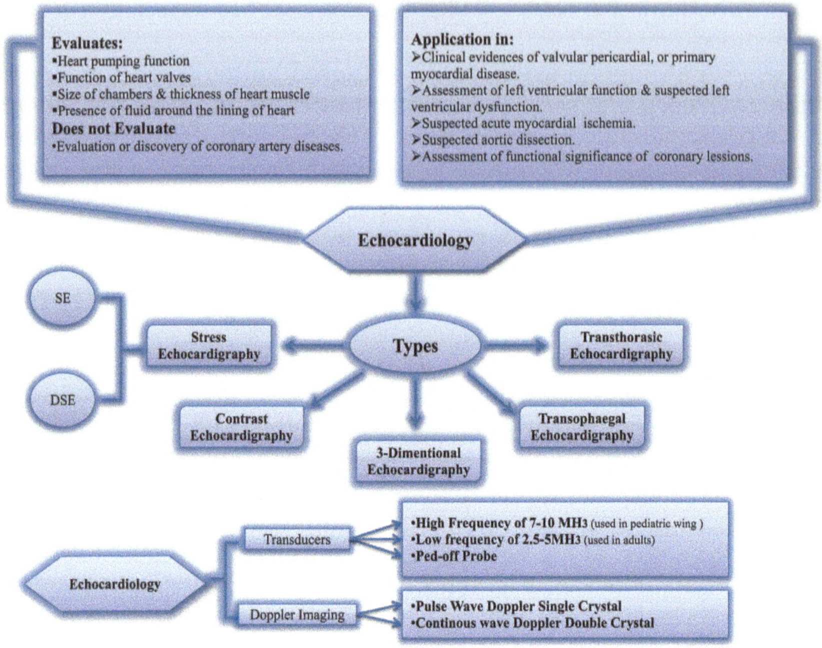

FIG. 1 Various types and applications of echocardiology used in clinical practice.

In 2-D imaging, phased array transducers are used. This transducer emits pulses of ultrasound in an ordered sequence and sequentially listens for returning echoes. This is referred as pulse-echo principle. Repetition of this sequence gives the moving images. The rate at which these pulses are emitted is called the pulse repetition frequency (PRF). Proper interpretation of returning signals is physically limited by the speed of sound in tissues in the depth of the tissues in interrogation which in turn gives the time taken by ultrasound signal to return to the transducer (Table 1).

2.1 Transthoracic echocardiography (TTE)

Transthoracic echocardiography uses a transducer that sends sound waves and picks up echoes that reflect to obtain the images of heart onto the different location over the chest and abdominal wall and can be used to identify anatomic abnormalities (e.g., valvular injury, pericardial effusion), ventricular dyskinesia, shunting of blood, and intracardiac thrombi [15]. This type of echocardiography provides necessary insights to patients who manifest hemodynamic instability or abnormal screening tests (ECG, cardiac enzymes) but has lower sensitivity in the patient whose condition is stable [16]. This transthoracic echocardiography has replaced invasive studies such as cardiac catheterization for the initial diagnosis

TABLE 1 List of various advantages and disadvantages offered by echocardiology.

Applications	Limitations
Regional wall motion abnormality (RWMA) for looking ischemia a. Hypokinesia b. Akinesia c. Dyskinesia d. Aneurysm	Operator based
Ejection fraction	Imaging artifacts
Valvular heart disease, identifying (i) Regurgitation (ii) Stenosis	Poor image quality leading to false-negative result for RWMA (means missing a wall motion abnormality, which is present)
Pericardial effusion—detection and quantification and dimensions	A patient free of chest pain at the time of imaging may not reveal RWMA
Detecting pericarditis and cardiomyopathy	Echo is insensitive for small areas of subendocardium or microvessel ischemia
Detecting Infective endocarditis Constrictive pericarditis Pulmonary hypertension Myocarditis Congenital heart diseases	Cannot find coronary artery diseases

of most forms of congenital heart diseases. A complete transthoracic echocardiographic examination usually entails combinations of M-mode two-dimensional (2-D) imaging, as well as pulsed, continuous, and color Doppler flow studies. Doppler tissue imaging and other new technologies provide a more quantitative assessment of ventricular function [17]. The 3-dimensional (3D) echocardiography provides valuable information regarding cardiac morphology.

Transthoracic echocardiography views:

- PLAX view—Patient should lie down in left lateral position and left hand above head. Put the transducer on third or fourth ICS (intercostals' space on left side chest and keeping the former toward the right shoulder.
- Short-axis view—From the PLAX view, rotate the transducer to 90 degree in a clockwise direction and tilt the former in superior to inferior to get the all short-axis views.
  - Aortic valve level
  - Mitral valve level
  - Apical.
- 4 "c" view—put the transducer in apex and keep toward the left and move laterally and inferiorly if required to get true apex.

- 5 "c" view—Tilt transducer anteriorly to obtain "5"c.
- 5. 2 "c" view and 3 "c" view—Rotate the transducer in 60–90 degree counterclockwise position from apex tilt the said to clockwise and posteriorly to open out descending aorta. Rotate transducer 30–45% anticlockwise for 3 "c" view.
- Suprasternal and subcostal—Make the patient supine and legs bent at knees. Put the transducer subxiphoid and midline and tilt anteriorly under sternum with groove toward patient's left for subcoastal 4 "c" view. Then, rotate 90 degrees clockwise to get the groove pointing anterosuperiorly to get views. You will have to sweep left to right to get apical, midventricular, and aortic valve levels.
- Suprasternal view—Patient supine, head-tilted backward. Put the transducer in a suprasternal notch with groove toward left, rotate to about 1 o'clock, parallel to trachea.
- IVC—for IVC, we put a scan plane directed toward the right midclavicular region with some counterclockwise rotation to open out the long axis of IVC. Color Doppler is used for IVC flow, PW is for hepatic veins [18].

2.2 Transesophageal echocardiography (TEE)

Transesophageal echocardiography (TEE) is a better method than TTE (transthoracic examination) where a small transducer is introduced into the patient's esophagus. In this method, a transducer is of higher frequency and remains in posterior and closer to the heart than can be achieved with TEE [19]. Transesophageal echocardiography is used routinely to monitor ventricular function in patients during difficult surgical procedures and can provide an immediate assessment of the result of surgical repair of congenital heart lesions.

Indications of transesophageal echocardiography (TEE) are as follows:

- Acute myocardial ischemic syndromes
- valvular heart disease
- prosthetic valves
- cardiomyopathy
- Hypotension (especially in the ICU)
- Aortic aneurysm
- Aortic dissection
- Intraoperative pericardial diseases
- Dyspnea (to differentiate cardiac vs noncardiac cause)
- Cardioembolic source
- In AF to find out LAA clot
- Trauma
- Congenital heart disease, etc.

Transesophageal echocardiography (TEE) has a useful modality in bedside, inpatient, and outpatient. It is useful in both the operative room and cardiac catheterization laboratory having applications in CVS. The first step of TEE is to look for dentures. It should be removed then insert IV line and blood pressure cuff in patient's arm. ECG lends to attach. A pulse oximeter should be removed for the arrangement of oxygen inhalation. Topical anesthesia should be applied to suppress gas reflex (use viscose 2% lidocaine), and if required IV sedation with midazolam, fentanyl, and sometimes with propofol is used. Patient is allowed to lie down facing the echo machine. The probe is inserted in two ways by either digital technique or rotary controls on the TEE probe [20].

Transducer positions are of four types:

1. upper esophageal
2. midesophageal
3. transgastric and
4. Deep transgastric.

Higher resolution, no acoustic shadowing by ribs, and especially identification of left arterial appendage clots used to guide cardioversion atrial fibrillation are the three advantages that transesophageal echocardiography have over transthoracic echocardiography.

Drawbacks of transesophageal echocardiography (TEE):

- Semi-invasive.
- Trabeculations in the atrial appendage may be wrongly studied as thrombi.
- Lipomatous hypertrophy of the interatrial septum can be incorrectly labeled as a mass.
- The transverse and oblique sinuses can be wrongly studied as abscess cavities.
- Lungs can bring reverberation artifacts, which can be wrongly interpreted as dissection flaps.
- Sometimes LV apex and right-sided structure may not be visible by TEE.
- Blind spot of acoustic shadowing where the trachea is interposed between the heart and esophagus.

2.3 Stress echocardiography

Stress echocardiography is the most commonly used technique for the assessment of coronary artery diseases by performing using either exercise on treadmill/bicycle or using the pharmacological stress with dobutamine hence corresponding to its two types [21]. The stress echo has the advantage of allowing acquisition of peak exercise images or repositioning the patient to obtain the accurate images [22]. Wall motion abnormalities in the ischemic segments are developed at the peak exercise during a stress echo and hibernating myocardium is associated with an improvement in wall motion on exercise which also provides a physiologically accurate assessment of the effects of exercise on cardiac function [23]. This is important in the setting of associated dyspnea, which may be caused by exercise-induced or exacerbated mitral regurgitation, left ventricular outflow tract obstruction, or pulmonary hypertension [24].

Dobutamine stress echocardiography (DSE) is the second kind of stress echocardiology that is performed on the patients unable to exercise; hence, the physiological assessment is difficult to interpret with this type of examination [25]. The protocol of the test involves the use of high-dose dobutamine with increasing increments to a maximum of 40mg/min and additionally atropine is used in the case of failure of achieving the required heart rate [26].

The advantage of this type of echocardiology is the assessment of viability more than the exercise echo by relying on the bimodal change in the wall motion which initially improves in function at low doses before the function worsens at high doses [27]. The reason behind this being dobutamine initially acting as inotropic agent, thereby improving myocardial contractility and at high doses acting at chronotropic agent because of which vasodilator effects become more prominent [28].

2.4 Three-dimensional echocardiography (3D Echo)

Three-dimensional echocardiography has some additional advantages that include faster acquisition which is due to the ability to obtain multiple planes in one single scan and more accurate assessment of ventricular volumes and functions. This 3D echo is considered more accurate than the 2D echo [29]. Recently, work is going on 3D echocardiography as is better used to study valvular heart disease and congenital heart disease and useful in interventions (surgical and percutaneous). It is also efficient in the quantification of LV and RV functions. It has reduced sensitivity in detecting anterior ischemia. Studies have demonstrated that there is a requirement to use intravenous contrast for adequate visualization of endocardium with this method [24].

This type of echo is severely used for the detection of RWMA which is of the following types:

1. Akinesia—Lack of movement or contraction of a region of the heart muscle.
2. Hypokinesia—Reduced movement or contraction of a segment of the heart muscle.
3. Dyskinesia—An abnormality bulging out in systole in place of contracting in systole.
4. Hyperkinesia—Increased contraction of heart muscle.

Echo should be used as the method of choice to evaluate serial changes in heart function, detect late side effects of treatment, and identify patients at risk of future decrease in LVEF. Decrease in LVEF suggests cardiotoxicity [30].

2.5 Contrast echocardiography

Contrast echocardiography has proven to be a valuable clinical tool for endocardial delineation in difficult to image patients, and accurate echocardiography perfusion estimates still remain a challenge and are not performed clinically [31]. Likewise, the technique, the anticipated option of targeted drug delivery through contrast agents, still remains to be under preclinical stage. The contrast agents for echo are composed of microbubbles of inert gas encapsulated within a lipid-protein or polymer coat that acts to scatter incident ultrasound at the gas-liquid interface in order to increase the strength of a returning echo signal [32].

In this modality of echocardiography, contrast agents are given as IV injection and the contrast agent is stabilized gas bubbles. Currently approved contrast agent is perfluorocarbon gases and this is used because of their quality of doing resistance to diffusion into the blood. If agitated saline is used, then larger bubbles are created [33]. This method gives the way to the assessment of myocardial microcirculation. Contrast agents are helpful in assessing aneurysm and diverticula, noncompaction cardiomyopathy, intracardiac mass, i.e., thrombi and tumor, and assessing their vascularity, MI complications like free wall rupture and pseudoaneurysms apical hypertrophy, and transient apical ballooning endomyocardial fibrosis. It is helpful in septal ablation for obstructive HCM [34].

These contrast agents are injected into the patient's arm and are transported throughout even to the small capillaries in order to enhance the blood pool signal which would without contrast appear black on the standard two-dimensional echo, this information due to the new enhanced signal helps the physician to analyze the parts of the patient's heart muscle which are perused poorly [35]. Nowadays in the coming future of medicine, the echocardiology is expanding its reach and offering advantages in the fetal monitoring in which fetal echocardiography is used on the expectant mothers during weeks 18 to 22 of pregnancy where the transducer is placed over the woman's abdomen to check for heart problems in the fetus. The test is considered safe for an unborn child because it does not use radiation, unlike an X-ray [36].

Echocardiology has become a routine practice in diagnosis, management, follow-up patients with any suspicion or known evidence of any heart disease as it can provide helpful information such as the size and shape of the heart (internal chamber size quantification), pumping capacity, and the location and extent of any tissue damage. The echocardiographic examination can be used to evaluate cardiac structure in congenital heart lesions; estimate intracardiac pressures and gradients across stenotic valves and vessels; quantify cardiac contractile function both systolic and diastolic; and determine the direction of flow across a defect, the presence of vegetation from endocarditic as well as the presence of pericardial fluid, cardiac tumors, and chamber thrombi. It can also give physicians other estimates of heart function, such as a calculation of the cardiac output, ejection fraction, and diastolic function (how well the heart relaxes) along with assessment of overall function of the heart. It can also be used to assist in the performance of interventional procedures, including pericardiocentesis, balloon atrial septostomy, atrial or ventricular septal defect closure and endocardial biopsy, and placement of flow-directed pulmonary artery monitoring catheters. It is a quite useful and informative test being used in the field of noninvasive cardiovascular imaging while some of the limitations posed by the test are taken for further investigation by the means of more specialized and stringent testing. Alternatives and more defined imaging options are explained further in this chapter.

3 Magnetic resonance imaging (MRI)

Cardiac magnetic resonance imaging is known for its immense abilities in the evaluation of cardiac structure and function as various important parameters such as delay gadolinium enhancement imaging, cardiac structure, function, abnormal ventricular wall motion, myocardial ischemia, infarction, and viability with stress perfusion that can be evaluated along with the additional advantages of being able to assess tissue characteristics, such as edema, fibrosis, and hemorrhage [37]. This technique can also display a reduction or deficiency of myocardial perfusion, and also, whole-heart-synchronized dynamic myocardial perfusion image can be obtained from high-resolution 3D MRI of the whole heart that improves the sensitivity of myocardial ischemia diagnosis and accurate quantitative analysis for myocardial ischemia [38].

This imaging technique has accuracy in diagnosing the patients with known or suspected coronary artery diseases as compared to the other diagnostic methods such as coronary angiography and generates 3D images of the cardiovascular system with good spatial resolution and contrast [39]. It is a noninvasive, X-ray-free imaging technique that is performed in two different methods, involving one method where perfusion imaging is used followed by the first pass of an intravenous bolus of gadolinium contrast, with the presence of perfusion defects (seen as subendocardial hypointense areas) in ischemic territories during peak hyperemia, typically defined as a 15 mmHg drop in systolic blood pressure and an increase in heart rate [40]. And, the second method of wall motion imaging is where stress is induced pharmacologically by the use of either dobutamine, dipyridamole, or adenosine, as physical exercise is difficult to perform within the magnet bore and often induces motion artifacts. Imaging of perfusion is commonly performed in three short-axis planes to obtain a 16-segment model, while wall motion abnormalities using dobutamine can also be imaged in long-axis planes [41]. The most important advantage that this technique has over others is that it is able to offer noninvasive imaging in any plane without any limitations due to lungs or ribs and avoiding nonionizing radiation [42].

An alternative to this cardiac MRI is the stress single-photon emission computed tomography and stress echocardiography but the advantage of this technique is that it does not have any radiation burden associated with it. Left and right ventricular dimensions, viability, and cardiac mass are also detected by the cardiac MRI [43]. Principally nuclei containing an odd number of protons and/or neutrons have characteristic motion or precession because nuclei are charged particles. This precession produces a small magnetic field. CMR scanners depend on the maintenance of a strong magnetic field that cannot be removed except in emergency situations [44]. Common implants hazardous to CMR scanning include cochlear implants, metal-containing ocular implants, pacing wires, metallic cerebral aneurysmal clips, hydrocephalus shunts, and neuron stimulators [45].

Contrast agents of cardiac MRI:

- Only gadolinium-based contrast agents (GBCAs) are used nowadays.
- GBCA use is safe but in some patients mild reactions such as nausea and mild skin rash occur, and severe experiences are extremely rare.
- All GBCAs are chelated to make them nontoxic and facilitate renal excretion.
- Patients with severe renal dysfunction and nephrogenic systemic fibrosis can develop.
- Judicious use of weight-based dosing and avoidance of GBCA use with eGFR less than 30 mL/min/1.73 mm2, NSF due to GBCA has near-zero incidences.

The use of gadolinium contrast also allows the assessment of myocardial fibrosis and scarring. Imaging performed 10–15 min after intravenous contrast injection, known as LGE imaging, can demonstrate the areas of myocardial infarct scar, and this can help in the differentiation between ischemic and nonischemic cardiomyopathy, while the transmural extent of the infarct scar can be used to determine myocardial viability [46] (Fig. 2).

The technique has a variety of applications in clinical arena such as detection of acute cardiac syndromes (ACSs) from noncoronary causes, detection and quantification of myocardial ischemia, and detection of hypertrophic cardiomyopathy [47]. It is found to be useful in arrhythmogenic right ventricular cardiomyopathy and beneficial in the detection of sarcoidosis [48]. Autopsy studies have shown that cardiac

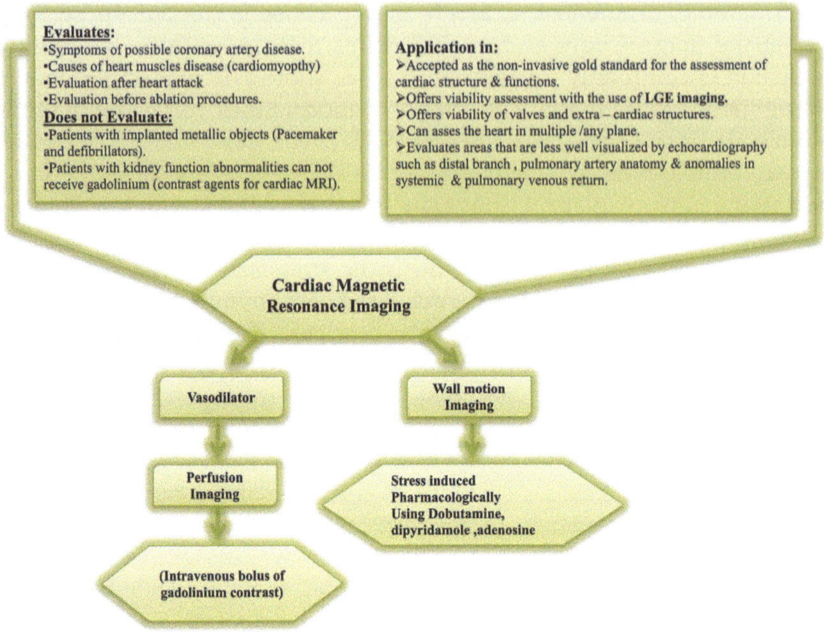

FIG. 2 Various types and applications of cardiac magnetic resonance imaging used in clinical practice.

involvement is found in 20%–30% of patients with sarcoidosis [49] to recognize cardiac involvement is important because sudden cardiac death is one of the most common causes of mortality in sarcoidosis patients [50]. This technique is very useful in making the differential diagnosis between ischemic (ICM) and nonischemic (NICM) etiology of cardiomyopathies, the site being essential: subendocardiac in ICM, with vascular distribution and varied pattern, without vascular distribution. Also, in patients affected by cardiomyopathies, LGE is a good predictor of adverse cardiovascular events [51]. Cardiac MRI has also been useful in detecting in amyloidosis which is a common cause of restrictive cardiomyopathy and is associated with poor prognosis and in detecting myocarditis [52]. In iron-overload cardiomyopathy, a cardiac MRI can be proven resourceful. The former can be inherited or acquired. In patients with transfusion-dependent thalassemia major, cardiac death as a result of myocardial toxicity occurs in 50% of patients. CMR quantifies myocardial iron [53]. It is helpful before planning electrophysiological procedures given its ability to identify the potential site of ablation or scar and to provide 3D volume mapping of the arterial or ventricles. It serves diagnostic purposes in patients with risk of sudden cardiac death

(SCD) by quantification of left ventricular ejection fraction (LVEF) and RV pathology, detection of myocardial scar, identification of anomalous coronary arteries, mapping of iron overload [54].

It is the current test of choice in differentiating constrictive pericarditis from restrictive cardiomyopathy [55]. It is done on the basis of assessing pericardial thickness and constrictive physiology from pericardial disease and the pattern of any myocardial infiltration from restrictive cardiomyopathy. Magnetic resonance imaging is proven to be helpful in detecting both loculated and circumferential pericardial effusion and in detecting cardiac masses and also in left ventricular thrombus, and also useful in Anderson-Fabry disease [56]. As this technique produces tomographic images of the heart in any projection with excellent contrast resolution of fat, myocardium, and lung as well as moving blood from blood vessel walls which makes it useful in evaluating areas that are less well visualized by echocardiography such as distal branch pulmonary artery anatomy and anomalies in systemic and pulmonary venous return [57].

Cardiac MRI has some limitations as it is relatively time-consuming, expensive, and not readily available in all centers. Also, not all patients are able to tolerate the scan, due to either claustrophobia or inability to breathe hold adequately and also cannot be performed in cases of renal impairment [58]. Gadolinium contrast should be used with caution only when strictly necessary in patients with an estimated glomerular filtration rate of <30 mL/min due to the risk of nephrogenic systemic fibrosis. Evidently, image quality may be degraded by arrhythmia, although techniques such as real-time imaging can be used to mitigate this [59]. Another limitation of the technique is that the parameters such as partial volume effect, presence of ventricular arrhythmias during acquisition, lack of adequate apnea, variability of the kinetics of the contrast agent can result in an inaccurate scarring characterization and hence can influence the imaging quality [60].

Cardiac magnetic resonance imaging is proven to be an efficient tool in the investigation of cardiovascular disorders as it is capable of providing the differential diagnosis between ischemic and nonischemic cardiomyopathy and of establishing the etiology in the nonischemic type, and also CMR has evolved in the last decade from an effective research tool into a clinically proven, safe, and comprehensive modality with a large spectrum of applications. The versatility and power of this technique to address key issues in the evaluation of therapies for many cardiovascular diseases provide the needed information to CMR specialists and others involved in such clinical trials which will ensure the best use of the information to evaluate promising new therapies and ultimately to deliver these therapies to patients with cardiovascular disease [61].

4 Cardiac computed tomography

Cardiac computed tomography or cardiac CT is the most traditional technique for coronary artery imaging and for the detection of aortic aneurysms, aortic dissections, pulmonary embolism, and pericardial diseases. It is also an X-ray-based technique that is used for rapid heart imaging and used with iodine-based contrast agents that help in clear visibility of the blood vessels [62]. The scanner used for the computed tomography consists of a track with an X-ray tube mounted opposite to the detector, both the tube and the detector are rotated during the scan, and as this tube rotates, it emits the X-rays that transport through the patient at the center of the scanner. The remaining X-rays which are not absorbed are picked up by the detector and the detected signal is then processed using the reconstruction algorithms which generate a 3D CT image. For the same purpose, ECG signal can also be used to trigger the image acquisition and for the same retrospective reconstruction [63].

This is done when other tests like X-ray, ECG, echocardiogram, or stress tests do not give them complete information about heart, and the main information that can be gathered from a cardiac CT includes the structure of the heart and its pumping ability, scarring of heart muscles, amount of plaque and narrowing of coronary arteries, fluid in the pericardial sac that covers the surface of the heart, and any abnormality in the large blood vessels leaving the heart [64].

As in similarity to the MRI, a contrast multidetector CT can also be used to reveal the cardiac regeneration after intramyocardial MSC transplantation. Computed tomography (CT) was invented by Sir Godfrey Hounsfield, an English Engineer, and Dr. Allan Cormack, an African Physicist, in 1971. First human CT was performed on a patient's brain in London in 1972. Both were given Nobel Prize for this. Thereafter technological advancement of CT systems has generated vast improvements in the detection and exclusion of anatomic and physiologic pathology in virtually all body systems [65]. One such improvement was the 64-detector row CT scanner in 2005, which afforded the necessary temporal and spatial resolution for capturing almost motion-free studies of the heart and coronary arteries that, when coupled with adequate voltage coverage, reduced breath-hold times such that a cardiac CT study was achievable by most patients. CT uses X-ray to make pictures of your heart. It is noninvasive test [66] (Fig. 3).

It is a noninvasive and predominantly anatomical technique that exists in two modalities, namely calcium scoring and CTCA which are carried out using ECG gating to minimize cardiac motion artifact, with images usually acquired in diastole where the heart is at its most still and breath-holding is also done to minimize respiratory motion artifact (Fig. 3) [67].

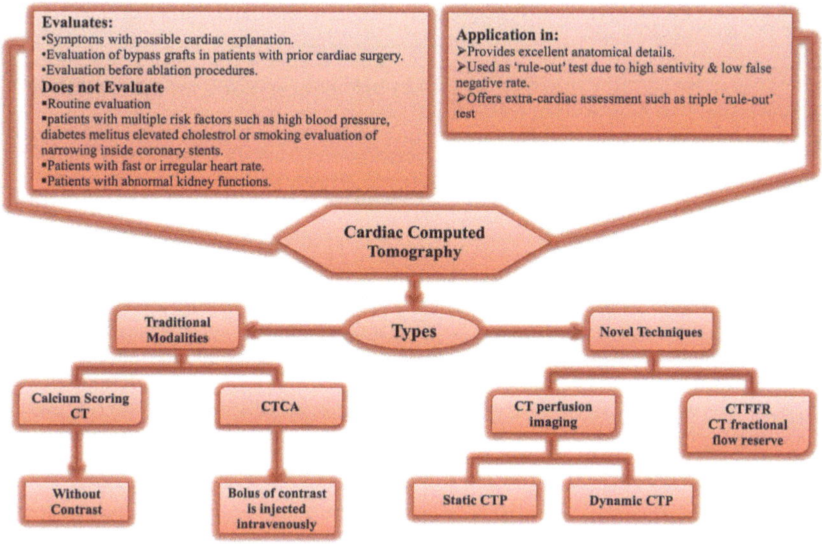

FIG. 3 Various types and applications of cardiac computed tomography used in clinical practice.

The first kind CT calcium scoring is done without contrast with modern 64-slice scanners, and imaging is usually acquired in one breath-hold. This is analyzed using the rapid postprocessing software to measure the amount of coronary calcium [68]. The most commonly used scoring method is the Agatston score which takes into account the size and radiographic density of the plaque to provide a score that reflects the coronary calcified plaque burden, though not all significant obstructive coronary lesions are calcified in fact heavily calcified coronary arteries can often be nonobstructed, hence calcium scoring alone cannot be used as a diagnostic tool for stable angina [69] (Table 2).

Due to this limitation of calcium scoring, CTCA is the more preferred method of CCT imaging where 64-slice CT scanners have reduced concerns about the radiation dose (radiation dose for one CCT is 1–2 mSv with appropriate scan optimization) [70]. For attaining such doses, the patients are required to have slow (<60 bpm), regular heart rates which can be achieved by oral and intravenous beta-blockade if necessary (calcium channel blockers and ivabradine can also be used). Here a bolus of contrast is injected intravenously and with the help of ECG gating a coronary angiogram is generated which provides anatomical detail and has shown excellent diagnostic accuracy in comparison with ICA [71]. This type of CT is recommended for the patients with low to intermediate risk of developing coronary artery diseases. This technique also has some limitations such as low specificity, ionizing radiation dose, and scan preparation, and

TABLE 2 Different characteristics of coronary artery calcium scoring scan and computed tomography coronary angiography (CTCA).

Coronary artery calcium scoring scan	Computed tomography coronary angiography (CTCA)
A calcium scan is a computed tomography of the heart that detects whether there are plaques that contain calcium involving the arteries of the heart. This test does not use any intravenous contrast	Computed tomography coronary angiography (CTCA) can provide information beyond a mere exclusion of obstructive coronary disease; characterization of atheromatous plaque can help in risk stratification
In selected individuals without known coronary artery disease, the presence/ amount of plaque can be used to determine future risk of cardiovascular events such as heart attacks	In the ischemia (initial invasive or conservative strategy for stable coronary disease) trial, after moderate to severe ischemia was diagnosed, CTCA is subsequently performed to rule out left main disease and nonobstructive coronary disease By ruling out significant coronary artery stenosis, CTCA potentially has a role in non-ST-segment elevation myocardial infarction
Advantages • Determine future risk of cardiovascular disease among individuals who have no symptoms • Useful only if the results of the test could influence treatment decisions	**Advantages** • Safe and quick procedure • Can be used for plaque characterization and risk stratification • Can reliably rule out obstructive coronary disease or left main stem disease • Can provide functional assessment when combined with CT-FFR • Can provide prognostic benefit • Can evolve a role in acute coronary syndrome
Disadvantages • Not useful for individuals with known coronary artery disease or who are already at high risk • Not recommended for individuals <40 y of age	**Disadvantages** • Increase in invasive coronary angiography • Image quality affected by various patient factors (i.e., heart rate, arrhythmias, BMI, coronary calcium, inability to hold breath) • Radiation exposure • Anatomical test and diagnostic test only (e.g., inability to intervene on lesions) • Risk of incidental findings exposing patients to further investigations (e.g., pulmonary nodules)
No preparation is required and needs to hold your breath for <10 s	The adverse effects of the technique include contrast-induced nephropathy, contrast allergy, side effects from beta-blockers and nitroglycerin (Anginine), and the possible adverse effects of incorrect treatment as a result of an overestimation of coronary artery disease due to the imperfect specificity of the test

the scan accuracy is found to get reduced in the presence of heavy calcification which results in **blooming artifact** which makes the study difficult to interpretation. Also, CTCA is not found accurate in the diagnosis of in-stent restenosis (particularly in stents <3 mm) and no functional information on stenosis severity hence is not recommended in low-intermediate risk patients and is known as a **rule-out technique** [72].

To overcome such limitations of this technique as a whole, novel techniques have been developed to add functional assessment to CTCA. The first one is **CT perfusion imaging** which is similar to cardiac MRI principle perfusion and relies on the passage of contrast (iodinated rather than gadolinium) from the myocardial blood pool into the myocardium and then areas of infraction or ischemia appear hypointense compared to normal myocardium. Again, imaging can be performed at rest and stress to determine the functional significance of lesions [73]. This procedure requires optimal timing, as continuous imaging is not performed to reduce radiation dose. It has not yet been adopted into routine clinical practice due to its reasonable diagnostic accuracy limited availability and radiation dose concerns [74].

Another promising technique emerged from cardiac CT is known as the **CT fractional flow reserve (CT-FFR)** where computational fluid dynamics are used to provide a prediction of the invasive FFR. An advantage of this technique is that it does not require any further imaging beyond a standard CTCA and does not involve any administration of adenosine to induce hyperemia as well. This technique shows good reproducibility and diagnostic accuracy [75] (Table 3).

It is an important method to evaluate coronary lesions noninvasively, providing the characteristics of calcification, plaque, and epicardial adipose tissue (EAT) with low-dose radiation and no contrast agent infection. There are the indications that correspond to arterial lesions and inflammatory activity which help in determining the obstructive coronary artery disease and help in deciding the further therapeutic guidance [78].

Before scanning, proper patient preparation is required for a high-quality cardiac scan. Low heart rate permits less motion contamination of images which brings improved diagnostic interpretation [79]. For getting a low heart rate, B-blocker is used. Metoprolol is a more frequently used drug in oral and IV form. On the night before the scan, Tab. metoprolol is used in oral form. In the CT scan, if the heart rate is less than 65 BPM, it is fine, but if not, IV metoprolol is given. Apart from metoprolol, atenolol is used. Calcium channel blockers are not much effective. Ivabradine can be used. Caffeine and nicotine should be avoided before scanning [80].

Due to contrast material, CKD, elevated serum creatinine levels, and increased risk of the contrast-induced nephropathy (CIN) should be carefully ascertained. For better visualization of the coronary artery lumen, 0.4 to 0.8 mg of S/L nitroglycerine is used to administer for its coronary

TABLE 3 Different characteristics of CT perfusion (CTP) imaging and CT fractional flow reserve (CT-FFR).

<table>
<tr><th colspan="2">CT perfusion (CTP) imaging [76]</th><th>CT fractional flow reserve (CT-FFR) [77]</th></tr>
<tr><td colspan="2">During CTP, radiopaque iodinated contrast is injected intravenously, transits the cardiac and pulmonary structures, and then perfuses through the coronary vasculature and myocardium. This "first pass" of contrast through the myocardial vasculature is directly proportional to myocardial blood flow</td><td>CCTA-derived fractional flow reserve (CT-FFR), a novel imaging modality, now permits the physiological assessment of coronary artery disease</td></tr>
<tr><td>Static perfusion imaging: On static perfusion imaging, one single image may be used to compare myocardial CTP from one region of the myocardium to other remote areas</td><td>Dynamic perfusion imaging: several sequential images can be acquired over time to determine both contrast perfusion into the myocardium and wash out and allows for semiquantitative and fully quantitative assessment of myocardial blood flow</td><td>The computational fluid dynamics enable the calculation of CCTA-derived FFR (FFRCT) from three-dimensional imaging anatomic models. Foremost, CT-FFR reflects a novel imaging modality with the ability to determine the hemodynamic significance of CAD without additional radiation exposure or contrast injection beyond the initial CCTA</td></tr>
<tr><td colspan="2">Stress CTP imaging uses vasodilator stress agents such as regadenoson, adenosine, and dipyridamole to evaluate for ischemia</td><td>CT-FFR has mostly focused on its diagnostic performance for the determination of ischemia</td></tr>
<tr><td colspan="2">Postprocessing of static CTP images is similar to nuclear imaging and generally analyzed visually where a myocardial segment with comparatively "normal" blood flow is required to visually detect segments with lower blood flow</td><td>On-going prospective studies designed to compare the diagnostic power of CT-FFR with other stress imaging tests will also offer clinicians an evidence-based alternative for detecting significant CAD, allowing for appropriate utilization of CT-FFR in the diagnostic purposes</td></tr>
<tr><td colspan="2">CTP can determine not just the presence or absence of ischemia, but also quantify the amount of ischemia, similar to the various semiquantitative scores that are used in nuclear cardiology but from a technical perspective, future studies should determine how to mitigate various artifacts, ranging from motion to beam hardening</td><td>While the technology has shown considerable promise, it is by no means the perfect diagnostic test. Under the current model of FFRCT analysis, where data have to be sent to the company for FFR calculation and sent back to the hospital, turnaround times are a concern, particularly for emergency department cases</td></tr>
</table>

vasodilatory effects at CT scanning. Contraindications to nitroglycerine are low BP or use of phosphodiesterase inhibitors (e.g., Sildenafil and tadalafil) that are used within 48h of scanning [81].

The cardiac CT has various derivatives in terms of modalities and a variety of functional applications in the field of cardiovascular imaging as along with the morphological assessment of coronary artery, it can evaluate and test the blood supply of the myocardium which makes it an apt technique to predict myocardial ischemia, proves to be a guide for its clinical treatment, and also monitors the clinical curative effect. Breading its different limitations which gives the scope for further continuous research this technique is valuable for accurate diagnosis of patients with suspected coronary artery diseases.

5 Radionucleotide imaging

This is the imaging technique in which a small dose of a radioactive isotope called a tracer is used to detect cancer, trauma, infection or CVS, and disorders. In this test, tracer is either injected or swallowed [82]. The chemical emits gamma rays. This test is a painless and noninvasive test. Tracer is injected into vein or swallowed. Tracer enters the body and travels through the bloodstream to the target organ such as heart, thyroid, and bones. Tracers get collected in different organs [83]. The tracer emits gamma rays that are in turn detected by a gamma camera and with the help of computer, it is analyzed and in turn image of target organ is formed. Sites of potential problems send out more intense gamma rays and appear as bright spots on the scan [84]. Examples of radionuclide scans are pet scans, gallium scans, and bone scans [85]. It is done to see the involvement of skeleton from cancer of prostate or breast or others. It is also used to check hidden fractures and infections of joints or bones. Gallium scan is used to identify sarcoidosis or pneumocystis carinii in the lungs or Crohn's disease in the abdomen [86].

This radionucleotide imaging gives important information for the diagnosis and prognosis of different diseases. In oncology, it is common to use pet scans to know the spread of cancer. The test is costly but frequently used in India for cancer patients to plan the line of treatment. This technique has a role in the detection of coronary artery disease, valvular disease, congenital cardiac disorder, cardiomyopathy, and other cardiac disorders [87]. This technique detects and evaluates the myocardial perfusion status and myocardial viability to provide the useful information on the coronary arteries. This test is particularly used for the imaging of heart in the diagnosis of chest pain when there exists an unknown cause as in if the coronary arteries are narrowed, radionucleotide imaging is used to learn how the narrowing is affecting the heart's blood supply and function along with this it also assesses the improvement in the blood supply to heart muscle after the bypass surgery or any similar procedure which may be helpful to determine the further prognosis after the patient has experienced a heart attack [88].

For the same purpose, different tracers are used which depends on what disorder is suspected for the evaluation of blood flow through heart muscle and these tracers are technetium-99 m sestamibi or thallium-201 usually while the person experiences an exercise stress test [89]. The amount of tracer absorbed by the heart muscles also depends on the blood flow as at the peak exercise an area of heart muscle that has an inadequate blood supply (ischemia) absorbs less tracer which produces a fainter image than the neighboring muscle with a normal supply [90]. For the patients who are unable to take the exercise test, an intravenous injection of a drug, such as dipyridamole, or adenosine, may be used to simulate the effects of exercise on blood flow. And after the person rests for a few hours, a repeat scan is done and the resultant image is compared with the original image obtained during the exercise [91]. The area where the heart has inadequate blood flow is distinguished between reversible by the narrowing of coronary arteries from areas where it is irreversible caused by scarring due to the previously occurred heart attack. The radioactive material that is used while the radioactive imaging remains in the body for the next few days [92].

6 Myocardial perfusion imaging

Myocardial perfusion imaging is a technique that makes qualitative evaluation of myocardial ischemia by observing myocardial perfusion defects through the visual inspection of the same [93]. This is the test that detects how blood flows through the heart muscle. It also suggests pumping status of heart. It is also known as a nuclear stress test. In this, a small amount of radioactive tracer will be injected through a small needle in a vein in the arm. This tracer through blood goes to the myocardium. In normal heart muscle, it goes with normal blood flow. In MPI, tracer produces the picture of heart [94].

This method first evaluates the contrast medium through the myocardium at rest and then during the corresponded pharmacological stress and then emphasizes the difference in the perfusion between the normal and ischemic myocardium which is done by two means of static and dynamic testings. Here static testing acquires data during a single phase of contrast medium through the myocardium and the dynamic acquires data at multiple phases [95].

A major advantage of this technique is that it can detect perfusion directly that is useful to assess the hemodynamic significance of coronary lesion, and combined with other techniques, it can obtain anatomic information about the coronary artery and also functional of myocardial perfusion in an optimized one-stop-shop pattern, compared with the conventional myocardial perfusion imaging (MPI) [96]. MPI tracers show the linear relationship between the peak stress myocardial blood flow and myocardial tracer concentration uptake but there exists a roll-off phenomenon where which increases in coronary flow beyond 1.5–2-fold are not accompanied by a directly proportional increase in tracer uptake which happens when the diameter of stenosis is in excess of ~50%–70% [97]. This technique involves progressive improvement of single-photon emission computed tomography (SPECT) technologies, new radiopharmaceuticals, and new software, hence also known as myocardial perfusion single-photon emission computed tomography (MPS) or nuclear stress test [98] (Fig. 4).

MPI is performed with ECG-gated SPECT for evaluation of both myocardial perfusion and cardiac function simultaneously, and for the same, several stress techniques are available, but the exercise stress test is the preferred method because it provides several important prognostic data in addition to increasing coronary blood flow but in case of contraindication to exercise stress testing, pharmacological stress is considered [45]. The main assessment efficiency of the MPS relies upon its diagnosis of

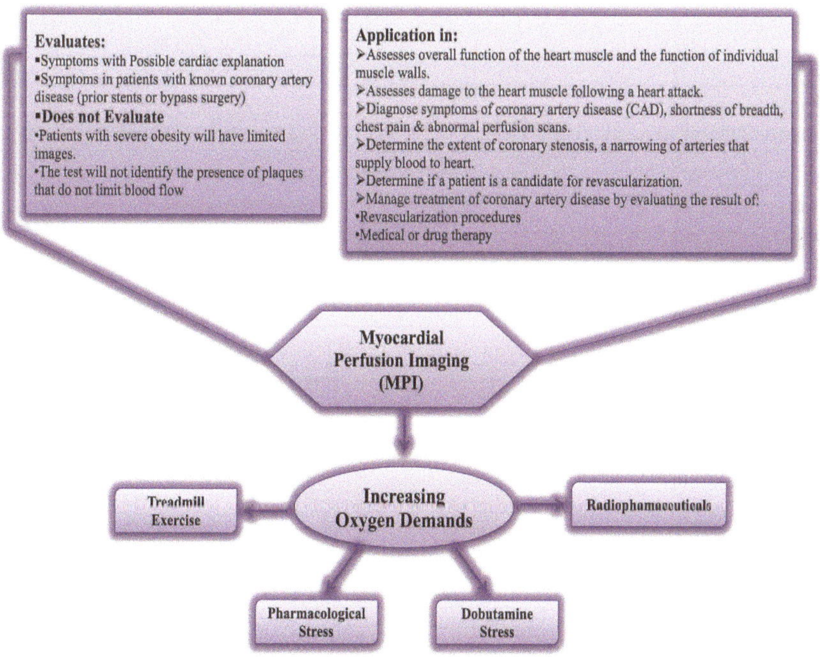

FIG. 4 Various types and applications of myocardial perfusion imaging used in clinical practice.

coronary artery disease (CAD), risk stratification in patients with known CAD, assessment of therapy and intervention, and myocardial viability before the percutaneous intervention or bypass surgery [99].

For the appropriate interpretation of the MPS results, a close collaboration between the both referring and intermediate physicians is crucial for the interpretation of the study, optimal patient preparation, and effective use of the test for further clinical management [100]. The patients are required to not have anything except medications and water before the test as this ensures interference with the evaluation of the inferior wall of the left ventricle (LV) by limiting the gut activity. If a vasodilator stress test is to be anticipated, then medications with methylxanthines or caffeine and food beverage with caffeine must be avoided for 12–24h [101].

Heterogeneity in the coronary flow is induced to assess the physiological significance of coronary stenosis and the resting flow is maintained until there is an approximately 90% reduction of coronary arterial flow, although the maximum flow termed as coronary flow reserve is impaired with approximately 50% coronary stenosis, and this increase can be achieved by increased oxygen demand with exercise by the means of treadmill or bicycle or b-adrenergic agonist (dobutamine) or by direct vasodilator (adenosine and dipyridamole) [102,103] (Table 4).

TABLE 4 Different means of increasing the oxygen demand for perfusion imaging and their corresponding agents commonly employed.

Means of increasing oxygen demand	Characteristics
Treadmill exercise [104]	Graded exercise testing is preferred over pharmacological testing in patients able to exercise adequately because it provides an assessment of well-validated prognostic markers such as workload, ST-segment depression, anginal symptoms, heart rate, and blood pressure response
Pharmacological stress [96]	Pharmacological vasodilatation with adenosine or dipyridamole is indicated for patients who are unable to exercise or unable to increase their heart rate (rate-limiting medications) and patients with left bundle branch block (LBBB) or paced ventricular rhythm
Dobutamine stress [105]	Dobutamine is a β-adrenergic agonist with both positive chronotropic and inotropic effects, resulting in coronary artery dilatation secondary to an increase in myocardial oxygen demand
Radiopharmaceuticals [97]	The injected radiopharmaceuticals are delivered to the myocardium in proportion to flow. There is reduced tracer uptake in the regions with reduced flow and incensed tracer uptake in the region with increased blood flow. The currently used radiotracers in MPS are 201Tl-chloride, 99mTc-sestamibi, and 99mTc-tetrofosmin

The interpretation of MPS images should be performed in a systematic and consistent order to include:

(1) evaluation of the raw images in cine mode to determine the presence of potential sources of image artifact and the distribution of extracardiac tracer activity;
(2) proper alignment of the poststress and rest images;
(3) interpretation of images with respect to the location, size, severity, and reversibility of perfusion defects as well as cardiac chamber sizes, and especially for Tl-201, presence or absence of increased pulmonary uptake;
(4) incorporation of the results of quantitative perfusion analysis;
(5) consideration of functional data obtained from the gated images; and
(6) Consideration of clinical factors that may influence the final interpretation of the study. All of these factors contribute to the production of a final clinical report [104].

There could be several artifacts and pitfalls that can affect the MPS imaging and limits its utilization in clinical therapeutics. This technique is a systematic approach for interpretation and reporting the blood flow to the heart muscle both under resting conditions and during stress which can be accomplished with either exercise or medications which can detect abnormal blood flow to the heart muscle from blockages or narrowing in the coronary arteries and the presence of prior heart attacks.

7 Summary

There are ample cardiovascular imaging techniques these days such as radionuclide cardiac imaging; echocardiography, cardiac computed tomography, and cardiac magnetic resonance imaging techniques that play an important role in the diagnosis of coronary artery disease, in the assessment of myocardial perfusion and viability, for the diagnosis of arrhythmogenic right ventricular dysplasia, and various contemporary imaging techniques, with either stress nuclear myocardial perfusion imaging or stress echocardiography, providing high sensitivity and specificity in the detection and risk assessment of coronary artery disease having incremental value over exercise electrocardiography and clinical variables.

There are some other techniques as well such as **magnetic resonance angiography (MRA)** that is extremely helpful in the diagnosis and management of patients with congenital heart disease that shows images in tomographic planes at different phases of cardiac cycle. Another technique like **magnetic resonance spectroscopy (MRS)** is a predominant research tool presently providing means of demonstrating relative concentrations of high-energy metabolites within the regions of the myocardium.

Radionucleotide angiography can also be used to detect and quantify shunts and to analyze the distribution of blood flow to each lung and heart to quantify the blood flow distribution. All these techniques have their own share of advantages and limitations but the employment of the correct technique depends on the conditions and testing requirement of the patient.

More than one technique out of these could be applied in combination to reach the correct diagnosis and to decide further prognosis for a patient, and also, the understanding of the patient condition and cooperation between the physicians and technicians also plays a great deal in reaching the correct clinical interpretation and implementation of therapeutic measures for the same.

References

[1] R. Blankstein, Introduction to noninvasive cardiac imaging, Circulation 125 (3) (2012) e267–e271.
[2] W.D. Gilson, D.L. Kraitchman, Noninvasive cardiovascular imaging techniques for basic science research: application to cellular therapeutics, Rev. Esp. Cardiol. 62 (8) (2009) 918–927.
[3] D. Shan, J. Yang, Y. Chen, Noninvasive cardiac imaging technologies in detecting coronary artery disease: from research to clinical practice, Cardiol. Plus 5 (1) (2020) 13.
[4] S. Dimmeler, A.M. Zeiher, Cell therapy of acute myocardial infarction: open questions, Cardiology 113 (3) (2009) 155–160.
[5] I.R. Mordi, et al., Efficacy of noninvasive cardiac imaging tests in diagnosis and management of stable coronary artery disease, Vasc. Health Risk Manag. 13 (2017) 427.
[6] J. Mieres, et al., Noninvasive cardiac imaging, Am. Fam. Physician 75 (8) (2007) 1219–1228.
[7] E.R. McVeigh, Emerging imaging techniques, Circ. Res. 98 (7) (2006) 879–886.
[8] Secretariat, Medical Advisory, Non-invasive cardiac imaging technologies for the diagnosis of coronary artery disease: a summary of evidence-based analyses, Ont. Health Technol. Assess. Ser. 10 (7) (2010) 1.
[9] R.J. Lederman, M.A. Guttman, D.C. Peters, R.B. Thompson, J.M. Sorger, A.J. Dick, et al., Catheter-based endomyocardial injection with real-time magnetic resonance imaging, Circulation 105 (2002) 1282–1284.
[10] F.E. Kloster, Echocardiographic examination techniques, in: Echocardiology, Springer, Dordrecht, 1977, https://doi.org/10.1007/978-94-010-1103-7_3.
[11] H. Hertz, Principles of ultrasound in medical diagnosis, in: Echocardiology, Springer, Dordrecht, 1977, https://doi.org/10.1007/978-94-010-1103-7_2.
[12] S.B. Malik, et al., Transthoracic echocardiography: pitfalls and limitations as delineated at cardiac CT and MR imaging, Radiographics 37 (2) (2017) 383–406.
[13] M.B. Jensen, et al., Transthoracic echocardiography for cardiopulmonary monitoring in intensive care, Eur. J. Anaesthesiol. 21 (9) (2004) 700–707.
[14] B.N. Shah, Echocardiography in the era of multimodality cardiovascular imaging, Biomed. Res. Int. 2013 (2013), 310483.
[15] D. Sobczyk, Echocardiography in acute aortic syndromes, in: New Approaches to Aortic Diseases from Valve to Abdominal Bifurcation, Academic Press, 2018, pp. 159–167.
[16] G.S. Hillis, P. Bloomfield, Basic transthoracic echocardiography, BMJ 330 (7505) (2005) 1432–1436.

[17] A.S. Nagre, Focus-assessed transthoracic echocardiography: implications in perioperative and intensive care, Ann. Card. Anaesth. 22 (3) (2019) 302.
[18] M. Galderisi, B. Cosyns, T. Edvardsen, N. Cardim, V. Delgado, G. Di Salvo, E. Donal, L.E. Sade, L. Ernande, M. Garbi, J. Grapsa, A. Hagendorff, O. Kamp, J. Magne, C. Santoro, A. Stefanidis, P. Lancellotti, B. Popescu, G. Habib, 2016–2018 EACVI Scientific Documents Committee, Standardization of adult transthoracic echocardiography reporting in agreement with recent chamber quantification, diastolic function, and heart valve disease recommendations: an expert consensus document of the European Association of Cardiovascular Imaging, Eur. Heart J. Cardiovasc. Imaging 18 (12) (2017) 1301–1310, https://doi.org/10.1093/ehjci/jex244. 29045589.
[19] W.G. Daniel, et al., Safety of transesophageal echocardiography. A multicenter survey of 10,419 examinations, Circulation 83 (3) (1991) 817–821.
[20] J.N. Hilberath, et al., Safety of transesophageal echocardiography, J. Am. Soc. Echocardiogr. 23 (11) (2010) 1115–1127.
[21] F.B. Sozzi, et al., Second harmonic imaging improves sensitivity of dobutamine stress echocardiography for the diagnosis of coronary artery disease, Am. Heart J. 142 (1) (2001) 153–159.
[22] P. Lancellotti, et al., Determinants and prognostic significance of exercise pulmonary hypertension in asymptomatic severe aortic stenosis, Circulation 126 (7) (2012) 851–859.
[23] D.G. Skolnick, et al., Enhanced endocardial visualization with noncontrast harmonic imaging during stress echocardiography, J. Am. Soc. Echocardiogr. 12 (7) (1999) 559–563.
[24] C. Aggeli, 3D STRESS ECHO: is it a fact or a fiction? Hell. J. Cardiol. 61 (5) (2020) 349–350, https://doi.org/10.1016/j.hjc.2019.12.009.
[25] S. Marechaux, et al., Usefulness of exercise-stress echocardiography for risk stratification of true asymptomatic patients with aortic valve stenosis, Eur. Heart J. 31 (11) (2010) 1390–1397.
[26] P. Lancellotti, et al., Prognostic importance of quantitative exercise Doppler echocardiography in asymptomatic valvular aortic stenosis, Circulation 112 (9_supplement) (2005) I-377.
[27] B.N. Shah, et al., Incremental diagnostic and prognostic value of contemporary stress echocardiography in a chest pain unit: mortality and morbidity outcomes from a real-world setting, Circ. Cardiovasc. Imaging 6 (2) (2013) 202–209.
[28] P. Lancellotti, P.A. Pellikka, W. Budts, F.A. Chaudhry, E. Donal, R. Dulgheru, T. Edvardsen, M. Garbi, J.W. Ha, G.C. Kane, J. Kreeger, L. Mertens, P. Pibarot, E. Picano, T. Ryan, J.M. Tsutsui, A. Varga, The clinical use of stress echocardiography in non-ischaemic heart disease: recommendations from the European Association of Cardiovascular Imaging and the American Society of Echocardiography, Eur. Heart J. Cardiovasc. Imaging 17 (11) (2016) 1191–1229, https://doi.org/10.1093/ehjci/jew190. Erratum in: Eur Heart J Cardiovasc Imaging. 2017 May 1; 18(8): 832 27880640.
[29] V. Ruddox, et al., Is 3D echocardiography superior to 2D echocardiography in general practice?: a systematic review of studies published between 2007 and 2012, Int. J. Cardiol. 168 (2) (2013) 1306–1315.
[30] K. Cheng, et al., 3D echocardiography: benefits and steps to wider implementation, Br. J. Cardiol. 25 (2018) 63–68.
[31] M.J. Stewart, Contrast echocardiography, Heart 89 (3) (2003) 342–348.
[32] R. Senior, et al., Contrast echocardiography: evidence-based recommendations by European Association of Echocardiography, Eur. J. Echocardiogr. 10 (2) (2009) 194–212.
[33] M. Turiel, P. Sarzi-Puttini, R. Cervera, Cardiac imaging techniques in systemic autoimmune diseases, Handb. Syst. Autoimmun. Dis. 1 (2003) 107–119.
[34] S.L. Mulvagh, et al., Contrast echocardiography: current and future applications, J. Am. Soc. Echocardiogr. 13 (4) (2000) 331–342.

[35] H. Becher, et al., Current indications for contrast echocardiography imaging, Eur. J. Echocardiogr. 6 (suppl_2) (2005) S1–S5.
[36] N.G. Chaubal, J. Chaubal, Fetal echocardiography, Indian J. Radiol. Imaging 19 (1) (2009) 60.
[37] M. Motwani, et al., Advanced cardiovascular magnetic resonance myocardial perfusion imaging: high-spatial resolution versus 3-dimensional whole-heart coverage, Circ. Cardiovasc. Imaging 6 (2) (2013) 339–348.
[38] M. Salerno, C.M. Kramer, Advances in parametric mapping with CMR imaging, JACC Cardiovasc. Imaging 6 (7) (2013) 806–822.
[39] A.R. Patel, et al., Assessment of advanced coronary artery disease: advantages of quantitative cardiac magnetic resonance perfusion analysis, J. Am. Coll. Cardiol. 56 (7) (2010) 561–569.
[40] C.M. Kramer, et al., Standardized cardiovascular magnetic resonance (CMR) protocols 2013 update, J. Cardiovasc. Magn. Reson. 15 (1) (2013) 1–10.
[41] C.J. Francois, Current state of the art cardiovascular MR imaging techniques for assessment of ischemic heart disease, Radiol. Clin. 53 (2) (2015) 335–344.
[42] S. Pflugi, et al., Accelerated cardiac MR stress perfusion with radial sampling after physical exercise with an MR-compatible supine bicycle ergometer, Magn. Reson. Med. 74 (2) (2015) 384–395.
[43] I. Mordi, et al., Comprehensive dobutamine stress CMR versus echocardiography in LBBB and suspected coronary artery disease, JACC Cardiovasc. Imaging 7 (5) (2014) 490–498.
[44] J.N. Dungu, et al., CMR-based differentiation of AL and ATTR cardiac amyloidosis, JACC Cardiovasc. Imaging 7 (2) (2014) 133–142.
[45] C. Jaarsma, et al., Diagnostic performance of noninvasive myocardial perfusion imaging using single-photon emission computed tomography, cardiac magnetic resonance, and positron emission tomography imaging for the detection of obstructive coronary artery disease: a meta-analysis, J. Am. Coll. Cardiol. 59 (19) (2012) 1719–1728.
[46] J. Romero, et al., CMR imaging assessing viability in patients with chronic ventricular dysfunction due to coronary artery disease: a meta-analysis of prospective trials, JACC Cardiovasc. Imaging 5 (5) (2012) 494–508.
[47] J. Schwitter, et al., MR-IMPACT: comparison of perfusion-cardiac magnetic resonance with single-photon emission computed tomography for the detection of coronary artery disease in a multicentre, multivendor, randomized trial, Eur. Heart J. 29 (4) (2008) 480–489.
[48] J. Zhang, et al., Cardiac magnetic resonance imaging for diagnosis of cardiac sarcoidosis: a meta-analysis, Can. Respir. J. 2018 (2018), 7457369.
[49] O. Vignaux, Cardiac sarcoidosis: spectrum of MRI features, Am. J. Roentgenol. 184 (1) (2005) 249–254.
[50] J. Schwitter, et al., Magnetic resonance imaging for myocardial perfusion assessment in coronary artery disease trial (MR-IMPACT II): a phase III multicenter, multivendor trial comparing perfusion cardiac magnetic resonance versus single photon emission computed tomography for the detection of coronary artery disease, Circulation 114 (2006) II_806.
[51] M.F. Jan, A. Jamil Tajik, Modern imaging techniques in cardiomyopathies, Circ. Res. 121 (7) (2017) 874–891.
[52] C. Schiau, et al., Cardiovascular magnetic resonance: contribution to the exploration of cardiomyopathies, Med. Pharm. Rep. 92 (4) (2019) 326.
[53] J.P. Greenwood, et al., Effect of care guided by cardiovascular magnetic resonance, myocardial perfusion scintigraphy, or NICE guidelines on subsequent unnecessary angiography rates: the CE-MARC 2 randomized clinical trial, JAMA 316 (10) (2016) 1051–1060.

[54] S.T. Hussain, et al., Design and rationale of the MR-INFORM study: stress perfusion cardiovascular magnetic resonance imaging to guide the management of patients with stable coronary artery disease, J. Cardiovasc. Magn. Reson. 14 (1) (2012) 1–10.
[55] F.E. Mordini, et al., Diagnostic accuracy of stress perfusion CMR in comparison with quantitative coronary angiography: fully quantitative, semiquantitative, and qualitative assessment, JACC Cardiovasc. Imaging 7 (1) (2014) 14–22.
[56] P. Ponikowski, et al., The Task Force for the diagnosis and treatment of acute and chronic heart failure of the European Society of Cardiology (ESC). 2016 ESC Guidelines for the diagnosis and treatment of acute and chronic heart failure, Eur. Heart J. 37 (2016) 2129–2200.
[57] M. Saeed, et al., Cardiac MR imaging: current status and future direction, Cardiovasc. Diagn. Ther. 5 (4) (2015) 290.
[58] E.L. Foster, et al., MR-compatible treadmill for exercise stress cardiac magnetic resonance imaging, Magn. Reson. Med. 67 (3) (2012) 880–889.
[59] S. Nazarian, et al., Magnetic resonance assessment of the substrate for inducible ventricular tachycardia in nonischemic cardiomyopathy, Circulation 112 (18) (2005) 2821–2825.
[60] I. Eitel, et al., Clinical characteristics and cardiovascular magnetic resonance findings in stress (takotsubo) cardiomyopathy, JAMA 306 (3) (2011) 277–286.
[61] M. Ugander, A.J. Oki, L.-Y. Hsu, Extracellular volume imaging by magnetic resonance imaging provides insights into overt and subclinical myocardial pathology, Eur. Heart J. 33 (2012) 1268–1278.
[62] X.-W. Tian, et al., Advances in cardiac computed tomography functional imaging technology, Cardiology 145 (10) (2020) 615–622.
[63] C.E. Rochitte, R.T. George, M.Y. Chen, A. Arbab-Zadeh, M. Dewey, J.M. Miller, et al., Computed tomography angiography and perfusion to assess coronary artery stenosis causing perfusion defects by single photon emission computed tomography: the CORE320 study, Eur. Heart J. 35 (2014) 1120–1130.
[64] J.M. Lee, G. Choi, B.K. Koo, D. Hwang, J. Park, J. Zhang, et al., Identification of high-risk plaques destined to cause acute coronary syndrome using coronary computed tomographic angiography and computational fluid dynamics, JACC Cardiovasc. Imaging 12 (2019) 1032–1043.
[65] J.D. Schuijf, et al., Cardiac imaging in coronary artery disease: differing modalities, Heart 91 (8) (2005) 1110–1117.
[66] K. Ueno, T. Anzai, M. Jinzaki, M. Yamada, Y. Jo, Y. Maekawa, et al., Increased epicardial fat volume quantified by 64-multidetector computed tomography is associated with coronary atherosclerosis and totally occlusive lesions, Circ. J. 73 (2009) 1927–1933.
[67] J.H. Lee, et al., Fractional flow reserve measurement by computed tomography: an alternative to the stress test, Interv. Cardiol. Rev. 11 (2) (2016) 105.
[68] Q. Yang, K. Li, X. Liu, et al., 3.0T whole-heart coronary magnetic resonance angiography performed with 32-channel cardiac coils: a single-center experience, Circ. Cardiovasc. Imaging 5 (2012) 573–579.
[69] A.S. Agatston, et al., Quantification of coronary artery calcium using ultrafast computed tomography, J. Am. Coll. Cardiol. 15 (4) (1990) 827–832.
[70] B.S. Ko, et al., Non-invasive CT-derived fractional flow reserve and static rest and stress CT myocardial perfusion imaging for detection of haemodynamically significant coronary stenosis, Int. J. Cardiovasc. Imaging 35 (11) (2019) 2103–2112.
[71] D. Andreini, et al., Comparison of feasibility and diagnostic accuracy of 64-slice multidetector computed tomographic coronary angiography versus invasive coronary angiography versus intravascular ultrasound for evaluation of in-stent restenosis, Am. J. Cardiol. 103 (10) (2009) 1349–1358.

[72] J.K. Min, et al., High-definition multidetector computed tomography for evaluation of coronary artery stents: comparison to standard-definition 64-detector row computed tomography, J. Cardiovasc. Comput. Tomogr. 3 (4) (2009) 246–251.
[73] P.S. Douglas, et al., Clinical outcomes of fractional flow reserve by computed tomographic angiography-guided diagnostic strategies vs. usual care in patients with suspected coronary artery disease: the prospective longitudinal trial of FFRCT: outcome and resource impacts study, Eur. Heart J. 36 (47) (2015) 3359–3367.
[74] I. Mordi, N. Tzemos, Incremental value of CT perfusion in the diagnosis of coronary artery disease, Eur. Heart J. Cardiovasc. Imaging 14 (5) (2013) 504.
[75] B.-K. Koo, et al., Diagnosis of ischemia-causing coronary stenoses by noninvasive fractional flow reserve computed from coronary computed tomographic angiograms: results from the prospective multicenter DISCOVER-FLOW (Diagnosis of Ischemia-Causing Stenoses Obtained via Noninvasive Fractional Flow Reserve) study, J. Am. Coll. Cardiol. 58 (19) (2011) 1989–1997.
[76] K.R. Branch, et al., Myocardial computed tomography perfusion, Cardiovasc. Diagn. Ther. 7 (5) (2017) 452.
[77] I. Danad, et al., Diagnostic performance of cardiac imaging methods to diagnose ischaemia-causing coronary artery disease when directly compared with fractional flow reserve as a reference standard: a meta-analysis, Eur. Heart J. 38 (13) (2017) 991–998.
[78] J. Terrovitis, K.F. Kwok, R. Lautamäki, J.M. Engles, A.S. Barth, E. Kizana, et al., Ectopic expression of the sodium-iodide symporter enables imaging of transplanted cardiac stem cells in vivo by single-photon emission computed tomography or positron emission tomography, J. Am. Coll. Cardiol. 52 (2008) 1652–1660.
[79] Y. Tanami, et al., Lack of association between epicardial fat volume and extent of coronary artery calcification, severity of coronary artery disease, or presence of myocardial perfusion abnormalities in a diverse, symptomatic patient population: results from the CORE320 multicenter study, Circ. Cardiovasc. Imaging 8 (3) (2015), e002676.
[80] D.A. Kedziorek, P. Walczak, Y. Fu, T. Ehtiati, A.B. Brost, N. Azene, et al., Precise targeting of x-ray-visible microencapsulated mesenchymal stem cells using c-arm CT for interrogation of a bioluminescent reporter gene probe, J. Nucl. Med. 50 (2009) 657.
[81] I.M. Bensenor, et al., Association between a healthy cardiovascular risk factor profile and coronary artery calcium score: results from the Brazilian Longitudinal Study of Adult Health (ELSA-Brasil), Am. Heart J. 174 (2016) 51–59.
[82] M.I. Travin, Cardiac radionuclide imaging to assess patients with heart failure, Semin. Nucl. Med. 44 (4) (2014) 294–313.
[83] A.E. Iskandrian, E.V. Garcia, Nuclear Cardiac Imaging: Principles and Applications, Oxford University Press, 2008.
[84] J.E. Udelson, C.D. Shafer, I. Carrió, Radionuclide imaging in heart failure: assessing etiology and outcomes and implications for management, J. Nucl. Cardiol. 9 (5) (2002) S40–S52.
[85] J. Stirrup, et al., Cardiac radionuclide imaging in stable coronary artery disease and acute coronary syndromes, Br. Med. Bull. 89 (1) (2009) 63–78.
[86] M.E. Harinstein, P. Soman, Radionuclide imaging applications in cardiomyopathies and heart failure, Curr. Cardiol. Rep. 18 (3) (2016) 1–9.
[87] P. Merlet, et al., Positron emission tomography with 11C CGP-12177 to assess beta-adrenergic receptor concentration in idiopathic dilated cardiomyopathy, Circulation 87 (4) (1993) 1169–1178.
[88] P. Soman, et al., I-123 MIBG cardiac imaging, J. Nucl. Cardiol. 22 (4) (2015) 677–685.
[89] A.J. Sinusas, et al., Biodistribution and radiation dosimetry of LMI1195: first-in-human study of a novel 18F-labeled tracer for imaging myocardial innervation, J. Nucl. Med. 55 (9) (2014) 1445–1451.
[90] R.A. Werner, et al., Retention kinetics of the 18F-labeled sympathetic nerve PET tracer LMI1195: comparison with 11C-hydroxyephedrine and 123I-MIBG, J. Nucl. Med. 56 (9) (2015) 1429–1433.

[91] C. Ayoub, et al., Advanced imaging of cardiac sarcoidosis, Curr. Cardiol. Rep. 17 (4) (2015) 17.
[92] R. Langah, et al., Effectiveness of prolonged fasting 18f-FDG PET-CT in the detection of cardiac sarcoidosis, J. Nucl. Cardiol. 16 (5) (2009) 801–810.
[93] R.C. Hendel, et al., The value and practice of attenuation correction for myocardial perfusion SPECT imaging: a joint position statement from the American Society of Nuclear Cardiology and the Society of Nuclear Medicine, J. Nucl. Med. 43 (2) (2002) 273–280.
[94] A. Coenen, A. Rossi, M.M. Lubbers, A. Kurata, A.K. Kono, R.G. Chelu, et al., Integrating CT myocardial perfusion and CT-FFR in the work-up of coronary artery disease, JACC Cardiovasc. Imaging 10 (2017) 760–770.
[95] M. Nahrendorf, D. Sosnovik, B. French, F. Swirski, F. Bengel, M. Sadeghi, et al., Multimodality cardiovascular molecular imaging—part II, Circ. Cardiovasc. Imaging 2 (2009) 56–70.
[96] P.C. Albro, et al., Noninvasive assessment of coronary stenoses by myocardial imaging during pharmacologic coronary vasodilatation. III. Clinical trial, Am. J. Cardiol. 42 (5) (1978) 751–760.
[97] G.A. Beller, D.D. Watson, Physiological basis of myocardial perfusion imaging with the technetium 99m agents, Semin. Nucl. Med. 21 (3) (1991) 173–181.
[98] S.S. Husain, Myocardial perfusion imaging protocols: is there an ideal protocol? J. Nucl. Med. Technol. 35 (1) (2007) 3–9.
[99] V.R. Taqueti, M.F. Di Carli, Radionuclide myocardial perfusion imaging for the evaluation of patients with known or suspected coronary artery disease in the era of multimodality cardiovascular imaging, Prog. Cardiovasc. Dis. 57 (6) (2015) 644–653.
[100] M. Ora, S. Gambhir, Myocardial perfusion imaging: a brief review of nuclear and nonnuclear techniques and comparative evaluation of recent advances, Indian J. Nucl. Med. 34 (4) (2019) 263.
[101] R. Hachamovitch, et al., Comparison of the short-term survival benefit associated with revascularization compared with medical therapy in patients with no prior coronary artery disease undergoing stress myocardial perfusion single photon emission computed tomography, Circulation 107 (23) (2003) 2900–2907.
[102] J.D. Schuijf, et al., Diagnostic and prognostic value of non-invasive imaging in known or suspected coronary artery disease, Eur. J. Nucl. Med. Mol. Imaging 33 (1) (2006) 93–104.
[103] R. Manka, L. Wissmann, R. Gebker, et al., Multicentre evaluation of dynamic three-dimensional magnetic resonance myocardial perfusion imaging for the detection of coronary artery disease defined by fractional flow reserve, Circ. Cardiovasc. Imaging 8 (2015), e003061.
[104] A. Fathala, Myocardial perfusion scintigraphy: techniques, interpretation, indications and reporting, Ann. Saudi Med. 31 (6) (2011) 625–634.
[105] E.H. Sonnenblick, W.H. Frishman, T.H. LeJemtel, Dobutamine: a new synthetic cardioactive sympathetic amine, N. Engl. J. Med. 300 (1) (1979) 17–22.

CHAPTER

9

Cardiovascular risk induced by radiotherapy

Mari Uyeda

AC Camargo Cancer Centre, São Paulo, Brazil

1 Introduction

In recent years, radiotherapy (RT) has made many advances, increasing the survival rate of patients undergoing cancer treatment [1]. An example would be patients with Hodgkin's and non-Hodgkin's lymphoma who now have an increased chance of cure to 80%, as well as patients with breast cancer with about 90% [2]. However, despite the increased longevity of these patients, the late side effects of this therapy appeared.

Cardiotoxicity induced by mediastinal radiation is one of the main concerns during treatment and should be defined in clinical oncology trials, through measurements of left ventricular ejection fraction (LVEF) [3]. The total cumulative dose of RT depends on the number of sessions and radiation dose to which patients are submitted. High doses of thoracic RT are generally used as adjunctive therapy or after conservative surgery in patients with breast cancer, adjuvant or definitive therapy in patients with esophageal or lung cancer, and a complementary therapy to the systemic treatment of lymphoma [3,4]. Radiation-induced heart disease (RIHD) can involve any cardiac structure including pericardium, myocardium, valves, conduction system, and coronary arteries. The cumulative prevalence of heart disease related to RT is estimated at 10 to 30% in 5 to 10 years after treatment and is dose-dependent [5]. In this way, the oncological patient is seen as a chronic disease carrier that, throughout its evolution, may present acute decompensations, such as cardiovascular manifestations.

Cardiovascular Toxicity and Therapeutic Modalities Targeting Cardio-oncology
https://doi.org/10.1016/B978-0-323-90461-2.00014-6

2 Cardiotoxicity

Cardiotoxicity is one of the most significant adverse events in cancer treatment and is responsible for considerable morbidity and mortality, presenting in an acute, subacute, or chronic form. Acute or subacute cardiotoxicity is characterized by sudden changes in ventricular repolarization, changes in the chemotherapy (CT) interval, supraventricular and ventricular arrhythmias, acute coronary syndromes, pericarditis, and myocarditis, usually seen from the beginning to 14 days after the end of treatment [6]. Chronic cardiotoxicity can be differentiated into two types, according to the onset of clinical symptoms. The first subtype occurs within 1 year after the end of CT, and the second usually occurs after 1 year after the end of CT. The most typical manifestation of chronic cardiotoxicity is systolic or diastolic ventricular dysfunction, which can lead to heart failure (HF) until cardiovascular death. In addition to not considering the clinical evaluation, different thresholds for the determination of cardiovascular toxicity have been used, which makes it difficult to establish the real incidence of cardiotoxicity over time [7]. To eliminate these difficulties, a common definition must be established, and preferably, it must incorporate clinical findings and not only complementary examinations. Some clinical manifestations of cardiotoxicity that can occur with cancer patients are shown in Table 1.

3 Risk factors

The volume of the irradiated cardiac area is the main determinant of RIHD and this relationship is very evident when comparing the effects of radiation in patients with cancer. Patients who receive high doses

TABLE 1 Clinical manifestations of cardiotoxicity in cancer patients.

Cardiotoxicity
Cardiac insufficiency
Ventricular and supraventricular arrhythmias
Acute myocardial ischemia with or without ST elevation
Asymptomatic left ventricular dysfunction
Systemic arterial hypertension
Pericardial disease
Thromboembolic events

ST: ST-segment of the electrocardiogram.

TABLE 2 Risk factors at RIHD.

Risk factors
Irradiation of anterior or left thorax
High cumulative dose (> 30 Gy)
Younger age (2 Gy per day)
Presence of tumor disease close to the heart or cardiac tumor
Lack of radiation field protection
Concomitant CT (mainly anthracyclines)
Cardiovascular risk factors (DM, smoking, obesity, high blood pressure, and dyslipidemia)
Preexisting cardiovascular disease

of irradiation show a threefold increase in death from cardiovascular causes, while patients who have undergone less cardiac exposure to radiation had a similar risk to the general population. The accumulated dose and its fractionation can determine the acute and chronic cardiac effects of RT. In the past, pericarditis was the most common side effect in patients receiving conventional RT for Hodgkin's lymphoma [8]. Dose restriction to 30 Gy with lower daily dose and subcarinal region block reduced the incidence of pericarditis from 20% to 2.5%. Radiation also increased the risk of cardiotoxic effects of some chemotherapeutic agents such as anthracyclines [9]. Younger patients at the beginning of RT also have a higher risk of developing heart disease, as well as smoking, arterial hypertension (AH), diabetes mellitus (DM), obesity, and dyslipidemia, as shown in Table 2.

4 Physiopathology

The progression of radiation-induced heart disease in the short and long term in each patient is still studied for better understanding. It is known that cardiac damage associated with RT is related to endothelial damage, affecting the micro- and macrovasculature [10]. Radiation increases oxidative stress and the formation of reactive oxygen species (ROS), consequently increasing the inflammatory response. Microvascular damage (drop-in capillary density resulting in ischemia) is associated with fibrosis, diastolic dysfunction, and HF. Primary radiation-related fibrosis does not occur as a direct effect, but as a repairing response of cardiac tissue to damage to the microvascular system. The macrovascular damage includes accelerated atherosclerosis and coronary strictures.

5 Heart failure (HF)

5.1 Incidence and clinical presentation

Among the adverse effects on the cardiovascular system, myocardial aggression with systolic ventricular dysfunction and HF stands out for determining the interruption of treatment, compromising the cure or the adequate control of cancer, with a worse prognosis than many neoplasms due to asymptomatic or symptomatic ventricular systolic and diastolic dysfunction ranging from 5% to 30% in clinical series, being more frequent in patients who have: extremes of age, previous ventricular dysfunction, AH, DM, use of combination CT, RT, and genetic susceptibility.

Cardiotoxic effects are cumulative and are related to dose, infusion rate, drug combination, and liver and kidney failure [11]. The mechanism responsible for the clinical symptoms and signs of HF may be due to systolic, diastolic, or both dysfunction, affecting one or both ventricles.

HF can be classified into evolutionary stages:

✓ **Stage A**—Includes patients at risk of developing HF, but still without noticeable structural disease and symptoms attributable to HF.
✓ **Stage B**—Patients who have acquired structural cardiac injury, but still have no symptoms attributable to HF.
✓ **Stage C**—Patients with structural cardiac injury and current or past HF symptoms.
✓ **Stage D**—Patients with symptoms refractory to conventional treatment, and who require specialized interventions or palliative care.

HF due to cardiotoxicity usually occurs in the first months after treatment and may still occur in the first weeks, and even later, years after treatment. However, acute conditions can still occur during treatment, especially in individuals with risk factors or when higher cumulative doses are used. The main symptom is paroxysmal nocturnal dyspnea, as well as orthopnea. Other symptoms include tiredness, fatigue, and digestive symptoms, such as anorexia, bloating, and diarrhea (in cases of ischemia or visceral congestion). Other signs seen include lower limb edema, hepatomegaly, ascites, and tachycardia. Typical signs of low cardiac output include hypotension, changes in the level of consciousness, oliguria, filiform pulse, and cold extremities. Pleural effusion is common in patients with decompensated HF.

5.2 Radiation toxicity

External radiation on the chest is associated with cardiovascular damage, including chronic pericarditis, pericardial effusion, constrictive pericarditis, restrictive cardiomyopathy, systolic dysfunction, valve disease, changes in the conduction system and accelerated coronary disease, with manifestations dependent on the radiation dose, the volume of the exposed heart and specific application techniques.

The spectrum of RT-induced heart disease includes systolic, diastolic dysfunction, and restrictive cardiomyopathy [12]. The predominant underlying pathophysiological mechanism in ischemic small vessel disease and fibrosis. Restrictive cardiomyopathy can be difficult to differentiate from constrictive pericarditis, and both can be present in the same patient. Valve disease is common, but it is often not serious. Either way, it can contribute to substantial morbidity that accompanies the irradiated heart. The most common injuries are tricuspid regurgitation, mitral regurgitation, and aortic regurgitation, which in some cases may be the cause of HF.

5.3 Monitoring of cardiotoxicity

Some examinations with a cardiovascular focus should be performed on patients who will undergo cardiotoxic therapies, such as a 12-lead electrocardiogram at rest and baseline assessment of left ventricular function by preferred echocardiography or radioisotopic ventriculography.

Patients undergoing cardiotoxic cancer therapies should be constantly monitored, especially for signs and symptoms of HF, and attention should be paid to early clinical manifestations of toxicity, which may present as a clinical picture of acute fulminant myocarditis and/or severe ventricular arrhythmias. Late manifestations can also occur and special attention should be paid to nonspecific symptoms such as tiredness, fatigue, and functional limitations for daily activities.

The widely accepted methods of assessing ventricular function are two-dimensional echocardiography and radioisotopic ventriculography. Both are capable of detecting baseline changes in cardiac function, being appropriate options for an initial assessment [13,14]. For the monitoring of sequential changes in ventricular function, it is suggested that the same method of analysis be maintained during follow-up, as the measurements obtained between different techniques are not interchangeable. Periodic monitoring of cardiotoxicity during infusion cycles is an essential strategy to prevent serious and irreversible myocardial injuries, although there are no prospective randomized studies that have tested such conduct. In general, we can classify the monitoring methods between those that identify global functional damage to the heart (traditionally assessed by LVEF) and those that seek to identify early damage to the cardiomyocyte (assessed by biomarkers) [15,16].

The evaluation of LVEF should be performed at different times of treatment:

(I) before the start of potentially cardiotoxic antineoplastic therapy,
(II) after administration of half of the total cumulative dose or after specific doses of anthracyclines or equivalent, and
(III) after each subsequent cycle of CT.

In the follow-up, after the completion of the cancer treatment, it is recommended to assess LVEF at variable intervals, according to the baseline risk of cardiotoxicity. The behavior of LVEF during follow-up has important therapeutic implications. Reduced LVEF greater than 10% and/or reduction to absolute values less than 50% are accepted and validated criteria for treatment suspension. The use of these criteria implied a substantial reduction in the risk of developing clinical HF. The disadvantage of this type of strategy, however, is that the diagnosis of cardiotoxicity is established only when the lesion is already established and, for the most part, it is irreversible.

5.4 Acute and late injuries

Acute injuries are difficult to diagnose, being less prevalent when compared to late injuries, which appear several years after treatment, and are difficult to distinguish and resemble diseases found in patients with heart disease of other etiologies. When the entire heart absorbs small doses of radiation, cardiac damage can manifest clinically in the first years after RT and continue to be important decades after exposure to radiation. The most common acute injuries are acute exudative pericarditis, usually as a reaction to necrosis and inflammation of tumors close to the heart. Pericarditis can also occur weeks after RT, presenting as an asymptomatic or symptomatic pericardial effusion. Cardiac tamponade is rare. As an acute injury, acute myocarditis can also occur, with inflammation, alteration of repolarization, and myocardial dysfunction.

When the doses prescribed for the clinical target in the mediastinum are typically higher ($\geq$ 40–50 Gy), smaller parts of the heart can absorb higher doses of radiation. Patients with esophageal cancer and lung cancer show that overdosing on the heart can put patients at a higher risk of noncancer-related death in the first years after the completion of RT [17].

5.5 Prevention of cardiotoxicity

The prevention of myocardiopathy involves the recognition of its risk factors, such as ischemic heart disease, ventricular dysfunction, valvular disease, uncontrolled AH and arrhythmias, cumulative anthracycline dose greater than 550 mg/m^2, age extremes, concomitant mediastinal RT, DM, obesity, and combination with other drugs, such as trastuzumab. The recognition of individuals at risk and the prevention of asymptomatic left ventricular dysfunction and HF are, therefore, important objectives in the management of these patients. Immunosuppressive therapy can predispose to an increase in the incidence of cancer, or even induce recurrence of the previous disease, being critical of the patient's careful evaluation for a better definition of treatment.

6 Myocardial ischemia

During cancer treatment, the patient can present any form of coronary disease, stable or unstable. Acute coronary syndrome (ACS) results from the interaction between the atherosclerotic lesion and the hematopoietic system, causing rupture of a vulnerable plaque or erosion of the plaque with the formation of a thrombus, which can partially occlude the vascular lumen (angina/acute myocardial infarction without supra-ST-segment elevation) or totally (acute myocardial infarction with ST-segment elevation). The clinical presentation of coronary heart disease in cancer patients is similar to that seen in the general population. Coronary heart disease involves an interaction between lipid metabolism, inflammation, and thrombosis. Cancer induces a prothrombotic state through several mechanisms, including platelet activation and aggregation, as well as an increase in procoagulant factors, the risk of which is increased with the use of chemotherapeutic agents [18]. Among the chemotherapeutic drugs most associated with myocardial ischemia, the following stand out:

✓ Antimetabolites (capecitabine and fluorouracil)
✓ Antimicrotubule (paclitaxel and docetaxel)
✓ Monoclonal antibodies (bevacizumab)
✓ Tyrosine kinase inhibitors (sorafenib and sunitinib)
✓ Vinca alkaloids (vincristine and vinorelbine).

The management of coronary heart disease in cancer patients must take into account the following aspects:

(a) the high incidence of noncardiac surgery in cancer patients.
(b) the increased potential for the occurrence of thrombocytopenia during evolution.
(c) predisposition to thrombosis.
(d) potential of drug interaction between drugs used in the management of coronary disease and CT.

For the management of myocardial ischemia, stable angina, and acute coronary syndrome, one must consider:

Stents: despite revolutionizing the management of coronary disease, it has thrombogenic potential, and antiplatelet therapy with drugs is essential until the stent is incorporated into the endothelium, that is, 4 weeks in patients without the use of CT. After endothelialization, antiplatelet inhibition with aspirin is recommended to prevent late thrombosis.

The impact of antineoplastic therapy on stent re-endothelialization could expose the patient to an even longer period of thrombotic risk, both with conventional or pharmacological stents and requiring prolonged double-antiplatelet therapy. Thus, angioplasty with a conventional stent is recommended as the first choice because of the lower risk of thrombosis.

Thrombocytopenia: platelets are a fundamental part of the pathogenesis of acute coronary syndrome (ACS), although this event is rare in patients with thrombocytopenia, for cancer patients it occurs in 30% of cases. Therapy involves medications such as acetylsalicylic acid, clopidogrel, heparin, thrombolytic, and percutaneous intervention. This therapy is associated with an increased risk of bleeding in patients with thrombocytopenia, and in the general population, thrombocytopenia usually results in a contraindication to these interventions [19]. The cause of coronary thrombosis in patients with thrombocytopenia is controversial and multifactorial. But the development and consequences of the occlusive thrombus are similar to atherosclerotic disease. Regardless of thrombocytopenia, cancer patients are predisposed to coronary thrombosis because the platelets are larger, more adherent to the vascular surface, and produce thrombogenic microparticles, promoting the formation of the hemostatic plug.

Treatment must be individualized and the benefit of treatment must be weighed against the risk of bleeding, especially considering other predisposing factors. In patients with thrombocytopenia, radial access is preferred when performing catheterization, as it allows for better local hemostasis.

7 Systemic arterial hypertension (SAH)

Systemic arterial hypertension (SAH) is a multifactorial clinical condition characterized by elevated and sustained levels of blood pressure (BP). It is frequently associated with functional and/or structural changes in the target organ (heart, brain, kidneys, and blood vessels) and metabolic changes, with a consequent increased risk of fatal and nonfatal cardiovascular events [20].

The prevalence of SAH in cancer patients before the introduction of angiogenesis inhibitors was similar to that of the adult population in general, between 32% and 48%. However, with the increased survival of cancer patients, and with the increased use of CT drugs related to SAH, the latter has been a more frequent diagnosis in these patients. The chemotherapeutics most related to SAH are angiogenesis inhibitors because they act on neurohormonal factors, such as renin, aldosterone, and catecholamines, inhibiting the growth receptor tyrosine kinase activity vascular endothelium, responsible for increased capillary permeability, nitric oxide production, migration, and proliferation of endothelial cells, resulting in vasoconstriction and sodium retention.

7.1 Clinical manifestations/diagnosis

SAH, in most cases, does not result in symptoms. It is an asymptomatic disease; however, when clinical manifestations occur, it causes complications, increasing morbidity and mortality, and worsening quality of life. It is manifested by hypertensive encephalopathy, stroke, acute pulmonary edema, ACS, malignant hypertension, and renal failure. The level of BP considered adequate in cancer patients is similar to that used in the adult population.

The prevention of SAH begins before the beginning of antineoplastic therapy, identifying the cardiovascular profile of the patient so that later it is decided on the therapeutic approach (hygienic dietary changes, selection of drugs, dose administered during each session, cumulative dose, route of administration, drugs, and drug sequence).

7.2 Treatment

According to risk stratification considerations, goals should always be pursued, as the control of SAH should allow patients to tolerate the maximum effective dose of treatment with CT and benefit from the control of tumor growth for a long period, improving quality and the duration of life. Hygienic dietary changes should always be encouraged, emphasizing the recommendation of a low-sodium diet and regular physical activity. However, nonpharmacological strategies are not always sufficient and adequate, due to the compromised performance status of many patients with advanced neoplasia. Thus, in this group, early pharmacological intervention should be considered. Every patient with stage 1 SAH ($\geq$ 140/90 mmHg) or an increase in diastolic BP by 20 mmHg or more should start antihypertensive therapy.

8 Arrhythmias

The incidence of arrhythmias in cancer patients varies according to treatment and may occur in several circumstances, such as infections, hydroelectrolytic disorders, in the postoperative environment, and during CT. Cancer by itself generates a pro-arrhythmogenic environment, generating, for example, atrial fibrillation that occurs twice as often in cancer patients, being responsible for morbidity after cancer surgery in up to 12.6. Arrhythmogenesis in cancer patients is related to fibrosis, chronic inflammation, and the direct effects of drugs on the conduction system [21].

8.1 Clinical manifestations/diagnosis

Bradycardia is typically defined as a hazard ratio (HR) < 60 bpm. However, several patients are asymptomatic with HR < 50 bpm. Some patients, however, may present associated symptoms, such as fatigue, limitation to physical activities, syncope, and dizziness.

Cancer patients are particularly prone to CT prolongation since 16% to 36% of cancer patients show abnormalities on the baseline electrocardiogram (ECG).

Cancer patients have a high prevalence of comorbidities, including structural heart disease, liver and kidney dysfunction, as well as the use of medications that prolong the CT interval used in the complications of CT and the progression of the neoplasia. Added to this is the fact that cancer patients frequently experience nausea, vomiting, diarrhea, and decreased oral intake, which can lead to hydro electrolytic disorders. The main agents associated with CT interval prolongation are histone deacetylase (vorinostat) inhibitors.

Tachycardia is defined by a heart rate above 100 bpm, and the diagnosis is electrocardiographic. In cancer patients, they can range from sinus tachycardia, in general, secondary to other comorbidities, to supraventricular (TSV) and ventricular tachycardias. Among supraventricular tachycardias, the most common is atrial fibrillation (AF), characterized by complete disorganization of atrial electrical activity. The AF is classified as follows:

✓ Initial (the first episode diagnosed),
✓ Paroxysmal (ends spontaneously without the use of drugs or electrical cardioversion, usually lasting less than 7 days),
✓ Persistent (the one that installs itself and does not interrupt) and
✓ Permanent (more than 1 year).

Atrial tachycardia shows an isoelectric line between atrial activations and depolarization of the atria is determined by P waves of a morphology different from that of sinus P.

Wide complex tachycardia (QRS) is defined as HR > 100 bpm and QRS > 120 milliseconds. Most of the time, it is of ventricular origin; however, TSV aberration may occur in patients with bundle branch block, with antidromic arrhythmias, and with ventricular preexcitation.

Ventricular tachycardias, in general, are associated with structural cardiomyopathies (coronary artery disease, dilated cardiomyopathies, right ventricular cardiopathies, congenital abnormalities, coronary spasm, and hypertrophic cardiomyopathy) [22]. These are divided into hemodynamically stable arrhythmias (asymptomatic or minimal symptoms such as palpitations) and hemodynamically unstable arrhythmias (presyncope, syncope, shock, cardiac collapse, and sudden death).

8.2 Treatment

It is essential to exclude or treat when hypovolemia, infection, hydroelectrolytic disorders, and thyroid disorders are present. In most cases of bradycardia or atrioventricular blocks, the patient is asymptomatic. In the specific treatment of arrhythmias, concomitance with cardiomyopathy and other comorbidities should always be assessed.

The goal of treating ventricular arrhythmias is to decrease morbidity and sudden death events, and it is essential to evaluate the triggering factors. Indicates whether pharmacological therapy for refractory and/or symptomatic cases. Thus, ventricular ectopic beats and nonsustained ventricular tachycardia (NSVT) monomorphic without symptoms in patients without structural heart disease should not be given antiarrhythmics and should only be followed up.

9 Thromboembolism

Venous thromboembolism (VTE) includes deep vein thrombosis (DVT) and pulmonary thromboembolism (PTE). It is a serious complication in cancer patients, being one of the main causes of death in these patients [23]. Neoplasms are associated with an increased risk of thrombosis by four times, and treatment with chemotherapy increases the risk by approximately six times.

The incidence of VTE in cancer patients has been increasing for several reasons. Some causes have been raised: antiangiogenic drugs, thalidomide, lenalidomide, bevacizumab, and the use of hormonal therapies (tamoxifen).

9.1 Diagnosis and risk factors

Cancer patients hospitalized with VTE have a higher mortality rate (an increase of two to eight times) and knowledge of risk factors is essential for the quality of treatment. Some risk factors are surgery, immobility, old age.

Although the association between cancer and VTE is clear, the risk is not the same for all patients and may vary with the location, treatment, and stage. Risk factors are:

- Women;
- Black ethnicity;
- Cancer sites and associated types: brain, pancreas, kidneys, ovary, stomach, lung, myeloma, and lymphoma;
- More advanced stage: VTE is more common in patients with metastatic disease when compared to patients without metastasis;
- Initial period after diagnosis: the risk of VTE is greater in the first 3 months after cancer diagnosis;
- Chemotherapy: is strongly associated with an increased risk of VTE.

9.2 Prophylaxis

Hospitalized cancer patients are at particularly high risk for VTE, although the risk can vary significantly between patient subgroups.

Mechanical prophylaxis: intermittent pneumatic compression (IPC) and graduated elastic compression stockings should always be considered in hospitalized patients diagnosed with cancer, in the absence of contraindications (open wounds arterial insufficiency), in association with pharmacological prophylaxis, when this is possible [24]. However, its isolated use should only be done in patients with contraindications to pharmacological prophylaxis.

Pharmacological prophylaxis: cancer patients, acutely ill, benefit from prophylaxis with low molecular weight heparin and fondaparinux (5%–15%). Pharmacological prophylaxis in cancer patients is always recommended due to the high risk of VTE.

Mechanical methods should be associated with pharmacological methods, but they should not be used as monotherapy unless there are contraindications to pharmacological therapy. Prophylaxis must be maintained for at least 7 to 10 postoperative days. Prophylaxis extended up to 4 weeks postoperatively should be considered in patients undergoing abdominal or pelvic oncology surgeries with high-risk characteristics, such as post-resection residual malignancy, obesity, or previous history of VTE.

10 Cardiotoxicity associated with radiotherapy (RT)

RT is used to treat various types of tumors. Thus, the cardiovascular complications of RT are an additional concern in the management of cancer patients, requiring early diagnosis and immediate intervention.

Coronary artery disease (CAD) is the most common cause of cardiac death in patients undergoing RT, with or without CT. Atherosclerotic lesions are similar to nonirradiated individuals and can be detected early, in the first 5 years after treatment, and the risk remains high, ≥ 15 years after RT and are characterized by intimal proliferation, accumulation of lipid-rich macrophages, and formation of plaque [25]. In radiation-related CAD, the middle vessel layer is destroyed with a loss of smooth muscle, and the adventitia is markedly thickened and fibrotic. The distribution of CAD correlates with the areas exposed to higher radiation doses. For example, in patients with left breast cancer, the anterior wall of the heart receives higher doses of radiation, and consequently, the anterior descending artery and diagonal distal branches are more involved. In patients with lymphoma undergoing mediastinal RT, the affected region is more commonly that of the internal mammary artery, and more basal cardiac structures and more ostial coronary lesions are seen in these patients.

The clinical presentation of coronary heart disease does not differ from the general population. It can present itself silently, in 50% of cases, in a typical anginal form or the form of sudden death. The silent form is more frequent, possibly because it also presents lesions in the nerve endings during RT. Patients may present with angina, dyspnea, HF, and even sudden death. The latter is a consequence of diffuse intimal hyperplasia, with significant stenosis of the anterior descending artery. The average time interval for the development of CAD after radiotherapy is 82 months. The management of these changes is similar to the treatment of atherosclerotic disease. RT also compromises the carotid arteries, resulting in extensive lesions in the bifurcations and atypical areas.

The treatment of patients with RT-related atherosclerosis can be performed with percutaneous interventions and surgical revascularization. In cases of surgical revascularization, care should be taken as the internal mammary arteries may also be damaged by radiation, preventing their use in grafts. The concomitant carotid disease may increase the perioperative risk of these patients, in addition to associated pulmonary fibrosis, porcelain aorta, and valve diseases, requiring joint surgical treatment.

In the pericardium, it can cause fibrous thickening that especially compromises the right ventricle. The most frequent clinical presentations are pericardial effusion and pericarditis. The interval between RT and the development of symptoms of pericardial disease varies from 2 to 145 months. In pericardial effusion, symptoms appear at the beginning of the disease, whereas in constrictive pericarditis, clinical manifestations are late, usually after 18 months.

Myocardial fibrosis is another side effect of RT. The fibrous thickening of the valves occurs, with the left side being more affected. However, the minority of patients have moderate to severe valve dysfunction, and the average time to onset of symptoms is approximately 98 months.

10.1 Pericardial diseases and cancer

The incidence of pericardial disease in cancer patients is around 7%–12%. On average, an average absorbed dose of 40 Gy in the mediastinum increases the prevalence of asymptomatic pericardial disease tenfold, but the clinically evident pericardial disease is rare [26]. This rate varies according to the type of neoplasia, being higher in patients with solid neoplasia, especially in the lung (19%–40%) and breast (10%–28%) tumors, followed by hematological tumors such as lymphoma and leukemias (9%–28%). The etiology is varied and may be due to the primary involvement of the pericardium by the neoplasm itself, resulting from complications of RT or side effects of some chemotherapeutic agents. Fortunately, the prevalence of complications related to the pericardium has been decreasing in recent years, probably due to advances in the early diagnosis of pericardial effusion, as well as in the treatment of cancer.

10.1.1 *Diagnosis*

The presence of pericardial effusion in patients with cancer is not always related to the underlying neoplasm, which can result from RT or other common clinical situations. The symptoms of pericardial diseases are mostly nonspecific, such as chest pain, asthenia, weakness, dyspnea, and fever. The diagnosis of pericardial effusion is confirmed by transthoracic echocardiography or chest tomography. In many cases, the pericardial effusion approach not only relieves symptoms but can also be the chance for the oncological diagnosis. The detection of neoplastic cells in the pericardial fluid can be done through cytological examination and pathological anatomy. With the help of immunohistochemistry and the measurement of some tumor markers such as CEA, CA19-9, CA125, NES, and CYFRA-21, the sensitivity and specificity of this diagnostic test increase significantly.

Another diagnostic option is pericardiotomy, a valuable diagnostic tool in some cases to elucidate the etiology of pericardial effusion. The main advantage of this test is the possibility of obtaining a fragment of the pericardium for histopathological analysis. If the analysis of the pericardial fluid is inconclusive, pericardiotomy should be indicated (Evidence IIa).

10.1.2 *Treatment*

The goal of treatment may be to relieve symptoms, prevent a long-term recurrence, or even to treat the tumor to increase patient survival. When the goal is only symptom relief, percutaneous or surgical drainage is appropriate. For the prevention of recurrences, several approaches have been proposed: pericardial window, local sclerosis, local or systemic CT, and RT.

(a) **Percutaneous drainage**: offers immediate relief of symptoms. The use of the echocardiogram to guide the procedure reduces the risk of accidental puncture of the right ventricle. The recurrence rate is high (40%) if no other treatment is performed, indicated for hemodynamically unstable patients, without surgical drainage conditions.

(b) **Surgical drainage**: a technique of choice for drainage of malignant pericardial effusion. It has a lower rate of recurrence than percutaneous drainage for two reasons: permanent communication through the pericardial window and inflammatory reaction that promotes adhesion of the parietal and visceral pericardium.

(c) **Sclerosis therapy**: prevents recurrence after drainage. The instilled agent promotes irritation, inflammation, and fibrosis of the pericardium. Side effects are pain, fever, and paroxysmal atrial fibrillation. Bleomycin and thiotepa have been used with a lower incidence of side effects. Most effective in breast cancer.

(d) **Local chemotherapy**: the agents used are especially platinum (for lung, ovarian, and mesothelioma cancer) and mitoxantrone (breast and other carcinomas). Treatment should be supplemented with systemic chemotherapy.
(e) **Radiotherapy**: indicated in radiosensitive tumors such as acute and chronic leukemias and breast tumors. It is effective, but involves radiation risks in addition to the high cost.

The treatment of terminally ill patients should aim only at relieving symptoms, while patients with a longer life expectancy must have effective control of pericardial effusion with measures that reduce recurrence, aiming at a greater and better quality of life.

10.2 Valves

Valve disease secondary to RT is seen in 81% of patients. The most affected valves are often aortic and mitral, saving tricuspid and pulmonary and more related to insufficiency than stenosis. Although common, they do not have clinical repercussions, with 70% of patients asymptomatic. Treatment is the same as in the general population, but in patients with extensive mediastinal fibrosis, surgical treatment can be complicated with increased perioperative morbidity and mortality. For clinically relevant valve disease, the latency of more than 10 years and a high risk of progressive valve deterioration have been described during the second and third decades after treatment. The prevalence increases with time after irradiation and appears to be independent of age.

10.3 Driving system

Changes in asymptomatic ventricular repolarization are common during RT. Conduction disorders after RT can appear several years later, making it difficult to establish cause and effect and their prevalence. The right bundle branch block is most commonly associated with mediastinal radiation, probably due to its proximity to the endocardium on the same side. Lesions to the right branch can occur directly by radiation or indirectly by myocardial fibrosis and ischemia. As the most serious manifestation, a total atrioventricular block may occur.

RT is also associated with prolongation of the corrected CT interval. The correlation of the conduction system disorder with radiotherapy may suggest a relationship between atrioventricular blocks and previous radiation:

✓ Total radiation dose to the heart > 40 Gy.
✓ 10 years or more after therapy.
✓ Changes such as right bundle branch block.
✓ Previous pericardial involvement.
✓ Associated cardiac or mediastinal injuries.

10.4 Cardiomyopathies

RT can directly affect myocytes, leading to interstitial fibrosis. Exposure to 250 mg/m^2 or more of anthracyclines and radiation greater than 1200 cGy to the heart increases the relative risk two to six times compared with nonirradiated survivors. Myocardial microvascular injury results in chronic ischemia, with myocardial fibrosis, typically manifesting itself as diastolic dysfunction.

Thoracic RT can lead to both systolic and diastolic dysfunction, as well as dilated or restrictive cardiomyopathy, especially when RT is associated with CT. The treatment does not differ from the treatment of other cardiomyopathies, with ACE inhibitors, angiotensin blockers, aldosterone antagonists, and beta-blockers.

10.5 Cardiotoxicity and thymus diseases

Thymomas and thymic carcinomas are rare thoracic cancers, most often located in the upper anterior mediastinum. The indication for radiation treatment is not common in this group of tumors, but when carefully considered in the postoperative scenario, the doses normally prescribed are in the range of 45 to 50 Gy for patients with clear or narrow surgical margins and up to 60–70 Gy for patients with gross or unresectable residual disease [27]. The relative 5-year survival rates for patients over 65 and under 25 years old were 60% and 78%, respectively.

Particle therapy plans, with protons and carbon ions, present lower doses for the lung, sinuses, heart, esophagus, and spinal cord, without compromising the coverage of the target planning volume (PTV) when compared to 3D-CRT, IMRT, VMAT, or helical tomotherapy, HT.

The DIBH technique is becoming globally available, saving the heart from exposure to radiation in high and low doses. The DIBH-proton combination achieves the lowest dose for the heart, lung, and sinuses, simultaneously with a higher target volume coverage. The dose for organs at risk (OAR) is reduced by 38%–83%.

10.6 Cardiotoxicity for malignant mesotheliomas

Radiotherapy for malignant pleural mesothelioma after extrapleural pneumonectomy, pleurectomy/decortication, or surgical biopsy is a real clinical challenge due to a large and circumferential PTV to encompass the pleural cavity [28]. As part of the multimodal treatment of malignant mesotheliomas in (neo) adjuvant environments, the prescribed doses are up to 50–60 Gy.

Viable toxicity profiles are used for adjuvant radiotherapy with helical tomotherapy, HT, IMRT, VMAT, and recently for proton techniques.

The incidence of grade 2 or higher cardiac events is in the range of 6 to 8%. IMPT has shown promise in results, particularly in limiting radiation exposure to thoracic OAR.

10.7 Cardiotoxicity after craniospinal irradiation

The total treatment dose prescribed for craniospinal irradiation (CSI) medulloblastoma ranges from 18 to 36 Gy given in 1.8 Gy fractions with a booster dose for intracranial or spinal metastases when indicated. It is essential to be aware of the possible late treatment-related toxicity, as the event-free 5-year survival rate in patients with medulloblastoma is high, reaching > 80%. In CSI, all structures of the heart are susceptible to heart damage due to the outgoing radiation dose.

Compared to conventional 2D-RT or 3D-CRT, advanced treatment techniques (i.e., IMRT, VMAT, HT, electron-based technique, and proton beam scanning) were all considered superior about dosimetric parameters to achieve better preservation of the heart and other OAR.

Proton beam CSI, compared to photon beam CSI techniques, confers lower risks of expected healthy tissue complications. Due to the Bragg peak, the output dose for OAR in the application of proton RT treatment is considerably reduced. The estimated risk reduction of the complication rate is based on different risk assessment models.

10.8 Cardiotoxicity after total body irradiation

Long-term survivors of hematopoietic cell transplantation have an increased risk of serious treatment-related chronic diseases due to intense QT to treatment with radiotherapy and the disease. The total body irradiation itself exposes long-term survivors to an increased risk of CVD due to the direct exposure of the heart to radiation and indirect disturbances metabolic, renal, pulmonary, neurological, and endocrine functions, causing late treatment sequelae, such as growth hormone deficiency, DM, dyslipidemia, metabolic syndrome, SAH, or renal failure.

11 Recommendations for follow-up after radiotherapy

All patients undergoing chest RT should be evaluated for the usual risk factors for cardiovascular disease. It is suggested that all be evaluated with a transthoracic echocardiogram before the start of treatment, to identify preexisting anomalies (Table 3).

In the follow-up of patients exposed to RT, one should seek to modify risk factors for cardiovascular diseases, such as encouraging healthy eating and weight control, blood pressure control, smoking cessation, stimulating physical exercise practices.

TABLE 3 Cardiac screening before and after radiotherapy.

Before radiotherapy, full screening and aggressive control of risk factors
Baseline transthoracic echocardiogram to detect structural changes
Annual follow-up cardiovascular history and physical examination
Transthoracic echocardiogram if murmurs detected careful investigation of symptoms
5-year follow-up
Transthoracic echocardiogram for high-risk asymptomatic patients
Noninvasive assessment of coronary heart disease with stress echocardiography or cardiac magnetic resonance imaging
Repeat transthoracic echocardiogram every 5 years
10-year follow-up
Transthoracic echocardiogram for asymptomatic nonhigh-risk patients
Repeat echocardiogram every 5 years

12 Modern radiotherapy techniques

In recent years, new radiotherapy techniques have been studied with less exposure to radiation without reducing the chance of a cure for the tumor. The three main strategies used to reduce cardiac toxicity are to reducing radiation dose; reducing radiation field and volume; and using new forms of application of RT, such as intensity-modulated radiotherapy (IMRT), volumetric modulated arc therapy (VMAT) or proton therapy, and deep inspiration breathing retention (DIBR). These ways reduce the radiation dose to the heart and its substructures.

New techniques include intensity-modulated RT, in which three-dimensional tomographic images are used coupled with computerized programming that controls the dynamic RT system. This strategy allows the doses of radiotherapy to be "sculpted" in three dimensions using exactly the design of the areas to be treated and protected. Another technique is image-guided radiotherapy (IGRT), which arose from the need to more precisely locate the tumor or internal organs, affected by the disease, at the time of treatment, so that a better correlation with reference images occurs. The purpose of the IGRT is to ensure as accurately as possible that the tumor will be within the irradiation field every day of the treatment since they can change positions between treatments, or even in the same treatment. This can occur due to respiratory movements, the filling or emptying of some organs, or even small changes in positioning from 1 day to the next. IGRT involves conformed radiotherapy guided by images, such as tomography, ultrasound, and X-rays, performed daily in the treatment room, before the radiotherapy procedure. The image obtained is compared with the image made in the planning process, and then, the necessary adjustments are made. This allows for greater precision in the treatment, saving the adjacent normal tissues. In some cases, doctors implant small markers in the tumor, helping to visualize the movement of the tumor or organs.

Proton therapy appears promising in reducing exposure of the OAR to radiation and early nonrandomized clinical studies, confirming the low toxicity rates are available, the expected benefit of proton therapy can vary from patient to patient. It is also associated with some disadvantages, namely possible uncertainties at the end of the field, movement management, unavailability, and cost of treatment. However, proton therapy should be reasonably considered in patients with appropriately selected lymphoma, when the dose for OAR can be significantly decreased. According to the recommendations of the International Lymphoma Radiation Group (ILROG), two groups of patients can benefit from proton therapy to reduce cardiotoxicity. The first group includes patients with mediastinal disease that extend below the origin of the left main coronary artery and are anterior, posterior, or on the left side of the heart, and the other includes intensely pretreated patients who are at a greater risk of toxicity related to radiation.

The cumulative incidences of cardiovascular disease after RT are low but should be considered in the context of global increases in the number of cancer survivors. Mediastinal RT with doses around 40 Gy with the 2D-RT technique increases the risk of cardiovascular death by approximately six times. The risk is the highest 10 years after treatment, increasing the inhomogeneity of the dose absorbed to the heart and the age at diagnosis.

Asymptomatic cardiovascular disease is more common and can be found in screening in most lymphoma survivors, although some of these abnormalities may be of less clinical relevance. The burden of cardiac toxicity is substantially reduced with modern RT techniques, compared to historical series. Every effort should be made to reduce the dose to the heart and its substructures as reasonably as possible, regardless of the patient's age or histology of the primary tumor. Technological developments, such as breathing control and improvements in treatment planning techniques, may further facilitate the reduction of cardiac mortality risks. Proton therapy is evolving rapidly and currently shows the best dosimetric advantages in the preservation of thoracic OAR when compared to advanced photon techniques in lymphoma, thymic neoplasms, malignant mesothelioma, and CSI.

References

[1] B. Rajamanickam, K.A. Lee, K.W. Yeo, Cancer and radiation therapy: current advances and future directions, Int. J. Med. Sci. 9 (3) (2012) 193–199, https://doi.org/10.7150/ijms.3635.

[2] S. Shanbhag, R.F. Ambinder, Hodgkin lymphoma: a review and update on recent progress, CA Cancer J. Clin. 68 (2) (2018) 116–132, https://doi.org/10.3322/caac.21438.
[3] S.H. Armenian, C. Lacchetti, A. Barac, J. Carver, L.S. Constine, N. Denduluri, D. Lenihan, Prevention and monitoring of cardiac dysfunction in survivors of adult cancers: American Society of Clinical Oncology Clinical Practice Guideline, J. Clin. Oncol. 35 (8) (2017) 893–911, https://doi.org/10.1200/JCO.2016.70.5400.
[4] A. Recht, S.E. Come, I.C. Henderson, R.S. Gelman, B. Silver, D.F. Hayes, J.R. Harris, The sequencing of chemotherapy and radiation therapy after conservative surgery for early-stage breast cancer, N. Engl. J. Med. 334 (21) (1996) 1356–1361, https://doi.org/10.1056/NEJM199605233342102.
[5] H. Wang, J. Wei, Q. Zheng, L. Meng, Y. Xin, X. Yin, X. Jiang, Radiation-induced heart disease: a review of classification, mechanism and prevention, Int. J. Biol. Sci. 15 (10) (2019) 2128, https://doi.org/10.7150/ijbs.35460.
[6] A. Hidron, N. Vogenthaler, J.I. Santos-Preciado, A.J. Rodriguez-Morales, C. Franco-Paredes, A. Rassi Jr., Cardiac involvement with parasitic infections, Clin. Microbiol. Rev. 23 (2) (2010) 324–349, https://doi.org/10.1128/CMR.00054-09.
[7] M. Volkova, R. Russell, Anthracycline cardiotoxicity: prevalence, pathogenesis and treatment, Curr. Cardiol. Rev. 7 (4) (2011) 214–220, https://doi.org/10.2174/157340311799960645.
[8] M. Witkowska, A. Majchrzak, P. Smolewski, The role of radiotherapy in Hodgkin's lymphoma: what has been achieved during the last 50 years? Biomed. Res. Int. 2015 (2015), https://doi.org/10.1155/2015/485071.
[9] E.T. Yeh, A.T. Tong, D.J. Lenihan, S.W. Yusuf, J. Swafford, C. Champion, M.S. Ewer, Cardiovascular complications of cancer therapy: diagnosis, pathogenesis, and management, Circulation 109 (25) (2004) 3122–3131, https://doi.org/10.1161/01.CIR.0000133187.74800.B9.
[10] J.D. Groarke, P.L. Nguyen, A. Nohria, R. Ferrari, S. Cheng, J. Moslehi, Cardiovascular complications of radiation therapy for thoracic malignancies: the role for non-invasive imaging for detection of cardiovascular disease, Eur. Heart J. 35 (10) (2014) 612–623, https://doi.org/10.1093/eurheartj/eht114.
[11] P. Mladěnka, L. Applová, J. Patočka, V.M. Costa, F. Remiao, J. Pourová, TOX-OER and CARDIOTOX Hradec Králové Researchers and Collaborators, Comprehensive review of cardiovascular toxicity of drugs and related agents, Med. Res. Rev. 38 (4) (2018) 1332–1403, https://doi.org/10.1002/med.21476.
[12] E. Muchtar, L.A. Blauwet, M.A. Gertz, Restrictive cardiomyopathy: genetics, pathogenesis, clinical manifestations, diagnosis, and therapy, Circ. Res. 121 (7) (2017) 819–837, https://doi.org/10.1161/CIRCRESAHA.117.310982.
[13] S. Kaul, C. Tei, J.M. Hopkins, P.M. Shah, Assessment of right ventricular function using two-dimensional echocardiography, Am. Heart J. 107 (3) (1984) 526–531, https://doi.org/10.1016/0002-8703(84)90095-4.
[14] L.F.L.D. Oliveira, J.L. O'Connell, E.E.V.D. Carvalho, É.C.C. Pulici, M.M.D. Romano, B.C. Maciel, M.V. Simões, Comparison between radionuclide ventriculography and echocardiography for quantification of left ventricular systolic function in rats exposed to doxorubicin, Arq. Bras. Cardiol. 108 (2017) 12–20, https://doi.org/10.5935/abc.20160194.
[15] R. Dhingra, R.S. Vasan, Biomarkers in cardiovascular disease: statistical assessment and section on key novel heart failure biomarkers, Trends Cardiovasc. Med. 27 (2) (2017) 123–133, https://doi.org/10.1016/j.tcm.2016.07.005.
[16] J. Wang, G.J. Tan, L.N. Han, Y.Y. Bai, M. He, H.B. Liu, Novel biomarkers for cardiovascular risk prediction, J. Geriatr. Cardiol. 14 (2) (2017) 135, https://doi.org/10.11909/j.issn.1671-5411.2017.02.008.

[17] I. Gayed, S. Gohar, Z. Liao, M. McAleer, R. Bassett, S.W. Yusuf, The clinical implications of myocardial perfusion abnormalities in patients with esophageal or lung cancer after chemoradiation therapy, Int. J. Cardiovasc. Imaging 25 (5) (2009) 487–495, https://doi.org/10.1007/s10554-009-9440-7.
[18] M.S. Hamza, S.A. Mousa, Cancer-associated thrombosis: risk factors, molecular mechanisms, future management, Clin. Appl. Thromb. Hemost. 26 (2020), https://doi.org/10.1177/1076029620954282, 1076029620954282.
[19] R. Gauer, M.M. Braun, Thrombocytopenia, Am. Fam. Physician 85 (6) (2012) 612–622. https://www.aafp.org/afp/2012/0315/p612.html.
[20] G.A. Mensah, J.B. Croft, W.H. Giles, The heart, kidney, and brain as target organs in hypertension, Cardiol. Clin. 20 (2) (2002) 225–247, https://doi.org/10.1016/S0733-8651(02)00004-8.
[21] H.M. Chang, R. Moudgil, T. Scarabelli, T.M. Okwuosa, E.T. Yeh, Cardiovascular complications of cancer therapy: best practices in diagnosis, prevention, and management: part 1, J. Am. Coll. Cardiol. 70 (20) (2017) 2536–2551, https://doi.org/10.1016/j.jacc.2017.09.1095.
[22] E.M. Lopez, R. Malhotra, Ventricular tachycardia in structural heart disease, J. Innov. Card. Rhythm Manag. 10 (8) (2019) 3762, https://doi.org/10.19102/icrm.2019.100801.
[23] J. Laryea, B. Champagne, Venous thromboembolism prophylaxis, Clin. Colon Rectal Surg. 26 (03) (2013) 153–159, https://doi.org/10.1055/s-0033-1351130.
[24] E. Hillegass, M. Puthoff, E.M. Frese, M. Thigpen, D.C. Sobush, B. Auten, Role of physical therapists in the management of individuals at risk for or diagnosed with venous thromboembolism: evidence-based clinical practice guideline, Phys. Ther. 96 (2) (2016) 143–166, https://doi.org/10.2522/ptj.20150264.
[25] A. Cassar, D.R. Holmes Jr., C.S. Rihal, B.J. Gersh, Chronic coronary artery disease: diagnosis and management, Mayo Clin. Proc. 84 (12) (2009) 1130–1146. Elsevier https://doi.org/10.4065/mcp.2009.0391.
[26] I. Ratosa, M.I. Pantar, Cardiotoxicity of mediastinal radiotherapy, Rep. Pract. Oncol. Radiother. 24 (6) (2019) 629–643, https://doi.org/10.1016/j.rpor.2019.09.002.
[27] R. Komaki, D. Gomez, Radiotherapy for thymic carcinoma: adjuvant, inductive, and definitive, Front. Oncol. 3 (2014) 330, https://doi.org/10.3389/fonc.2013.00330.
[28] R.M. Flores, H.I. Pass, V.E. Seshan, J. Dycoco, M. Zakowski, M. Carbone, V.W. Rusch, Extrapleural pneumonectomy versus pleurectomy/decortication in the surgical management of malignant pleural mesothelioma: results in 663 patients, J. Thorac. Cardiovasc. Surg. 135 (3) (2008) 620–626, https://doi.org/10.1016/j.jtcvs.2007.10.054.

CHAPTER

10

Cardiovascular complication from cancer therapy

Navin Kumar, Swagata Chatterjee, and Bhavna Solanki

School of Biotechnology, Gautam Buddha University, Greater Noida, Uttar Pradesh, India

1 Introduction

Cancer is the result of uncontrolled proliferation of cells with the potential to disseminate or invade the distant parts of the body. According to the Global Cancer Observatory (GCO) (2019), almost 18 million new cancer cases are reported annually worldwide and about 8.8 million people are died due to cancer each year as per the data given by Global Burden of Disease Study 2015 (GBD). The occurrence of cancer is not limited to the people of a particular age group, but it is more prevalent in the elder population [1]. The common practice for the treatment of cancer patients is chemotherapy along with radiation therapy, immunotherapy, and surgical interventions. In recent years, a tremendous advancement in cancer therapy is made so far to minimize the subsequent health consequences of the therapy at both levels: surgical operations and targeted medications. It is believed that clinical advances in early diagnosis, periodic surveillance, and well-coordinated therapeutic decision-makings can transform cancer into a manageable disease. But one of the biggest challenges is to limit the comorbid diseases and to combat subsequent therapeutic side effects. Several clinical observations reported that many of the cancer survivors particularly those having breast cancer, Hodgkin's lymphoma, prostate, bladder, lung cancer, and testicular cancers are potentially associated with cardiovascular complications. The situation is more critical in the patients who are diagnosed with cancer at a later age (< 60 years). The highest rate of mortality in cancer survivors is associated with cardiovascular complications. Recently, standardized mortality ratios were calculated to compare the general United States (US) population to their cancer survivors

Cardiovascular Toxicity and Therapeutic Modalities Targeting Cardio-oncology
https://doi.org/10.1016/B978-0-323-90461-2.00005-5

(3234 256) between 1973 and 2012. Among them, 38% of patients died from cancer and 11% of patients died from cardiovascular diseases (CVD). CVD mortality risk was reported highest in cancer survivors at <35 years of age within the first year of cancer diagnosis [2]. The off-target effects of drug-targeted therapy in cancer patients are associated with hypertension, thrombosis, QT prolongation, atrial fibrillation, and heart failure. Radiation therapy is often associated with the induction of atherosclerosis, coronary artery diseases, pericardial disease, and damage to the heart valve which may take months or years to evident clinically [3,4]. Severity of cardiotoxicity against cancer therapy depends on several parameters of treatment like dose and treatment time of the drugs, radiation dose and exposure time, health status and age of patients, and area or type of the cancer. Anticancer drugs like anthracyclines and several VSP (VEGF signaling pathway) inhibitors are associated with unexpected cardiovascular complications including heart failure, hypertension, thrombosis, and cardiomyopathy [5]. Similarly, a retrospective study was performed on 2232 Hodgkin's lymphoma patients who received the dose of radiation >30 Gy resulted 3.5-fold greater risk for cardiac death in comparison with control population [6]. The impact of cancer therapeutic approaches remains for years and impacts significantly on patient outcomes. Therefore, a deep understanding of health complications of cancer therapy and coordinated therapeutic treatment may contribute to the effective management of cancer diseases. In this chapter, cardiovascular complications including vascular conditions, cardiac structural problems, cardiac dysfunctions, and heart failure from cancer therapy are highlighted to become familiar with serious off-target outcomes and management of the cancer diseases.

2 Vascular conditions

2.1 Atherosclerosis

Atherosclerosis is a lipid-mediated chronic inflammatory disease that results in the plaque formation in the endothelial arteries. Standard endothelial cells are highly sensitive to detect changes in cellular microenvironment and regulate their functions accordingly. The development of atherosclerosis is a multistage process that involves disruption to the endothelial lining of the arteries, permeating plasma lipoproteins (i.e., low-density lipoprotein) in the arterial intima, formation of lipid-laden macrophages, migration of smooth muscle cells (SMCs) from the media into the intima, synthesis of connective tissue constituents, and formation of fibrous cap [7]. Levels of plasma cholesterol, LDL, and apolipoproteins (e.g., apolipoprotein B) are associated with the onset of clinical atherosclerosis. Importantly, activation of immune cells and subsequent production of free reactive oxygen species (ROS) significantly promote the formation of atherosclerosis. Briefly, once the endothelial cell of coronary arteries is injured, damage results permeation of free-circulating plasma LDL from endothelial lumen to the intima. Followed by circulating, monocytes adhere to the surface receptor of activated endothelium and then migrate to intima. Accumulation of LDL and migrated monocytes/macrophages produces ROS that oxidized LDL and lost the nitric oxide (NO) production in the intima. Oxidised LDL (ox-LDL) acts as an exogenous antigen for monocyte leads to the differentiation into macrophage which engulf ox-LDL. Continuous engulfment of ox-LDL by macrophages produces ox-LDL-laden foam cells. Foam cells release more cytokines and chemokines that recruit more inflammatory cells including T cells in the intima. Meanwhile, cytokine and chemokine activate SMCs present in the media and allow them to migrate into intimal layer. Activated SMCs also uptake free ox-LDL and act as foam cells. Hence, foam cells develop a flat fatty streak, a lipid-rich core that is the precursor lesion of atheromatous plaques. Overeating of ox-LDL by foam cells and production of more ROS induces cell death via both apoptosis and narcosis. Simultaneously, activated SMCs proliferate, synthesize, and secrete extracellular matrix and lead to the formation of a fibrous cap, a collagen-rich cap matrix, around the lipid core that stabilizes the mature atherosclerotic plaque (Fig. 1).

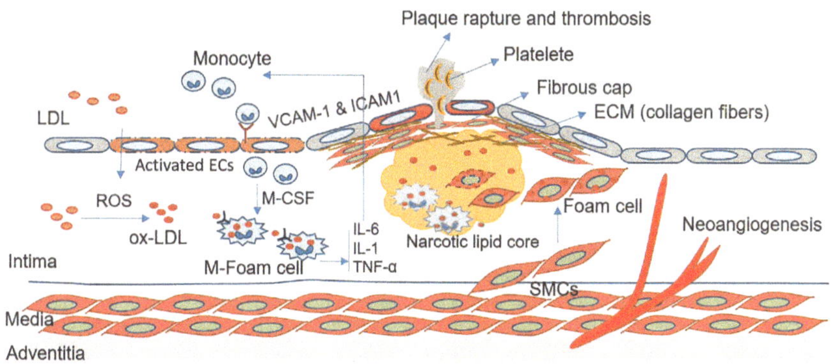

FIG. 1 Integrated pathogenesis of atherosclerosis. In response to the extracellular environment, endothelial cells (ECs) in the endothelium monolayer of arteries get activated and allow migration of circulating lipoprotein (LDL) into the arterial intima. Followed by circulating, monocytes adhere with overexpressed intercellular adhesion molecule-1 (ICAM1) and vascular cell adhesion molecule 1 (VCAM-1) on the surface of ECs. Migrated monocyte then differentiated into macrophage in the presence of macrophage colony-stimulating factor (M-CSF). Oxygenated environment and inflammatory stress produce reactive oxygen species (ROS) that oxidize free circulating LDL in the intima. Macrophages engulf oxidized LDL (ox-LDL) and continuous engulfment by a single macrophage gives rise to macrophage foam cell (M-foam cell). Activated smooth muscle cells (SMCs) migrate to intima from media and act as foam cells as well as secrete an extracellular matrix (ECM) to form a collagen-rich, fibrous cap around the lipid core. Apoptosis and secondary necrosis of M-foam cells and SMC-foam cells develop a calcified lipid-rich necrotic core. Finally, the proteolytic enzymes from SMCs may rupture or release the plaque into the coronary arteries and can lead to the development of thrombosis.

The thick fibrous cap and activation of macrophage result in increased oxygen demand in the atherosclerotic plaque that induces neovascularization, which is the formation of new vessels in the arterial wall. The production and activation of vascular endothelial growth factor (VEGF) signal to endothelial cells and SMCs to develop new vessels. Neovascularization may provide oxygen and nutrients which contribute to the progression of atherosclerotic plaque. Cytokines like IFN-γ, interleukin-1 (IL-1), and tumor necrosis factor-α (TNF-α) within the fibrous cap activate SMCs to release a variety of proteolytic enzymes including matrix metalloproteinases (MMPs) lead to plaque rupture. Consequently, vascular structures, necrotic core components, and aggregation of platelets further lead to the progression of coronary thrombosis. Hyperlipidemia, hypertension, diabetes, smoking, and medications including cancer therapy are associated with the induction and development of atherosclerotic lesions, arterial thrombosis in the coronary arteries, and other vascular complications including hypertension, venous thromboembolism as discussed later.

3 Cancer therapy vs atherosclerosis

Cancer therapy is used to kill the cancer cells. Multiple therapeutic approaches including surgery, chemotherapy, and radiation therapy are commonly employed to cure the cancer. The other options of cancer therapy are drug-targeted therapy, immunotherapy, and hormone therapy which are not only used for effective treatment of the cancer but also minimize the subsequent side effects/toxicity of the therapy. It is well documented that several anticancer agents and radiation therapy are directly or indirectly associated with the occurrence of cardiovascular complications which may cause transient or permanent functional disorders of heart and vascular systems, including atherosclerosis in cancer survivors. Usually, cardiovascular complications are associated with delayed or late effect of cancer therapy while it is completely absent or undetectable during cancer therapy.

Conventional anticancer agents may have acute and long-term effects on the endothelial functions. Treatment with anthracyclines induces vascular endothelial dysfunction in children having acute lymphoblastic leukemia [8]. Likewise, one of the potent chemotherapeutic drugs, cisplatin, is associated with atherosclerosis and may also contribute to induce acute thrombosis and vasospasm. Long-term cancer survivors treated with cisplatin-based therapy have shown a high incidence of endothelial cell damage, dysfunction, and increased levels of circulating endothelial cell even two decades after the chemotherapy. Similarly, the presence of circulating cisplatin is found detectable more than a decade after the cisplatin treatment in the cancer survivor [9]. Of note, cisplatin-based therapy (bleomycin, etoposide, and cisplatin) induces about a twofold higher risk of cardiovascular complication than expected [10]. Similarly, platinum-based drugs specifically at higher dose >3000 mg are thought to be associated with CAD.

Targeted therapy requires precision medicine that targets proteins which essentially regulate the growth, proliferation, and spread of the cancer cells. VEGF is the molecule that is found to be overexpressed in the cancer and involved in angiogenesis, cell division, and malignant disease progression. VEGF plays a critical role in leukemia, lymphoma, and a variety of solid malignant cancers [11]. Drug-targeting VEGF signaling significantly reduced the malignancy but at the same time induced endothelial dysfunction like inhibition nitric oxide production, thereby associated with induction or progression of atherosclerosis and plaque neovascularization. It is evident that pan-VEGF inhibition elicits experimental atherosclerosis and alters the endothelial homeostasis in a rodent model even though there is no change in plaque vulnerability. In another study, VEGF-targeting bevacizumab therapy is found to be associated with the inhibition of neovascularization and growth of the established plaque [12]. Moreover, it is also documented that the effect of VEGF-targeting therapy on endothelial function and subsequent cardiovascular therapy is inconsistent. The overall effect of the therapy is very difficult to predict, depending on the several parameters including age of the patient, dose, duration, stage of atherosclerosis, or VEGF inhibition may not always be synonymous all the time with anti-VEGF signaling. Moreover, meta-analysis showed that tyrosine kinase inhibitors (TKIs), nilotinib and ponatinib, that target BCR-ABL-1 and widely used to treat chronic myeloid leukemia are associated with cardiotoxicity. Additionally, the effect of both the drugs on vascular system is more aggressive when supplemented in combination. It is hypothesized that both nilotinib and ponatinib downregulate the VEGF signaling pathway [13,14]. Another potent drug 5-fluorouracil (5-FU), a pyrimidine analogue, used to treat a variety of multiple solid tumor, exhibits a wide spectrum of cardiotoxicity including myocardial ischemia, dilated cardiomyopathy, and prothrombotic state. However, the incidence of cardiotoxicity with 5-FU therapy largely varies 0%–35% that might be depending on dose, preexisting cardiac comorbidity, schedule of chemotherapy, and variability in multiple-drug combinations [15,16]. The exact mechanism of 5-FU-induced cardiotoxicity is still undefined though it is reported that this drug may induce endothelial injury, oxidative stress, coronary artery vasospasm, hypercoagulation, vasoconstriction of arteries, increased endothelin-1 levels, etc. [16].

Immunotherapy is another advanced and effective approach to chemotherapy that boosts host immunity to kill cancer cell. It is a well-known fact that microenvironment of cancer cells is able to avoid detrimental effects of immune responses, usually, via inactivating immune checkpoint proteins (e.g., PD-1 or CTLA-4) on immune cells, particularly T cells. Activated cytotoxic T cells recognize cancer cells as an exogenous antigen and kill them. Anti-PD-1 monoclonal antibodies (e.g., atezolizumab, durvalumab, and avelumab) and anti-CTLA-4 monoclonal antibodies (e.g., ipilimumab and tremelimumab) are supplemented to inhibit PD1/PDL-1 and CTLA-4/B7 interaction, respectively, which in turn activate T cell responses. However, at the same time activated T cells elicit inflammation, oxidative stress, and cytokine level. Elevated levels of cytokines (e.g., IL-1, IL-6, TNF-α, and IFN-γ) may accelerate atherosclerosis [17,18]. It is evident that $Pd1^{-/-}Ldlr^{-/-}$ mice (double-gene knockout mice) developed larger atherosclerosis and inflammation in comparison with $Ldlr^{-/-}$ control mice. Similarly, anti-PD-1 monoclonal antibody-treated $Ldlr^{-/-}$ mice showed induced proatherogenic immune responses [17,19].

3.1 Cancer therapy vs hypertension

Among the complicacies arises due to cancer therapy, hypertension is one of the most common cardiovascular anomalies. Blood pressure is a hydrostatic pressure generated by heart to move the blood through the arteries. In a healthy human body, blood pressure is well regulated via a number of physiological pathways. Being a multifactorial disease, hypertension can be linked to any influencer that alters the blood pressure homeostasis. For cancer therapy recipients, the etiology of the increased blood pressure cannot be one-dimensional. The specific medications, stage of malignancy, and several patient factors like age and preexisting comorbidities are considered to be major decisive parameters for developing hypertension in cancer survivors. Chemotherapeutic agents like angiogenesis inhibitors, protease inhibitors, mitotic inhibitors, calcineurin inhibitors, anthracyclines, antimetabolite agents, alkylating agents, and steroids have been identified as profound systemic hypertension inducers as discussed as follows.

Angiogenesis inhibitors: In the natural course of cancer, tumor growth and metastasis essentially require angiogenesis. So, angiogenesis inhibitors are widely used to hinder the vascular endothelial growth factor (VEGF) signaling pathways. Bevacizumab and ramucirumab directly inhibit VEGF ligand's ability to bind to its target receptor and cause systemic hypertension in 4% to 35% of the bevacizumab-treated patients (Table 1) [4,20]. A successful VEGF ligand-receptor binding eventually turns on the activity of tyrosine kinase. Tyrosine kinase inhibiting agents like sorafenib and sunitinib are accountable for an elevated blood pressure in 7% to 43% and 5%to 24% cases, respectively [21,22]. To understand the pathophysiology of onset of a new or worsening hypertension in angiogenesis-restricted cancer patients, the VEGF-mediated cell signaling mechanism should be considered. Activation of VEGF receptor 2 (VEGFR-2) which are

TABLE 1 List of anticancer agents associated with cardiovascular complications.

Anticancer agents	Generic name of commonly used drugs	Used to treat cancer	Associated cardiac complications	Mechanism of cardiotoxicity
Anthracyclines	Doxorubicin and epirubicin	Leukemia, lymphoma, breast, stomach, uterine, bladder, and lung cancers	Atherosclerosis, hypertension, valvular heart disease, arrhythmia, and myocarditis	Oxidative stress and topoisomerase II poisoning
Alkylating and alkylating-like agents	Cisplatin, dasatinib, ponatinib, cyclophosphamide, and busulfan	Leukemia, lymphoma, multiple myeloma, sarcoma, Hodgkin disease, lung, breast, and ovary cancers	Hypertension, Arterial thrombosis, venous thromboembolism, Atherosclerosis, and pericardial effusion	Nephrotoxicity, SRC inhibition, impaired VEGF and c-AB1 signaling pathways, cell apoptosis, and release of procoagulant endothelial microparticles
Angiogenesis inhibitors	Bevacizumab, ramucirumab, sorafenib, and sunitinib	Colon, breast, lung, and bladder cancers and renal cell carcinoma	Hypertension, atherosclerosis, LVSD, heart failure, conduction abnormalities, and QT prolongation	Tyrosine kinase inhibitors, suppress eNOS activity, increased endothelin-1 levels, rarefaction of capillaries, drop in LVEF, impaired cardiac hypertrophy, loss of contractile function, and microvascular plasticity disruption
Proteasome inhibitors	carfilzomib, bortezomib, and ixazomib	Multiple myeloma, cardio-oncology	Heart failure, hypertension, and arrhythmias	Impaired NO production and subsequent vasoconstriction
Mitotic inhibitors	Docetaxel and paclitaxel	Breast and ovarian cancers	Hypertension and congestive heart failure	Impaired cell division
HER2 inhibitors	Trastuzumab	Metastatic breast and stomach cancers	Symptomatic heart failure and valvular heart disease	Inhibit HER2 signaling
Radiotherapy		mediastinal lymphoma and breast, lung, and oesophageal cancers	Pericardial effusion, valvular heart disease, and heart failure	Oxidative stress, Baroreceptor failure, microvascular changes, fibrotic changes of valves, morphological changes of pericardium, and alteration in micro-RNA expression
Others	5-Fluorouracil, arsenic trioxide, and bleomycin	Various solid tumors and head, neck, and testicular malignancy	Myocardial ischemia and dilated cardiomyopathy	Endothelial injury, oxidative stress, coronary artery vasospasm, hypercoagulation, vasoconstriction of arteries, increased endothelin-1 levels

predominantly expressed in endothelial cells, upon binding with VEGFA, initiates signaling through tyrosine kinases that stimulate many pathways, including phosphoinositide 3-kinase/AKT/protein kinase B-mammalian target of rapamycin, endothelial NO synthase, and prostacyclin, that regulate vasodilation and inflammatory responses [23,24]. VEGFIs eventually suppress the endothelial nitric oxide synthase (eNOS) activity and resulted in a reduction in bioavailability of NO. Further, an increase in plasma endothelin 1, a potent vasoconstrictor as opposed to vasodilator NO, is also reported [25]. So when VEGFIs are used to control the metastasis via impaired angiogenesis, elevation in blood pressure is quite inevitable. Ironically, VEGF signaling inhibitor-induced rise in blood pressure is an indication of effectiveness of the anticancer therapy by showing target toxicity [26].

Proteasome inhibitors: Inhibition of proteasome has been strategically used to promote programmed cell death in multiple myeloma cells. Among the various complex and cascading pathways involved in apoptosis of neoplastic cells, the activation and stabilization of tumor suppressor protein p53 by proteasome inhibitors can be linked to hypertension. The Kruppel-like factor (KLF) 2, a pivotal endothelial factor that is important to NO production, is inhibited by p53 [27]. Also, Qin Wei et al. reported that eNOS activity is markedly reduced in proteasome-inhibited cells [28]. Thus, proteasome inhibiting drugs like carfilzomib, bortezomib, ixazomib, marizomib, oprozomib, and delanzomib may cause hypertension possibly due to impaired NO production and subsequent vasoconstriction (Table 1).

Alkylating and alkylating-like agents: These groups of anticancer drugs are used since the early days of invention of therapies against cancer. The mode of action of this oldest class of drugs is to simply damage the DNA of cancer cells by attaching an alkyl group to the nitrogenous base of the DNA and thereby arresting the cell proliferation. The mechanism of developing hypertension by alkylating agents like cyclophosphamide, chlorambucil, and busulfan is not much clear, but alkylating-like agents such as cisplatin-induced high blood pressure may be an outcome of nephrotoxicity [29]. Chemotherapeutic agent dasatinib, a novel multitargeted tyrosine kinase inhibitor (TKI), is reported to cause pulmonary hypertension. The exact mechanism of dasatinib-induced increase in pulmonary arterial pressure is unknown. However, it is already reported that these TKI drugs also target SRC kinases that are important in vascular regulation, including vasodilation and vascular proliferation [30]. SRC inhibition by dasatinib leads to pulmonary hypertension [31]. However, few reports suggest that some pathways of dasatinib-induced pulmonary hypertension may not be related to SRC inhibition [32]. Another chemotherapeutic drug notoriously associated with pulmonary hypertension is bleomycin [33].

Radiotherapy may disrupt the blood pressure homeostasis, possibly due to baroreceptor failure [34].

Steroids: Combination of steroid therapy and chemotherapy is frequently used in cancer treatment. Unfortunately, it is well established that steroids have a contribution in increasing blood pressure. Corticosteroid-induced elevated blood pressure is due to the overstimulation of mineralocorticoid receptor, resulting sodium retention in kidney. The high sodium retention leads to volume expansion and subsequent increment in blood pressure [35]. An altered vascular reactivity is considered as the central reason behind glucocorticoid-induced hypertension [36].

Prior to start of anticancer therapy, blood pressure monitoring is a must. However, hypertension may be turned up even after 10 years of the treatment. A persistent BP $\geq$ 140/90 enhances the chances of mortality from congestive heart failure, myocardial infarction, stroke, and renal insufficiency. The treatment of hypertension cannot be unified for all and also sometimes a combination treatment is required to control the rise in BP. Effective management of hypertension as early as possible is very crucial for cancer therapy to avoid any acute complication as well as withholding the anticancer drug in an individual. As it is already discussed, the use of VEGF inhibitors as chemotherapeutic agents has an adverse effect of elevated blood pressure. In such prognosis, angiotensin-converting enzyme inhibitors (ACEIs), angiotensin receptor blockers (ARBs), and dihydropyridine calcium channel blockers (CCBs) are preferred first-line treatment. ACEIs impose a halt in the functioning of the angiotensin-converting enzyme. Angiotensin-converting enzyme converts angiotensin I to angiotensin II, which is directly responsible for vasoconstriction of precapillary arterioles and postcapillary venules, inhibition of the norepinephrine reuptake, stimulation of the catecholamine release from the adrenal medulla, reduced urinary excretion of sodium and water, biosynthesis of aldosterone and collagen type I and type III in fibroblasts, and hypertrophy of both vascular smooth muscle cells and cardiac myocytes [37]. All these events are recognized as a favorable precondition of hypertension. So decrease in angiotensin II production alone can reduce the blood pressure in multifaceted way. Apart from blocking angiotensin II production, ACEIs also reported to stimulate the release of endothelial NO and minimize the catabolism of bradykinin, a potent vasodilator [38]. Although ARBs also target the renin-angiotensin pathway to reduce high blood pressure, this class of drugs are more effective in lowering down the blood pressure. It overcomes the shortcomings of ACEIs as it replaces angiotensin II from angiotensin receptor site and helps in complete loss of angiotensin function, whereas CCBs inhibit the influx of calcium ions in vascular smooth muscle cells and cardiac muscle cells. Because muscle contraction is largely dependent upon influx of calcium, its inhibition causes relaxation of vascular and arterial smooth muscle cells resulting in arterial vasodilation and thereby useful in hypertension treatment [39]. The detrimental drug-drug interaction should be well taken care of in choosing antihypertensive drugs when oncogenic therapy is going on as, for example, drugs like diltiazem and verapamil should be avoided when sorafenib is used in anticancer therapy. These antihypertensive drugs block sorafenib metabolization via CYP3A4 isoenzyme inhibition [4].

3.2 Arterial thrombosis

The risk of arterial thrombosis, blood clot in the artery, is paying attention as an important cardiac complication of malignancy and of novel anticancer agents. The development of arterial thrombosis is the result of the complex interaction including preexisting cardiovascular risks, type of cancer, and selected cancer therapy. The basic mechanism of arterial thrombosis is associated with disruption of fibrous cap (Fig. 1), and subsequently, formation of blood clot or "thrombus" around the atheromatous plaque results in blocking the arteries primarily aorta, coronary, and cerebral arteries. A gray-white thrombus is a complex structure, consisting of platelets and fibrins meshwork, thrombin, red blood cells, tissue factors, and degenerative leukocytes. When a large fragment of thrombus travels through arteries, it restricts or blocks blood flow, termed as arterial thromboembolism (ATE). The clinical manifestation of ATE includes myocardial infarction, stroke, peripheral artery disease, and critical limb ischemia.

The development of AT in cancer patients is associated with a four- to fivefold increased risk of death [40,41]. Similarly, a recent prospective observational cohort study reported that the occurrence of ATE is associated with a 3.2-fold higher risk of all the death. Importantly, the risk of ATE was found greater in the lung and kidney cancer patients [42]. It is evident that thrombosis and bleeding are associated with increased morbidity and mortality in cancer patients. The increased risk of ATE in the cancer patient is found to be associated with multiple cancer therapeutic practices including platinum-based agents [43,44], VEGF signaling pathway inhibitors [45–47], Bcr-Abl TKI [48], tamoxifen, erythropoietin, and radiation therapy [49]. Of note, the pathophysiology of ATE greatly varies during cancer therapy and being high at the initial 6 to 12 months after diagnosis and then starts to decline [41,50]. Cisplatin and ponatinib are reported as potent inducers of ATE [44,48]. VEGF inhibitor-mediated hypertension may increase more risk of cardiovascular complications [51]. Bcr-Abl TKIs are commonly used to treat hematologic malignancy but are the potent inducer of vascular complications. Both VEGF and c-Abl signaling pathways operate in synergy, are important to growth, and repair cardiovascular structures.

However, it is difficult to know the real culprit for induction of ATE whether it is due to cancer or treatment. Antiplatelet therapy or anticoagulant agents are common and effective treatments of ATE in the patients. But, here it is important to note that bleeding is the limiting factor of antiplatelet therapy, particularly in the cancer patients because thrombocytopenia is a common clinical manifestation in cancer patients. McCarthy et al. suggested to adopt more conservative approach to treat cancer patients if platelet count is <50,000 [52]. Therefore, a collaborative approach between cardiologist and oncologist is required to provide an evidence-based and personalized care to overcome the risk of cancer morbidity and mortality.

3.3 Pulmonary embolus

Venous thromboembolism (VTE) is differentiated into two closely related diseases, namely deep vein thrombosis (DVT) and pulmonary embolism (PE). DVT is the formation of blood clots at the end of the deep veins, mostly in the leg. However, if this blood clot detaches from its site and eventually lodges in the lung arteries via travelling through right heart, is termed as PE. Cancer patients have a five- to sevenfold higher risk for developing VTE [53]. The incidence of VTE greatly varies (4%–20%) among cancer patients depending on cancer site, stage, comorbidity, selected therapy, and prothrombotic state including platelet count, leukocyte, tissue factor, and P-selectin [54]. Disruption in the balance between the blood coagulation and natural coagulant or fibrinolytic mechanisms gives rise to thrombosis. Fibrin, red blood cells, and platelets are the major components of venous thrombosis. The precise mechanism of induction and progression of VTE via cancer patients or treatment is not clear. However, there are two basic reasons for the induction of thrombosis are suggested: firstly, the release of procoagulants (i.e., tissue factors, ADP, thrombin, and podoplanin) and cytokines (IL-1β and TNF-α) from cancer cells and, secondly, reduction of natural anticoagulants and activation of vascular endothelium. Most cancer patients with blood coagulation test abnormalities show symptoms of accelerated platelet activation and aggregation, accelerated proteolysis, and the upregulation of the coagulation cascade. Some of the most common causes of such activation are comprised of activation of the host cells (e.g., monocytes or endothelium cells); discharge of cancer procoagulants; overexpression of plasminogen activator inhibitor-1(PAI-1); and altered expression or activity of proteins produced by the liver including protein C, protein S, and antithrombin [55].

Cancer is associated with hypercoagulant state and is often activated by cancer therapy including surgery, chemotherapy, and radiation therapy. VTE is a common vascular complication of major surgery in both patients with or without cancer. PE remains one of the most common causes of postoperative death. Histologically, it is reported that the risk of postoperative DVT and PE in cancer patients is increased twofold and >threefold, respectively, in comparison with noncancer patients [56].

Chemotherapy-mediated VTE is reported predominant in stomach and pancreas cancer than other cancer types [53]. There is a high probability of occurrence of VTE during the first year after cancer diagnosis and also soon after the commencement of the treatment [57]. Cisplatin-based therapy is associated with both developing VTE and ATE [58]. Cisplatin-oxaliplatin combination increased about a twofold risk of thrombotic complication in esophagogastric cancer [59]. Lechner et al. observed that cisplatin treatment on human endothelial cells induces apoptosis as well as promotes the release of procoagulant endothelial microparticles that produce thrombin independent of tissue factor activity [60]. Similarly, antiangiogensis drug bevacizumab, targeting VEGF, is associated with vascular thrombosis [61].

4 Cardiac structural problem

4.1 Valvular heart disease

The four heart valves, such as mitral (between left atrium and left ventricle), aortic (between left atrium and aorta), tricuspid (between right atrium and right ventricle), and pulmonary (between right ventricle and pulmonary vein), maintain the unidirectional flow of blood among the four chambers of heart. Each valve has flaps that open and close once in each heartbeat. In valvular heart disease (VHD), one or more valves do not open or close appropriately, thereby causing disruption of blood flow inside the heart. Regurgitation (leaking valve), stenosis (narrowed valve), and atresia (presence of sheet of solid tissue instead of a functional valve), these three structural anomalies are generally found to be the root cause behind the irregular functioning of heart valves.

The prevalence of VHD is estimated as 10% of the treated patients and more common among young survivors who received thoracic radiation therapy for certain malignancies, such as Hodgkin's lymphoma and breast cancer [62]. Although it is considered as a late (>10 years following radiation) cardiotoxic effect of radiation treatment, recent radiation techniques are a safer option than chemotherapy in reducing the risk of VHD [63–65]. As valves are largely avascular, the radiation-induced valve damages cannot be explained by microvascular changes that are happened in the case of other radiation-induced heart diseases. The mechanism of developing valvular disease due to radiation exposure is not completely understood. The only concrete information we have is that flaps of the valves undergo fibrotic changes with or without calcification upon radiation exposure. Left-sided valves are more prone to damage by radiation, probably because higher systemic pressure has an additional effect on the pathogenesis [66,67].

Some drugs, to be precise, ergot alkaloids such as methysergide and ergotamine, ergot-derived dopaminergic agonists such as pergolide and the recreational drug ecstasy, and drugs metabolized into norfenfluramine such as fenfluramine and dexfenfluramine are reported to have direct effects on valvular dysfunction. These agents stimulate a specific serotonin receptor in valve tissues. The possible outcomes of this stimulation are fibroblast proliferation and collagen synthesis which may lead to VHD [68]. However, chemotherapy-induced valvular diseases are hardly reported. Only few case reports are known where anthracyclines in combination with radiation therapy or independently can cause VHD [69–71]. The serotonin receptor-based mechanism is not established in chemotherapy-induced VHD. Rather cellular toxicity in mature cells like myocardium is an adverse effect of chemotherapy. Likewise, valvular endothelium can be damaged as well, leading to scarring, leaflet retraction, and thickening, which could cause regurgitation and eventually stenosis. Some suggest that anthracyclines may cause papillary dysfunction that impairs lateral shortening between the papillary muscles that tethers the leaflet edges, leading to incomplete leaflet closure and significant regurgitation even in the absence of LV and mitral annular dilation [72]. Valvular heart disease is considered as the only early indication of trastuzumab-induced cardiotoxicity (TIC), which is occasionally life-threatening [73]. Often valve regurgitation is considered as a secondary effect of anthracycline-treated Hodgkin's lymphoma survivors and appears much before the onset of symptomatic cardiomyopathy.

Mostly valvular dysfunction is mild to moderate but sometimes it becomes severe, and this type of cardiac malfunction should get proper attention. The radiation therapy has been improved a lot to minimize the collateral damage. In the initial days of radiation treatment, the primary objective was only to kill the malignant cells and to achieve that a uniform radiation field with a broad coverage area was being used. Successive progress in technology and growing awareness on radiation-induced heart diseases enable us to use more precise radiation field that excludes a greater portion of heart. Radiation therapy passed a long technologically evolved journey, started from mantle radiation to the most recent proton radiotherapy [74,75].

Though there is a high risk involved in the surgery of radiation-induced dysfunctional cardiac valve, surgical intervention becomes unavoidable in case of severe valvular abnormalities. Percutaneous placement of transcatheter aortic valves or MitraClip has been successfully performed as better alternatives to traditional surgery [76,77].

4.2 Pericardial effusion

Anatomically pericardium is the fibrous sac where the heart and beginning of the great vessels are safely encased. Functionally, it has quite a major role in physiological cardiac function. There are two layers in the pericardium, the outer fibrous pericardium and the inner serous pericardium. In between these two layers, a lubricating fluid is present. The optimal amount (normally 50 mL) of this fluid actually assists in the movements of the heart during each heartbeat, but a little excess of the fluid can be detrimental if not treated timely. Pericardial effusion is a condition of heart when there is an accumulation of excess fluid between the layers of pericardium. Accumulated surplus fluid that exerts an increased intrapericardial pressure affects diastolic feeling of the heart, leading to decreased cardiac output. Although it shows mild or moderate symptoms, cardiac tamponade resulting from pericardial effusion can be life-threatening and is a medical emergency, requiring urgent drainage of the fluid.

Pericardial effusion is a common symptom of cardiotoxicity manifested due to anticancer therapy. Pericardial effusion is 70%–90% prevalent among patients who received mediastinal radiotherapy. The pericardium of radiation-hit heart becomes more porous and tortuous. This change in the morphology resulting in a neutrophilic infiltrates and accumulation of high protein exudate [78]. Ischemia and late fibrosis used to happen as irradiated pericardium is prone to neovascularization. Due to fibrosis of venous and lymphatic channels, the drainage of extracellular fluid is somehow compromised. Consequently, the fibrinous exudate buildup in excess gives rise to the condition of pericardial effusion [79]. Chemotherapy-related effusions are occasionally observed (up to 1%–2% incidence) following exposure to busulfan, cytarabine, cyclophosphamide or tretinoin, especially when these drugs are used at high doses or in combination for the treatment of hematological malignancies ([80]). Even acute myopericarditis is observed in some cases. This collateral damage is often registered with anthracycline-treated cancer patients [81]. Myopericarditis is such a cardiotoxicity where both pericardium and myocardium are inflamed. It is speculated that the elevated level of troponin can be a triggering factor of myopericarditis. Even bleomycin has been directly linked to life-threatening pericarditis [82]. To treat locally advanced esophageal cancer, definitive concurrent chemoradiotherapy (CCRT) is adopted and here also pericardial effusion has been reported as a hazardous side effect of CCRT [83].

Pericardial effusion is not only a common complication of cancer therapy. It can also come out as a primary or secondary disease of any cancer, especially breast, lung, and Hodgkin's lymphoma, when these cancers metastasize to the pericardium. Therefore, it is of utmost importance to identify the actual cause behind pericardial effusion. Electrocardiography is used in general to detect pericardial effusion. In chronic cases, pericardiocentesis, and thereafter cytological examinations, flow cytometry, etc., has to be done to establish the etiology. The first line of treatment includes a trial of more intensive chemotherapy or steroid. If that does not work out, then less invasive pericardiocentesis, prolonged pericardial drainage, or surgical procedures such as subxiphoid pericardiostomy, with or without intrapericardial instillation of sclerosing or cytotoxic agents, percutaneous balloon pericardiotomy, and pericardial window are the other options to be considered [84,85]. However, before proceeding to the surgical procedure, the high risk of potential morbidity has to be evaluated well. At the same time, if the fluid accumulation is too fast to be drained out by pericardiocentesis or frequent relapsing of the disease, ultimately culminating in tamponade, extensive surgical management with both visceral and parietal pericardial stripping become a necessity [86].

4.3 Cardiac conduction diseases

Cardiac conduction system ensures that an electric impulse that is generated by pacemaker cells in the sinoatrial node can travel through the entire ventricular and atrial myocardium which thus creates a timely contraction and generates the impulse. The conduction cardiac system comprises the atrioventricular node, the sinoatrial node, the bundle of His, and the Purkinje network. Being a physiologic system, the cardiac conduction system is stimulated without any requirement of an external stimulus. In the so-called heart cell, the cardiac myocytes initiate the contraction via an electrical impulse known as the cardiac impulse that travels really through the atrioventricular myocardium. This phenomenon occurs due to electrically coupled junctions known as gap junctions. Every single myocyte can conduct cardiac impulses. During a normal activation, an impulse is originated in the pacemaker located in the sinoatrial node which then spreads uniformly in the entire myocardium. This impulse then travels via the AV node through internodal pathways and spreads in the entire ventricles and ventricular myocardium via the Purkinje fibers. The irregularity if any can cause cardiac arrhythmias leading to an abnormal ECG. Sodium, potassium, and calcium are the electrolytes responsible for the contraction of the cardiac myocyte. Electrolyte imbalance alters cardiac electrical impulses and results in arrhythmia development.

Several chemotherapeutic agents including novel targeted drugs are reported to have adverse effects on electrophysiological changes in cardiac system result bradycardia, arrhythmia, QT prolongation, myocardial ischemia, and heart failure [3] (Fig. 2). Anthracycline antibiotics like doxorubicin and epirubicin are often associated with primary arrhythmia as well as a secondary clinical manifestation of cardiomyopathy. This anthracycline-induced arrhythmia is eventually shown up in abnormal ECG indicating ST-T segment changes, QRS voltage lowering, T-wave flattening, and QTc interval prolongation in a dose-independent manner [87]. Arsenic trioxide is currently applied in cases of intractable hematological malignancies. Among the arsenic trioxide-receiving patients, conduction abnormalities are demonstrated through ECG signs including

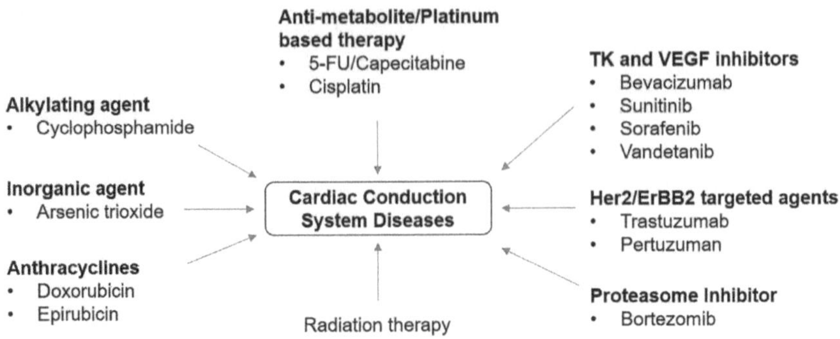

FIG. 2 Anticancer agents involved in the development of arrhythmia.

QRS widening, prolonged QT interval, ST-depression, and T-wave flattening. Sinus tachycardia, premature ventricular contractions (PVCs), and nonsustained ventricular tachycardia (NSVT) [88]. Acute cardiotoxicity is a confirmed side effect of high-dose alkylating agents like cyclophosphamide. In a study where cyclophosphamide is used before bone marrow transplantation, 33% of patients are diagnosed with arrhythmia among which 27% developed an early decrease in total QRS voltage [89]. Radiotherapy can also be the causative agent for the conduction abnormalities including varying degrees of AV block, sick sinus syndrome, prolonged QT, supraventricular arrhythmias, and ventricular tachycardia [90,91]. Conduction disorders may be treated with medicines, pacemakers, implantable cardioverter-defibrillators (ICDs), and surgical procedures.

5 Cardiac dysfunction and heart failure

Cardiac dysfunction is the precursor of heart failure where heart is unable to supply sufficient oxygen-rich blood to vital organs as per their metabolic and functional requirements. The underlying condition of improper contraction and relaxation of heart chambers results in a depressed cardiac output that is broadly described as heart failure. A number of cardiovascular abnormalities such as hypertension, coronary artery diseases, myocardial infarction, diabetes, atherosclerosis, valvular diseases may end up to heart failure. Here, the three most widely used anticancer therapies, namely radiation therapy, antiangiogenic therapy, and trastuzumab, are scrutinized in the perspective of cardiac dysfunction and heart failure.

5.1 Radiation therapy

Radiation therapy is considered one of the successful tools to enhance the survival rate among patients with different malignancies. Despite being a very crucial part of multimodality cancer therapy, radiation therapy comes with its own price tag. As mentioned earlier, radiation-induced heart diseases (RIHDs) are the concerning areas in case of application of radiotherapy to subside the cancerous growth, particularly for mediastinal lymphoma, breast, lung, and esophageal cancers. The development of RIHD gets aggravated when it is combined with chemotherapy. The excess mortality has been reported in irradiated breast cancer patients along a timeline of 10 to 15 years of completion of radiation treatment, and unfortunately, these deaths are mostly due to inefficient cardiac functionalities [92].; [93]). Similarly, there is always a higher risk of developing cardiac problems among the mediastinal radiation (>40 Gy)-treated Hodgkin's lymphoma survivors [94]. Radiation-induced heart diseases ultimately lead to cardiac dysfunction and heart failure. Depending upon the doses and field area, the ionizing radiation can severely damage any tissue inside the heart. The systolic and diastolic dysfunctions are often associated with myocardial fibrosis and remodeling and are listed in the clinical symptoms of heart failure. Coronary artery disease (CAD) is another independent predictor of heart failure. After a significant latency, radiation-induced pericarditis may turn up as congestive heart failure (CHF) [95].

A number of interlinked mechanisms involving many factors like endothelial cell injury, inflammatory reaction, OS, mitochondria and endoplasmic reticulum injury, various cytokines, calcium overload, and micro-RNAs are found to be present behind the pathophysiology of radiation-induced cardiac dysfunction and heart failure [67,96,97]. Radiation-induced fibrosis is established by an acute proinflammatory environment where profibrotic cytokines such as platelet-derived growth factor (PDGF), transforming growth factor β (TGF-β), basic fibroblast growth factor (bFGF), insulin-like growth factor (IGF), connective tissue growth factor (CTGF), and various interleukins are released [98]. Here it is worthy to be mentioned that TGF-β has a pivotal role in tissue-irradiated fibrosis by altering the balance of extracellular matrix remodeling to induce collagen synthesis, decrease the production of collagenase and other proteases, and increase the production of protease inhibitors [99]. The increased collagen deposition thickens the vessel wall and stenosis of lumen and reduces the elasticity of cardiac muscles. All these changes worsen the blood flow to the myocardium. Ionizing radiation is capable of decomposing cellular water and production of reactive oxygen species (ROS) and reactive nitrogen species (RNS) [100]. Oxidative stress due to the overproduction of ROS has a significant role in the secretion of proinflammatory cytokines and chemokines [101]. The basic cause of CAD is endothelial damage which may be attributed to the overproduction of ROS, cytokines, or other inflammatory mediators [102]. Further, irradiated cells exhibit a vicious cycle of mitochondrial ROS production and Ca + release from endoplasmic reticulum calcium pull and subsequently experience premature aging and apoptosis [103]. Several recent studies also suggest that alteration in micro-RNA (miRNA) expression is a prime factor in radiation-induced cardiac hypertrophy and heart failure [104].

The first step to manage radiation-induced cardiac dysfunction and heart failure should be the optimized application of radiation. There has to be a fine balance between the radiotherapy for malignancies and RIHD. Application of modernized automated three-dimensional conformal radiotherapy, intensity-modulated radiation therapy, and volumetric arc therapy, etc., helps us to minimize the off-target exposure [105–107]. However, some drugs including statins, ACEIs, and antioxidants are known to be effective in the treatment of RHID [108–110].

5.2 Antiangiogenic therapy

Angiogenesis is defined as the formation of blood vessels from the preexisting ones. The abnormal growth of cancer cells is somehow supported by angiogenesis that provides nutrients to the tumor cells. So to cut off the supply of blood to the malignant cells, antiangiogenic therapy became a better choice over the traditional chemotherapies in treating cancers of the colon, breast, lung, and bladder as well as renal cell carcinoma and non-small cell lung cancer (NSCLC). The potential targets of antiangiogenic agents are generally the modulating factors of angiogenesis. In most of the cases of clinical practices, VEGF inhibitors, VEGF receptor inhibitors, and small-molecule tyrosine kinase inhibitors (TKIs) are used as monotherapy or in combination with other chemotherapeutic agents or radiotherapy. Despite having quite a straightforward approach, antiangiogenic agents have certain snags with regard to patients' cardiac health.

So far, five FDA-approved antiangiogenic drugs are more regularly used in the management of solid tumors. Bevacizumab, sunitinib, sorafenib, pazopanib, and vandetanib are VEGF signaling inhibitors that block tumor angiogenesis. Although initial studies did not report any major toxicity related to antiangiogenic therapy, later on adequate clinical data arrived confirming the adverse effects of cardiotoxicity including hypertension, left ventricular systolic dysfunction (LVSD) and heart failure, conduction abnormalities, and QT prolongation (with the risk of torsades de pointe and sudden death) [111]. Advance cases of breast, lung, and colorectal carcinomas are often taken care of by administering humanized monoclonal antibody raised against VEGF-A [112,113]. Bevacizumab-associated heart failure hovers around 1.7% to 3%; risk is more when used after chemotherapy [114]. Except for bevacizumab, the other four are small-molecule tyrosine kinase inhibitors and are multitargeted as they are capable to deter tumor angiogenesis (via VEGF and PDGF pathway inhibition) as well as cancer metastasis.

In general, all these antiangiogenic agents inactivate VEGF signaling pathway that is crucial for maintaining the cardiovascular homeostasis. This obstruction gives rise to some pathophysiological events like rarefaction of capillaries, drop-in LVEF, impaired cardiac hypertrophy, and loss of contractile function after pressure overload. Another aspect of cardiotoxicity induced by these drugs is via disrupting the microvascular plasticity that allows adaptation of the vascular network and thus oxygen supply to enhanced metabolic demand due to pressure overload [115].

An important feature of VSP inhibitor-induced cardiotoxicity is its reversibility in most but not all (60–80%) of the patients upon withdrawal [116]. Decline in cardiac functions due to antiangiogenic therapy initiates with elevated blood pressure. Therefore, it is wiser to control the blood pressure with standard antihypersensitive medication with a close monitoring of blood pressure. These antihypersensitive drugs should be chosen considering there is no cross-reactivity on antiangiogenic therapy [117]. ACE inhibitors can also be used as a first line of therapy as they can prevent proteinuria and plasminogen activator inhibitor-1 expression [118]. Thus, more pieces of evidence regarding the pathomechanisms of antiangiogenic agent-induced cardiotoxicity will allow treating patients with VSP inhibitor therapy with a more precise understanding of the risk and mitigation strategies for cardiotoxicity.

5.3 Trastuzumab

Trastuzumab (Herceptin) is a humanized monoclonal antibody directed against human epidermal growth factor receptor 2 (HER2) which is a protooncogene. This targeted therapy has been approved by FDA long back in 1998 as a breakthrough in the treatment of metastatic breast cancer where HER2 protein is overexpressed, eight years after the molecular engineering of the humanized antibody [119]. Later on, it also got approval for the adjuvant treatment of early-stage (localized) HER2-positive node-positive breast cancer and for HER2-positive metastatic stomach cancer [120,121]. Due to upregulation of HER2 in tumor cells, there is a hyperactivation in signaling pathways, in particular Ras/Raf/mitogen-activated protein kinase (MAPK), the phosphoinositide 3-kinase/Akt, and the phospholipase Cγ (PLCγ)/protein kinase C (PKC) pathways involved in cell growth and cell cycle progression [122]. HER2 has an extracellular ligand-binding domain, a transmembrane region, and an intracellular or cytoplasmic tyrosine kinase domain. The active form of this receptor is a homodimer or heterodimer trastuzumab binds specifically to the extracellular domain of HER2 and subsequently prevents the homodimerization or heterodimerization of HER2. Trastuzumab-bound inactive HER2 is unable to receive oncogenic growth signals, thereby exerting a tumor-suppressive effect. Trastuzumab also reverses inhibition of B cell lymphoma 2-associated agonist of cell death (BAD), which leads to BCL-2-associated X protein (BAX) oligomerization at the mitochondrial membrane, release of cytochrome *c* (Cyt c), and caspase activation to cause apoptosis of tumor cells [123]. Now, HER2 has a vital role in cardiac development and that became evident when HER2 conditional mutant mouse model exhibits multiple independent parameters of dilated cardiomyopathy, including chamber dilation, wall thinning, and decreased contractility within a short time span after birth [124]. In adult cardiomyocytes except for HER2, two other epidermal growth factors, namely HER3 and HER4, are also expressed and HER2 acts as a coreceptor for HER4. It has been also reported that neuregulin1 (NRG-1), a ligand of ErbB3 and ErbB4, demonstrates its cardioprotective nature via NRG-1/ErbB2 and ErbB4 signaling pathway [125]. HER2 knockout mice develop age-related dilated cardiomyopathy and have decreased survival after pressure overload stress, which mimics severe hypertension [126]. Thus, trastuzumab-inhibited HER2 signaling causes a hindrance in embryonic cardiomyocyte proliferation and indicates as the central mechanism for cardiotoxicity. However, possible altered mitochondrial integrity due to trastuzumab-mediated inhibition leads to ATP depletion and contractile dysfunction [127]. The drug-induced cardiotoxicity might be linked to an accumulation of toxic reactive oxygen species (ROS) in human cardiocytes due to suppression of autophagy as a consequence of trastuzumab-inhibited HER2 signaling [128].

It has been well documented that trastuzumab-induced cardiotoxicity increased many folds when used in combination with anthracycline. To manage this adverse situation, different strategies like larger intervals between the administration of anthracyclines and trastuzumab or to diminish the duration of trastuzumab adjuvant therapy are adopted. The good thing is that unlike anthracycline, trastuzumab-induced cardiotoxicity is often reversible. Calculating the risk over benefit, it has come up that 1.7%–4.1% of trastuzumab (exclusively)-administered patient develops heart failure, whereas it can reduce the 3-year risk of recurrence of breast cancer by about one-half [3,4,120]. The standard medication of dexrazoxane, ACE inhibitors, and β blockers can be considered to prevent heart failure. To continue uninterrupted trastuzumab therapy, it is recommended to start the heart failure treatment at subclinical level. Nevertheless, in asymptomatic patients, if LVEF declines more than 20 points from baseline or LVEF < 30%, and in symptomatic patients, if LVEF reduces more than 30 points from baseline, trastuzumab withdrawal is proposed by some authors [129]. However, every effort should be made to avoid the cessation of cancer therapy.

References

[1] A. Jemal, F. Bray, M.M. Center, J. Ferlay, E. Ward, D. Forman, Global cancer statistics, CA Cancer J. Clin. 61 (2) (2011) 69–90, https://doi.org/10.3322/caac.20107.

[2] K.M. Sturgeon, L. Deng, S.M. Bluethmann, S. Zhou, D.M. Trifiletti, C. Jiang, S.P. Kelly, N.G. Zaorsky, A population-based study of cardiovascular disease mortality risk in US cancer patients, Eur. Heart J. 40 (48) (2019) 3889–3897, https://doi.org/10.1093/eurheartj/ehz766.

[3] H.M. Chang, R. Moudgil, T. Scarabelli, T.M. Okwuosa, E. Yeh, Cardiovascular complications of cancer therapy: best practices in diagnosis, prevention, and management: part 1, J. Am. Coll. Cardiol. 70 (20) (2017) 2536–2551, https://doi.org/10.1016/j.jacc.2017.09.1096.

[4] H.-M. Chang, T.M. Okwuosa, T. Scarabelli, R. Moudgil, E.T.H. Yeh, Cardiovascular complications of cancer therapy, J. Am. Coll. Cardiol. 0735-1097, 70 (20) (2017), https://doi.org/10.1016/j.jacc.2017.09.1095.

[5] M.S. Ewer, S.M. Ewer, Cardiotoxicity of anticancer treatments, Nat. Rev. Cardiol. 12 (9) (2015) 547–558, https://doi.org/10.1038/nrcardio.2015.65.

[6] S.L. Hancock, M.A. Tucker, R.T. Hoppe, Factors affecting late mortality from heart disease after treatment of Hodgkin's disease, JAMA 270 (16) (1993) 1949–1955.

[7] R. Ross, The pathogenesis of atherosclerosis: a perspective for the 1990s, Nature 362 (1993) 801–809, https://doi.org/10.1038/362801a0.

[8] M. Jang, S.S. Kim, J. Lee, Cancer cell metabolism: implications for therapeutic targets, Exp. Mol. Med. 45 (10) (2013), https://doi.org/10.1038/emm.2013.85, e45.

[9] J.A. Gietema, M.T. Meinardi, J. Messerschmidt, T. Gelevert, F. Alt, D.R. Uges, D.T. Sleijfer, Circulating plasma platinum more than 10 years after cisplatin treatment for testicular cancer, Lancet (London, England) 355 (9209) (2000) 1075–1076, https://doi.org/10.1016/s0140-6736(00)02044-4.

[10] A.W. van den Belt-Dusebout, J. Nuver, R. de Wit, J.A. Gietema, W.W. ten Bokkel Huinink, P.T. Rodrigus, E.C. Schimmel, B.M. Aleman, F.E. van Leeuwen, Long-term risk of cardiovascular disease in 5-year survivors of testicular cancer, J. Clin. Oncol. Off. J. Am. Soc. Clin. Oncol. 24 (3) (2006) 467–475, https://doi.org/10.1200/JCO.2005.02.7193.

[11] F. Larcher, R. Murillas, M. Bolontrade, C.J. Conti, J.L. Jorcano, VEGF/VPF overexpression in skin of transgenic mice induces angiogenesis, vascular hyperpermeability and accelerated tumor development, Oncogene 17 (3) (1998) 303–311, https://doi.org/10.1038/sj.onc.1201928.

[12] R.M. Touyz, S. Herrmann, J. Herrmann, Vascular toxicities with VEGF inhibitor therapies-focus on hypertension and arterial thrombotic events, J. Am. Soc. Hypertens. 12 (6) (2018) 409–425, https://doi.org/10.1016/j.jash.2018.03.008.

[13] A. Gover-Proaktor, G. Granot, S. Shapira, O. Raz, O. Pasvolsky, A. Nagler, D.L. Lev, A. Inbal, I. Lubin, P. Raanani, A. Leader, Ponatinib reduces viability, migration, and functionality of human endothelial cells, Leuk. Lymphoma 58 (6) (2017) 1455–1467, https://doi.org/10.1080/10428194.2016.1239258.

[14] E. Hadzijusufovic, K. Albrecht-Schgoer, K. Huber, G. Hoermann, F. Grebien, G. Eisenwort, W. Schgoer, S. Herndlhofer, C. Kaun, M. Theurl, W.R. Sperr, U. Rix, I. Sadovnik, B. Jilma, G.H. Schernthaner, J. Wojta, D. Wolf, G. Superti-Furga, R. Kirchmair, P. Valent, Nilotinib-induced vasculopathy: identification of vascular endothelial cells as a primary target site, Leukemia 31 (11) (2017) 2388–2397, https://doi.org/10.1038/leu.2017.245.

[15] C. Kosmas, M.S. Kallistratos, P. Kopterides, J. Syrios, H. Skopelitis, N. Mylonakis, A. Karabelis, N. Tsavaris, Cardiotoxicity of fluoropyrimidines in different schedules of administration: a prospective study, J. Cancer Res. Clin. Oncol. 134 (1) (2008) 75–82, https://doi.org/10.1007/s00432-007-0250-9.

[16] A. Polk, M. Vaage-Nilsen, K. Vistisen, D.L. Nielsen, Cardiotoxicity in cancer patients treated with 5-fluorouracil or capecitabine: a systematic review of incidence,

manifestations and predisposing factors, Cancer Treat. Rev. 39 (8) (2013) 974–984, https://doi.org/10.1016/j.ctrv.2013.03.005.

[17] I. Gotsman, N. Grabie, R. Dacosta, G. Sukhova, A. Sharpe, A.H. Lichtman, Proatherogenic immune responses are regulated by the PD-1/PD-L pathway in mice, J. Clin. Invest. 117 (10) (2007) 2974–2982, https://doi.org/10.1172/JCI31344.

[18] H. Schuett, R. Oestreich, G.H. Waetzig, W. Annema, M. Luchtefeld, A. Hillmer, U. Bavendiek, J. von Felden, D. Divchev, T. Kempf, K.C. Wollert, D. Seegert, S. Rose-John, U.J. Tietge, B. Schieffer, K. Grote, Transsignaling of interleukin-6 crucially contributes to atherosclerosis in mice, Arterioscler. Thromb. Vasc. Biol. 32 (2) (2012) 281–290, https://doi.org/10.1161/ATVBAHA.111.229435.

[19] C. Cochain, S.M. Chaudhari, M. Koch, H. Wiendl, H.H. Eckstein, A. Zernecke, Programmed cell death-1 deficiency exacerbates T cell activation and atherogenesis despite expansion of regulatory T cells in atherosclerosis-prone mice, PLoS One 9 (4) (2014), https://doi.org/10.1371/journal.pone.0093280, e93280.

[20] H. Essa, R. Dobson, D. Wright, G.Y.H. Lip, Hypertension management in cardio-oncology, J. Hum. Hypertens. 34 (2020) 673–681, https://doi.org/10.1038/s41371-020-0391-8.

[21] G.D. Demetri, A.T. van Oosterom, C.R. Garrett, et al., Efficacy and safety of sunitinib in patientswith advanced gastrointestinal stromal tumour after failure of imatinib: a randomised controlledtrial, Lancet 368 (2006) 1329–1338.

[22] M.J. Ratain, T. Eisen, W.M. Stadler, et al., Phase IIplacebo-controlled randomized discontinuationtrial of sorafenib in patients with metastatic renalcell carcinoma, J. Clin. Oncol. 24 (2006) 2505–2512.

[23] C. Heiss, A. Schanz, N. Amabile, et al., Nitric oxide synthase expression and functional response to nitric oxide are both important modulators of circulating angiogenic cell response to angiogenic stimuli, Arterioscler. Thromb. Vasc. Biol. 30 (2010) 2212–2218.

[24] T.W. Hein, R.H. Rosa, Y. Ren, X. Wenjuan, L. Kuo, VEGF receptor-2–linked PI3K/calpain/SIRT1 activation mediates retinal arteriolar dilations to VEGF and shear stress, Invest. Ophthalmol. Vis. Sci. 56 (9) (2015) 5381–5389, https://doi.org/10.1167/iovs15-16950.

[25] M.H. Kappers, J.H. van Esch, W. Sluiter, S. Sleijfer, A.H. Danser, A.H. van den Meiracker, Hypertension induced by the tyrosine kinase inhibitor sunitinib is associated with increased circulating endothelin-1 levels, Hypertension 56 (2010) 675–681.

[26] M. Simons, A. Eichmann, On-target cardiac effects of anticancer drugs: lessons from new biology, J. Am. Coll. Cardiol. 60 (2012) 626–627.

[27] A. Kumar, C.S. Kim, T.A. Hoffman, A. Naqvi, J. Dericco, S.B. Jung, Z. Lin, M.K. Jain, K. Irani, p53 impairs endothelial function by transcriptionally repressing Kruppel-Like Factor 2, Arterioscler. Thromb. Vasc. Biol. 31 (1) (2011) 133–141, https://doi.org/10.1161/ATVBAHA.110.215061.

[28] Q. Wei, Y. Xia, Proteasome inhibition down-regulates endothelial nitric-oxide synthase phosphorylation and function, J. Biol. Chem. 281 (31) (2006) 21652–21659, https://doi.org/10.1074/jbc.M602105200.

[29] E.C. Kooijmans, A. Bökenkamp, N.S. Tjahjadi, J.M. Tettero, B.E. van Dulmen-den, H.J. van der Pal, et al., Early and late adverse renal effects after potentially nephrotoxic treatment for childhood cancer, Cochrane Database Syst. Rev. 3 (2019), CD008944.

[30] D. Montani, E. Bergot, S. Günther, L. Savale, A. Bergeron, A. Bourdin, H. Bouvaist, M. Canuet, C. Pison, M. Macro, P. Poubeau, B. Girerd, D. Natali, C. Guignabert, F. Perros, D.S. O'Callaghan, X. Jaïs, P. Tubert-Bitter, G. Zalcman, O. Sitbon, G. Simonneau, M. Humbert, Pulmonary arterial hypertension in patients treated by dasatinib, Circulation 125 (17) (2012) 2128–2137.

[31] D. Dumitrescu, C. Seck, H. ten Freyhaus, F. Gerhardt, E. Erdmann, S. Rosenkranz, Fully reversible pulmonary arterial hypertension associated with dasatinib treatment for chronic myeloid leukaemia, Eur. Respir. J. 38 (1) (2011) 218–220.

[32] C. Guignabert, C. Phan, A. Seferian, A. Huertas, L. Tu, R. Thuillet, C. Sattler, M. Le Hiress, Y. Tamura, E.M. Jutant, M.C. Chaumais, S. Bouchet, B. Manéglier, M. Molimard, P. Rousselot, O. Sitbon, G. Simonneau, D. Montani, M.J. Humbert, Dasatinib induces lung vascular toxicity and predisposes to pulmonary hypertension, J. Clin. Investig. 126 (9) (2016) 3207–3218.

[33] T. Reinert, C.S.R. Baldotto, F.A.P. Nunes, A.A.S. Scheliga, Bleomycininduced lung injury, J. Cancer Res. 2013 (2013) 9.

[34] A. Leibowitz, E. Grossman, A. Berkovitch, M. Levartovski, S. Appel, Y. Sharabi, I. Gluck, The effect of head and neck radiotherapy on blood pressure and orthostatic hypotension in patients with head and neck tumors, Am. J. Hypertens. 31 (2) (2013) 235–239, https://doi.org/10.1093/ajh/hpx158.

[35] E.B. Clyburn, D.J. DiPette, Hypertension induced by drugs and other substances, Semin. Nephrol. 15 (2) (1995) 72–86.

[36] F. Mantero, M. Boscaro, Glucocorticoid-dependent hypertension, J. Steroid Biochem. Mol. Biol. 43 (5) (1992) 409–413, https://doi.org/10.1016/0960-0760(92)90077-v.

[37] F. Fyhrquist, K. Metsärinne, I. Tikkanen, Role of angiotensin II in blood pressure regulation and in the pathophysiology of cardiovascular disorders, J. Hum. Hypertens. 9 (Suppl 5) (1995) S19–S24.

[38] M.L. Maitland, G.L. Bakris, H.R. Black, et al., Initial assessment, surveillance, and management of blood pressure in patients receiving vascular endothelial growth factor signaling pathway inhibitors, J. Natl. Cancer Inst. 102 (2010) 596–604.

[39] Y. Lin, L. Ma, Blood pressure lowering effect of calcium channel blockers on perioperative hypertension: a systematic review and meta-analysis, Medicine (Baltimore) 97 (48) (2018), e13152.

[40] A.A. Khorana, C.W. Francis, E. Culakova, R.I. Fisher, N.M. Kuderer, G.H. Lyman, Thromboembolism in hospitalized neutropenic cancer patients, J. Clin. Oncol. 24 (3) (2006) 484–490.

[41] B.B. Navi, A.S. Reiner, H. Kamel, Risk of arterial thromboembolism in patients with cancer, J. Am. Coll. Cardiol. 70 (8) (2017) 926–938.

[42] E. Grilz, O. Königsbrügge, F. Posch, M. Schmidinger, R. Pirker, I.M. Lang, I. Pabinger, C. Ay, Frequency, risk factors, and impact on mortality of arterial thromboembolism in patients with cancer, Haematologica 103 (9) (2018) 1549–1556, https://doi.org/10.3324/haematol.2018.192419. 29794142. PMC6119137.

[43] R. Barceló, A.L.-V.G. Muñoz, Prospective evaluation of major vascular events in patients with nonsmall cell lung carcinoma treated with cisplatin and gemcitabine, Cancer 104 (2005) 1110–1111.

[44] R.A. Moore, N. Adel, E. Riedel, M. Bhutani, D.R. Feldman, N.E. Tabbara, High incidence of thromboembolic events in patients treated with cisplatin-based chemotherapy: a large retrospective analysis, J. Clin. Oncol. 29 (2011) 3466–3473.

[45] T.K. Choueiri, F.A. Schutz, Y. Je, J.E. Rosenberg, J. Bellmunt, Risk of arterial thromboembolic events with sunitinib and sorafenib: a systematic review and meta-analysis of clinical trials, J. Clin. Oncol. 28 (2010) 2280–2285.

[46] J. Herrmann, E.H. Yang, C.A. Iliescu, M. Cilingiroglu, K. Charitakis, A. Hakeem, Vascular toxicities of cancer therapies: the old and the new—an evolving avenue, Circulation 133 (2016) 1272–1289.

[47] V. Ranpura, S. Hapani, J.W.S. Chuang, Risk of cardiac ischemia and arterial thromboembolic events with the angiogenesis inhibitor bevacizumab in cancer patients: a meta-analysis of randomized controlled trials, Acta Oncol. 49 (2010) 287–297.

[48] FDA, Ponatinib, 2014, Available at: https://www.accessdata.fda.gov/drugsatfda_docs/label/2014/203469s009lbl.pdf.

[49] V. De Stefano, Arterial thrombosis and cancer: the neglected side of the coin of Trousseau syndrome, Haematologica 103 (9) (2018) 1419–1421, https://doi.org/10.3324/haematol.2018.197814. 30171017. PMC6119149.

[50] B. Zöller, J. Ji, J. Sundquist, K. Sundquist, Risk of coronary heart disease in patients with cancer: a nationwide follow-up study from Sweden, Eur. J. Cancer 48 (1) (2012) 121–128.
[51] A.K. Pandey, E.K. Singhi, J.P. Arroyo, T.A. Ikizler, E.R. Gould, J. Brown, Mechanisms of VEGF (vascular endothelial growth factor) inhibitor-associated hypertension and vascular disease, Hypertension 71 (2018) e1–e8.
[52] C.P. McCarthy, G. Steg, D.L. Bhatt, The management of anti-platelet therapy in acute coronary syndrome patients with thrombocytopenia: a clinical conundrum, Eur. Heart J. 38 (2017) 3488–3492.
[53] N.B. Abdol Razak, G. Jones, M. Bhandari, M.C. Berndt, P. Metharom, Cancer-associated thrombosis: an overview of mechanisms, risk factors, and treatment, Cancers (Basel) 10 (10) (2018) 380, https://doi.org/10.3390/cancers10100380. 30314362. PMC6209883.
[54] R. Sud, A.A. Khorana, Cancer-associated thrombosis: risk factors, candidate biomarkers and a risk model, Thromb. Res. 123 (Suppl 4) (2009) S18–S21, https://doi.org/10.1016/S0049-3848(09)70137-9. 19303497.
[55] A.I. Al Diab, Cancer-related venous thromboembolism: insight into underestimated risk factors, Hematol. Oncol. Stem Cell Ther. 3 (4) (2010) 191–195, https://doi.org/10.5144/1658-3876.2010.191. 21150239.
[56] G. Agnelli, G. Bolis, L. Capussotti, R.M. Scarpa, F. Tonelli, E. Bonizzoni, M. Moia, F. Parazzini, R. Rossi, F. Sonaglia, B. Valarani, C. Bianchini, G. Gussoni, A clinical outcome-based prospective study on venous thromboembolism after cancer surgery: the @RISTOS project, Ann. Surg. 243 (1) (2006) 89–95, https://doi.org/10.1097/01.sla.0000193959.44677.48. 16371741. PMC1449979.
[57] P.D. Stein, A. Beemath, F.A. Meyers, E. Skaf, J. Sanchez, R.E. Olson, Incidence of venous thromboembolism in patients hospitalized with cancer, Am. J. Med. 119 (1) (2006) 60–68, https://doi.org/10.1016/j.amjmed.2005.06.058. 16431186.
[58] P.M. Czaykowski, M.J. Moore, I.F. Tannock, High risk of vascular events in patients with urothelial transitional cell carcinoma treated with cisplatin based chemotherapy, J. Urol. 160 (6 Pt 1) (1998) 2021–2024, https://doi.org/10.1097/00005392-199812010-00022. 9817314.
[59] D. Cunningham, N. Starling, S. Rao, T. Iveson, M. Nicolson, F. Coxon, G. Middleton, F. Daniel, J. Oates, A.R. Norman, Upper Gastrointestinal Clinical Studies Group of the National Cancer Research Institute of the United Kingdom, Capecitabine and oxaliplatin for advanced esophagogastric cancer, N. Engl. J. Med. 358 (1) (2008) 36–46, https://doi.org/10.1056/NEJMoa073149. 18172173.
[60] D. Lechner, M. Kollars, A. Gleiss, P.A. Kyrle, A. Weltermann, Chemotherapy-induced thrombin generation via procoagulant endothelial microparticles is independent of tissue factor activity, J. Thromb. Haemost. 5 (12) (2007) 2445–2452, https://doi.org/10.1111/j.1538-7836.2007.02788.x. Epub 2007 Oct 8 17922809.
[61] S.R. Nalluri, D. Chu, R. Keresztes, X. Zhu, S. Wu, Risk of venous thromboembolism with the angiogenesis inhibitor bevacizumab in cancer patients: a meta-analysis, JAMA 300 (19) (2008) 2277–2285, https://doi.org/10.1001/jama.2008.656. 19017914.
[62] J.L. Zamorano, P. Lancellotti, D. Rodriguez Muñoz, V. Aboyans, R. Asteggiano, M. Galderisi, G. Habib, D.J. Lenihan, G. Lip, A.R. Lyon, T. Lopez Fernandez, D. Mohty, M.F. Piepoli, J. Tamargo, A. Torbicki, T.M. Suter, ESC Scientific Document Group, 2016 ESC position paper on cancer treatments and cardiovascular toxicity developed under the auspices of the ESC Committee for Practice Guidelines: the Task Force for cancer treatments and cardiovascular toxicity of the European Society of Cardiology (ESC), Eur. Heart J. 37 (36) (2016) 2768–2801, https://doi.org/10.1093/eurheartj/ehw211.
[63] J.R. Carver, C.L. Shapiro, A. Ng, L. Jacobs, C. Schwartz, K.S. Virgo, K.L. Hagerty, M.R. Somerfield, D.J. Vaughn, ASCO Cancer Survivorship Expert Panel, American Society of Clinical Oncology clinical evidence review on the ongoing care of adult cancer

survivors: cardiac and pulmonary late effects, J. Clin. Oncol. Off. J. Am. Soc. Clin. Oncol. 25 (25) (2007) 3991–4008.
[64] C.G. Lenneman, D.B. Sawyer, Cardio-oncology: an update on cardiotoxicity of cancer-related treatment, Circ. Res. 118 (6) (2016) 1008–1020.
[65] L.S. Mehta, K.E. Watson, A. Barac, T.M. Beckie, V. Bittner, S. Cruz-Flores, S. Dent, L. Kondapalli, B. Ky, T. Okwuosa, I.L. Piña, A.S. Volgman, American Heart Association Cardiovascular Disease in Women and Special Populations Committee of the Council on Clinical Cardiology, Council on Cardiovascular and Stroke Nursing, Council on Quality of Care and Outcomes Research, Cardiovascular disease and breast cancer: where these entities intersect: a scientific statement from the American Heart Association, Circulation 137 (8) (2018) e30–e66, https://doi.org/10.1161/CIR.0000000000000556.
[66] R.G. Carlson, W.R. Mayfield, S. Normann, J.A. Alexander, Radiation-associated valvular disease, Chest 99 (3) (1991) 538–545.
[67] J.P. Veinot, W.D. Edwards, Pathology of radiation-induced heart disease: a surgical and autopsy study of 27 cases, Hum. Pathol. 27 (8) (1996) 766–773, https://doi.org/10.1016/s0046-8177(96)90447-5.
[68] M. Andrejak, C. Tribouilloy, Drug-induced valvular heart disease: an update, Arch. Cardiovasc. Dis. 5 (2013) 333–339.
[69] J. Allen, J.D. Thomson, I.J. Lewis, J.L. Gibbs, Mitral regurgitation after anthracycline treatment for childhood malignancy, Heart 85 (4) (2001) 430–432.
[70] K. Murbraech, T. Wethal, K.B. Smeland, H. Holte, J.H. Loge, E. Holte, A. Rösner, H. Dalen, C.E. Kiserud, S. Aakhus, Valvular dysfunction in lymphoma survivors treated with autologous stem cell transplantation: a national cross-sectional study, J. Am. Coll. Cardiol. Img. 9 (3) (2016) 230–239, https://doi.org/10.1016/j.jcmg.2015.06.028.
[71] S. Myrehaug, M. Pintilie, R. Tsang, R. Mackenzie, M. Crump, Z. Chen, A. Sun, D.C. Hodgson, Cardiac morbidity following mod-ern treatment for Hodgkin lymphoma: supra-additive cardiotoxicity of doxorubicin and radiation therapy, Leuk. Lymphoma 49 (8) (2008) 1486–1493.
[72] K. Kalra, Q. Wang, B.V. McIver, W. Shi, R.A. Guyton, W. Sun, E.L. Sarin, V.H. Thourani, M. Padala, Temporal changes in interpapillary muscle dynamics as an active indicator of mitral valve and left ventricular interaction in ischemic mitral regurgitation, J. Am. Coll. Cardiol. 64 (18) (2014) 1867–1879, https://doi.org/10.1016/j.jacc.2014.07.988.
[73] A. Sato, A. Yoshihisa, M. Miyata-Tatsumi, M. Oikawa, A. Kobayashi, T. Ishida, Y. Takeishi, Valvular heart disease as a possible predictor of trastuzumab-induced cardiotoxicity in patients with breast cancer, Mol. Clin. Oncol. 10 (2019) 37–42, https://doi.org/10.3892/mco.2018.1764.
[74] L.L. Lin, S. Vennarini, A. Dimofte, D. Ravanelli, K. Shillington, S. Batra, Z. Tochner, S. Both, G. Freedman, Proton beam versus photon beam dose to the heart and left anterior descending artery for left-sided breast cancer, Acta Oncol. 54 (7) (2015) 1032–1039, https://doi.org/10.3109/0284186X.2015.1011756.
[75] M.V. Maraldo, N.P. Brodin, I.R. Vogelius, M.C. Aznar, P. Munck Af Rosenschöld, P.M. Petersen, L. Specht, Risk of developing cardiovascular disease after involved node radiotherapy versus mantle field for Hodgkin lymphoma, Int. J. Radiat. Oncol. Biol. Phys. 83 (4) (2012) 1232–1237, https://doi.org/10.1016/j.ijrobp.2011.09.020. Epub 2012 Jan 21 22270170.
[76] N. Beohar, et al., Trends in complications and outcomesof patients undergoing transfemoral transcatheteraortic valve replacement: experience from the PART-NER continued access registry, JACC Cardiovasc. Interv. 9 (4) (2016) 355–363.
[77] O. Franzen, et al., Acute outcomes of MitraClip therapyfor mitral regurgitation in high-surgical-risk patients: emphasis on adverse valve morphology and severe left ventricular dysfunction, Eur. Heart J. 31 (11) (2010) 1373–1381.

[78] L.F. Fajardo, J.R. Stewart, K.E. Cohn, Morphology of radiation-induced heart disease, Arch. Pathol. 86 (5) (1968) 512–519.
[79] D. Raghunathan, M.I. Khilji, S.A. Hassan, S.W. Yusuf, Radiation-induced cardiovascular disease, Curr. Atheroscler. Rep. 19 (5) (2017) 22, https://doi.org/10.1007/s11883-017-0658-x.
[80] G. Griguolo, V. Guarneri, Cardiac complications of cancer and anticancer treatment, in: ESMO Handbook of Oncological Emergencies, 2016.
[81] A.K. Ghosh, T. Crake, C. Manisty, M. Westwood, Pericardial disease in cancer patients, Curr. Treat. Options Cardiovasc. Med. 20 (7) (2018) 60, https://doi.org/10.1007/s11936-018-0654-7.
[82] J.D. Floyd, D.T. Nguyen, R.L. Lobins, Q. Bashir, D.C. Doll, M.C. Perry, Cardiotoxicity of cancer therapy, J. Clin. Oncol. Off. J. Am. Soc. Clin. Oncol. 23 (30) (2005) 7685–7696, https://doi.org/10.1200/JCO.2005.08.789.
[83] A. Konski, T. Li, M. Christensen, J.D. Cheng, J.Q. Yu, K. Crawford, et al., Symptomatic cardiac toxicity is predicted by dosimetric and patient factors rather than changes in 18F-FDG PET determination of myocardial activity after chemoradiotherapy for esophageal cancer, Radiother. Oncol. 104 (1) (2012) 72–77.
[84] S. Çelik, C. Lestuzzi, E. Cervesato, D. Dequanter, P. Piotti, M. De Biasio, M. Imazio, Systemic chemotherapy in combination with pericardial window has better outcomes in malignant pericardial effusions, J. Thorac. Cardiovasc. Surg. 148 (5) (2014) 2288–2293.
[85] J.M. McDonald, B.F. Meyers, T.J. Guthrie, R.J. Battafarano, J.D. Cooper, G.A. Patterson, Comparison of open subxiphoid pericardial drainage with percutaneous catheter drainage for symptomatic pericardial effusion, Ann. Thorac. Surg. 76 (3) (2003) 811–816, https://doi.org/10.1016/s0003-4975(03)00665-9.
[86] J. Sagristà-Sauleda, J. Angel, A. Sánchez, G. Permanyer-Miralda, J. Soler-Soler, Effusive-constrictive pericarditis, N. Engl. J. Med. 350 (5) (2004) 469–475, https://doi.org/10.1056/NEJMoa035630.
[87] J.M. Horacek, M. Jakl, J. Horackova, R. Pudil, L. Jebavy, J. Maly, Assessment of anthracycline-induced cardiotoxicity with electrocardiography, Exp. Oncol. 31 (2) (2009) 115–117.
[88] D. Unnikrishnan, J.P. Dutcher, S. Garl, N. Varshneya, R. Lucariello, P.H. Wiernik, Cardiac monitoring of patients receiving arsenic trioxide therapy, Br. J. Haematol. 124 (5) (2004) 610–617, https://doi.org/10.1111/j.1365-2141.2003.04817.x.
[89] M. Kupari, L. Volin, A. Suokas, T. Timonen, P. Hekali, T. Ruutu, Cardiac involvement in bone marrow transplantation: electrocardiographic changes, arrhythmias, heart failure and autopsy findings, Bone Marrow Transplant. 5 (2) (1990) 91–98.
[90] P.A. Heidenreich, J.R. Kapoor, Radiation induced heart disease: systemic disorders in heart disease, Heart 95 (3) (2009) 252–258, https://doi.org/10.1136/hrt.2008.149088.
[91] U. Nellessen, M. Zingel, H. Hecker, J. Bahnsen, D. Borschke, Effects of radiation therapy on myocardial cell integrity and pump function: which role for cardiac biomarkers? Chemotherapy 56 (2) (2010) 147–152, https://doi.org/10.1159/000313528.
[92] J. Cuzick, H. Stewart, R. Peto, M. Baum, B. Fisher, H. Host, J.P. Lythgoe, G. Ribeiro, H. Scheurlen, A. Wallgren, Overview of randomized trials of postoperative adjuvant radiotherapy in breast cancer, Cancer Treat. Rep. 71 (1) (1987) 15–29.
[93] M. Clarke, R. Collins, S. Darby, C. Davies, P. Elphinstone, V. Evans, J. Godwin, R. Gray, C. Hicks, S. James, E. MacKinnon, P. McGale, T. McHugh, R. Peto, C. Taylor, Y. Wang, Early Breast Cancer Trialists' Collaborative Group (EBCTCG), Effects of radiotherapy and of differences in the extent of surgery for early breast cancer on local recurrence and 15-year survival: an overview of the randomised trials, Lancet (London, England) 366 (9503) (2005) 2087–2106, https://doi.org/10.1016/S0140-6736(05)67887-7).
[94] H. Wang, J. Wei, Q. Zheng, L. Meng, Y. Xin, X. Yin, X. Jiang, Radiation-induced heart disease: a review of classification, mechanism and prevention, Int. J. Biol. Sci. 15 (10) (2019) 2128–2138, https://doi.org/10.7150/ijbs.35460.

[95] R. Madan, R. Benson, D.N. Sharma, P.K. Julka, G.K. Rath, Radiation induced heart disease: pathogenesis, management and review literature, J. Egypt. Natl. Canc. Inst. 27 (4) (2015) 187–193, https://doi.org/10.1016/j.jnci.2015.07.005.
[96] N.K. Taunk, B.G. Haffty, J.B. Kostis, S. Goyal, Radiation-induced heart disease: pathologic abnormalities and putative mechanisms, Front. Oncol. 5 (2015) 39, https://doi.org/10.3389/fonc.2015.00039.
[97] L.K. Liu, W. Ouyang, X. Zhao, S. Su, Y. Yang, W.J. Ding, D.X. Luo, Z. He, B. Lu, Pathogenesis and prevention of radiation-induced myocardial fibrosis, Asian Pac. J. Cancer Prev. 18 (3) (2017) 583–587, https://doi.org/10.22034/APJCP.2017.18.3.583.
[98] J. Yarnold, M.C. Brotons, Pathogenetic mechanisms in radiation fibrosis, Radiother. Oncol. 97 (1) (2010) 149–161, https://doi.org/10.1016/j.radonc.2010.09.002.
[99] M. Martin, J. Lefaix, S. Delanian, TGF-beta1 and radiation fibrosis: a master switch and a specific therapeutic target? Int. J. Radiat. Oncol. Biol. Phys. 47 (2) (2000) 277–290, https://doi.org/10.1016/s0360-3016(00)00435-1.
[100] E. Vorotnikova, R.A. Rosenthal, M. Tries, S.R. Doctrow, S.J. Braunhut, Novel synthetic SOD/catalase mimetics can mitigate capillary endothelial cell apoptosis caused by ionizing radiation, Radiat. Res. 173 (6) (2010) 748–759, https://doi.org/10.1667/RR1948.1.
[101] C. Moro, M.G. Jouan, A. Rakotovao, M.C. Toufektsian, O. Ormezzano, N. Nagy, A. Tosaki, J. de Leiris, F. Boucher, Delayed expression of cytokines after reperfused myocardial infarction: possible trigger for cardiac dysfunction and ventricular remodeling, Am. J. Physiol. Heart Circ. Physiol. 293 (5) (2007) H3014–H3019, https://doi.org/10.1152/ajpheart.00797.2007.
[102] S.W. Yusuf, S. Sami, I.N. Daher, Radiation-induced heart disease: a clinical update, Cardiol. Res. Pract. 2011 (2011), https://doi.org/10.4061/2011/317659, 317659.
[103] T. Yamamori, H. Yasui, M. Yamazumi, Y. Wada, Y. Nakamura, H. Nakamura, O. Inanami, Ionizing radiation induces mitochondrial reactive oxygen species production accompanied by upregulation of mitochondrial electron transport chain function and mitochondrial content under control of the cell cycle checkpoint, Free Radic. Biol. Med. 53 (2) (2012) 260–270, https://doi.org/10.1016/j.freeradbiomed.2012.04.033.
[104] B. Kura, C. Yin, K. Frimmel, J. Krizak, L. Okruhlicova, R.C. Kukreja, J. Slezak, Changes of microRNA-1, −15b and −21 levels in irradiated rat hearts after treatment with potentially radioprotective drugs, Physiol. Res. 65 (Suppl 1) (2016) S129–S137, https://doi.org/10.33549/physiolres.933399.
[105] M.V. Maraldo, N.P. Brodin, M.C. Aznar, I.R. Vogelius, P. Munck af Rosenschöld, P.M. Petersen, L. Specht, Estimated risk of cardiovascular disease and secondary cancers with modern highly conformal radiotherapy for early-stage mediastinal Hodgkin lymphoma, Ann. Oncol. 24 (8) (2013) 2113–2118, https://doi.org/10.1093/annonc/mdt156.
[106] K.A. Reardon, P.W. Read, M.M. Morris, M.A. Reardon, C. Geesey, K. Wijesooriya, A comparative analysis of 3D conformal deep inspiratory-breath hold and free-breathing intensity-modulated radiation therapy for left-sided breast cancer, Med. Dosim. 38 (2) (2013) 190–195, https://doi.org/10.1016/j.meddos.2013.01.002.
[107] D.N. Yeboa, S.B. Evans, Contemporary breast radiotherapy and cardiac toxicity, Semin. Radiat. Oncol. 26 (1) (2016) 71–78, https://doi.org/10.1016/j.semradonc.2015.09.003.
[108] J.J. Kruse, E.G. Strootman, J. Wondergem, Effects of amifostine on radiation-induced cardiac damage, Acta Oncol. 42 (1) (2003) 4–9, https://doi.org/10.1080/0891060310002168.
[109] S.J. van der Veen, G. Ghobadi, R.A. de Boer, H. Faber, M.V. Cannon, P.W. Nagle, S. Brandenburg, J.A. Langendijk, P. van Luijk, R.P. Coppes, ACE inhibition attenuates radiation-induced cardiopulmonary damage, Radiother. Oncol. 114 (1) (2015) 96–103, https://doi.org/10.1016/j.radonc.2014.11.017.
[110] K. Zhang, X. He, Y. Zhou, L. Gao, Z. Qi, J. Chen, X. Gao, Atorvastatin ameliorates radiation-induced cardiac fibrosis in rats, Radiat. Res. 184 (6) (2015) 611–620, https://doi.org/10.1667/RR14075.1.

[111] E.T.H. Yeh, A.T. Tong, D.J. Lenihan, S.W. Yusuf, et al., Cardiovascular complications of cancer therapy. Diagnosis, pathogenesis, and management, Circulation 109 (2004) 3122–3131.
[112] H. Hurwitz, L. Fehrenbacher, W. Novotny, T. Cartwright, J. Hainsworth, W. Heim, J. Berlin, A. Baron, S. Griffing, E. Holmgren, N. Ferrara, G. Fyfe, B. Rogers, R. Ross, F. Kabbinavar, Bevacizumab plus irinotecan, fluorouracil, and leucovorin for metastatic colorectal cancer, N. Engl. J. Med. 350 (23) (2004) 2335–2342, https://doi.org/10.1056/NEJMoa032691.
[113] A. Sandler, R. Gray, M.C. Perry, J. Brahmer, J.H. Schiller, A. Dowlati, R. Lilenbaum, D.H. Johnson, Paclitaxel-carboplatin alone or with bevacizumab for non-small-cell lung cancer, N. Engl. J. Med. 355 (24) (2006) 2542–2550, https://doi.org/10.1056/NEJMoa061884.
[114] K.D. Miller, L.I. Chap, F.A. Holmes, M.A. Cobleigh, P.K. Marcom, L. Fehrenbacher, M. Dickler, B.A. Overmoyer, J.D. Reimann, A.P. Sing, V. Langmuir, H.S. Rugo, Randomized phase III trial of capecitabine compared with bevacizumab plus capecitabine in patients with previously treated metastatic breast cancer, J. Clin. Oncol. Off. J. Am. Soc. Clin. Oncol. 23 (4) (2005) 792–799, https://doi.org/10.1200/JCO.2005.05.098.
[115] B.I. Lévy, Microvascular plasticity and experimental heart failure, Hypertension 47 (5) (2006) 827–829.
[116] M.S. Ewer, T.M. Suter, D.J. Lenihan, L. Niculescu, A. Breazna, G.D. Demetri, R.J. Motzer, Cardiovascular events among 1090 cancer patients treated with sunitinib, interferon, or placebo: a comprehensive adjudicated database analysis demonstrating clinically meaningful reversibility of cardiac events, Eur. J. Cancer 50 (12) (2014) 2162–2170.
[117] S. Wu, J.J. Chen, A. Kudelka, J. Lu, X. Zhu, Incidence and risk of hypertension with sorafenib in patients with cancer: a systematic review and meta-analysis, Lancet Oncol. 9 (2) (2008) 117–123, https://doi.org/10.1016/S1470-2045(08)70003-2.
[118] M. Dincer, K. Altundag, Angiotensin-converting enzyme inhibitors for bevacizumab-induced hypertension, Ann. Pharmacother. 40 (12) (2006) 2278–2279, https://doi.org/10.1345/aph.1H244.
[119] H.M. Shepard, P. Jin, D.J. Slamon, Z. Pirot, D.C. Maneval, Herceptin, Handb. Exp. Pharmacol. (181) (2008) 183–219, https://doi.org/10.1007/978-3-540-73259-4_9.
[120] J. Baselga, E.A. Perez, T. Pienkowski, R. Bell, Adjuvant trastuzumab: a milestone in the treatment of HER-2-positive early breast cancer, Oncologist 11 (Suppl 1) (2006) 4–12, https://doi.org/10.1634/theoncologist.11-90001-4.
[121] K.S. Gunturu, Y. Woo, N. Beaubier, H.E. Remotti, M.W. Saif, Gastric cancer and trastuzumab: first biologic therapy in gastric cancer, Ther. Adv. Med. Oncol. 5 (2) (2013) 143–151.
[122] J. Albanell, J. Codony, A. Rovira, B. Mellado, P. Gascón, Mechanism of action of anti-HER2 monoclonal antibodies: scientific update on trastuzumab and 2C4, Adv. Exp. Med. Biol. 532 (2003) 253–268.
[123] S. Gorini, A. De Angelis, L. Berrino, N. Malara, G. Rosano, E. Ferraro, Chemotherapeutic drugs and mitochondrial dysfunction: focus on doxorubicin, trastuzumab, and sunitinib, Oxid. Med. Cell. Longev. 2018 (2018) 7582730, https://doi.org/10.1155/2018/7582730.
[124] C. Ozcelik, B. Erdmann, B. Pilz, N. Wettschureck, S. Britsch, N. Hübner, K.R. Chien, C. Birchmeier, A.N. Garratt, Conditional mutation of the ErbB2 (HER2) receptor in cardiomyocytes leads to dilated cardiomyopathy, Proc. Natl. Acad. Sci. U. S. A. 99 (13) (2002) 8880–8885, https://doi.org/10.1073/pnas.122249299.
[125] K.L. Carraway 3rd, Involvement of the neuregulins and their receptors in cardiac and neural development, Bioessays 18 (4) (1996) 263–266, https://doi.org/10.1002/bies.950180403.
[126] S.A. Crone, Y.Y. Zhao, L. Fan, Y. Gu, S. Minamisawa, Y. Liu, K.L. Peterson, J. Chen, R. Kahn, G. Condorelli, J. Ross Jr., K.R. Chien, K.F. Lee, ErbB2 is essential in the prevention of dilated cardiomyopathy, Nat. Med. 8 (5) (2002) 459–465, https://doi.org/10.1038/nm0502-459.

[127] T. Force, D.S. Krause, R.A. Van Etten, Molecular mechanisms of cardiotoxicity of tyrosine kinase inhibition, Nat. Rev. Cancer 7 (5) (2007) 332–344, https://doi.org/10.1038/nrc2106.
[128] N. Mohan, Y. Shen, Y. Endo, M.K. ElZarrad, W.J. Wu, Trastuzumab, but not pertuzumab, dysregulates HER2 signaling to mediate inhibition of autophagy and increase in reactive oxygen species production in human cardiomyocytes, Mol. Cancer Ther. 15 (6) (2016) 1321–1331, https://doi.org/10.1158/1535-7163.MCT-15-0741.
[129] E.T.H. Yeh, MD Anderson Practices in Onco-Cardiology, Primedia E-launch LLC, Texas, 2016. 1944785949, 9781944785949.

CHAPTER

11

Novel anticancer drugs related to cardiotoxicity

Pallabi Banerjee and Imteyaz Qamar
School of Biotechnology, Gautam Buddha University, Greater Noida, Uttar Pradesh, India

1 Introduction

Cardiotoxicity is defined as the advent of cardiac muscle dysfunction, such as left ventricular ejection fraction (LVEF) due to exposure to antineoplastic therapy, which may progress to heart failure (HF). During the previous couple of decades, anticancer treatment has accomplished a remarkable advancement in enhancing the quality of life and survival rates of cancer patients. In parallel with the increase in the development of novel cancer therapeutics, the literature pertaining to cancer metastasis has grown enormously.

As a result of recent developments, drugs designed to target the activity of tyrosine kinases or tumor receptors have been documented to cause unwanted effects on the cardiovascular system [1]. In various studies, it has been proposed that anthracyclines (ANTs) cause irreversible type 1 CTX by producing reactive oxygen species (ROS) and reactive nitrogen species (RNS) [2]. Moreover, they target topoisomerase II (Top2), which binds to both DNA and Top2 causing apoptosis. Targeted therapy presents a surprisingly promising method for treating cancer. It involves using drugs to inhibit the dysregulated genes that cause malignancy. A wide range of cancer therapies is currently available, from monoclonal antibodies (mAbs) that target growth factor receptors to small molecules that inhibit kinases (mostly tyrosine kinase inhibitors or TKIs). Cardiotoxicity as a result of highly targeted therapies can be divided into two categories: on-target and off-target toxicity. On-target cardiotoxicity is a form of toxicity that is caused by the inhibition of a specific kinase responsible for the transformation of malignant cells, which significantly influences the

Cardiovascular Toxicity and Therapeutic Modalities Targeting Cardio-oncology
https://doi.org/10.1016/B978-0-323-90461-2.00006-7

maintenance of normal heart function. This therapy is therefore associated with a risk of cardiotoxicity. Trastuzumab, a humanized monoclonal antibody that inhibits the extracellular region of the HER2/ErbB2 receptor, is an example of on-target toxicity [3]. Nevertheless, anti-HER2 therapy can lead to complications such as systemic heart failure, characterized by an increase in the risk of asymptomatic decreased left ventricular ejection fraction (LVEF). Researchers examined 11,882 patients from 10 randomized trials and found that adding trastuzumab to anthracycline-based chemotherapy significantly increased the risk of cardiac dysfunction (relative risk 4.27; CI, 2.75–6.61, $P < 0.00001$) [4]. Over the past 10 years, cancer immunotherapy employing inhibitory effects on immune checkpoint receptors has emerged as a highly effective modality for the treatment of several malignancies, and it has proved highly promising in the cardio-oncology domain. Through this new method of cutting-edge treatment, anticancer treatments have been revolutionized [5]. In particular, anticytotoxic T lymphocyte-associated antigen 4 (CTLA-4) and antiprogrammed cell death protein-1 (PD-1) are effective in the treatment of melanoma [6]. Cancer cells usually use these receptors to avoid destruction by immune system T cells by binding to CLTA-4 and reducing T cell activation, or by expressing the cell death protein ligand-1 (PD-L1) that binds to PD-1 and inhibits T cell activity and apoptosis [7]. Specifically, ipilimumab (anti-CTLA-4 monoclonal antibody), pembrolizumab, and nivolumab (anti-PD-1) bind to the receptor site and block the inhibitory effects, increasing the effectiveness of the cytotoxic immune response to the cancer cells [5].

2 Clinically used cardioprotective drugs against cardiotoxicity

Cardioprotective medicines can help treat patients at risk for or with heart disease (CVD). A number of cardioprotective therapies have been found to be effective in treating anticancer-induced cardiotoxicity.

2.1 Dexrazoxane

In addition to being a highly effective anthracycline antagonist, dexrazoxane was found to be a catalytic inhibitor of the enzyme topoisomerase II at high concentrations administered by the Food and Drug Administration (FDA) [8]. It is a cyclic derivative of EDTA that acts as a chelator for iron, thereby reducing the production of reactive oxygen species and free radicals. Dexrazoxane was designated as an orphan drug by the FDA for "prevention of cardiomyopathy in children and adolescents treated with anthracyclines" in August 2014 [9]. In general, dexrazoxane is normally prescribed for patients with metastatic breast cancer who have received cumulative anthracycline doses above 300 mg/m^2. Moreover, due to its potential to reduce the anticancer effect of anthracyclines, dexrazoxane is not recommended in the beginning of treatment plan [10]. It is unclear whether dexrazoxane actually improves cardiac health, but some studies have shown it to be beneficial. In contrast, dexrazoxane has been shown to worsen thrombocytopenia and granulocytopenia. Clinical trials demonstrated a reduction in myocardial injury after treatment with dexrazoxane for children with acute lymphoblastic leukemia treated with doxorubicin, as shown by a decrease in serum troponin T levels [11].

2.2 Antioxidants

The glutathione precursor and analog *N*-acetyl-L-cysteine (NAC), both excellent antioxidants, can be given to patients at relatively high dosage levels with little side effects. Using murine models, NAC presented protective and synergistic antitumor [12] and antimetastatic [13] effects, but was unable to reverse long-standing doxorubicin-induced cardiotoxicity when tested in 19 patients with sarcoma [14]. In addition to many antioxidants and free radical scavengers, some vitamins, micronutrients, enzymes, and hormones, such as selenium, zinc, Q10 coenzyme, and melatonin, exhibit an enhanced cardiotoxic potential, as do anthracyclines, when added to trastuzumab, with the potential to reduce the likelihood of drug-induced cardiotoxicity. Resveratrol, a polyphenolic compound, is also known to reverse the effects of DOX-induced apoptosis and fibrosis in rat myocardium, both prophylactically and therapeutically [15]. Mice treated with baicalein, a bioflavonoid, has been proven to reduce cardiotoxicity [16]. While antioxidant drugs do eliminate ROS and RNS, this may result in detrimental health effects and even adverse physiological cellular functions [17]. Experiments have shown that several plant derivatives, such as extracts from Gingko biloba [18] and grape seed [19] as well as polyphenols, previously studied for their antiangiogenic, antiinflammatory, and anticancer activities [20,21], possess cardioprotective properties.

2.3 Statins

Patients suffering from atherosclerosis can lower low-density lipoprotein (LDL) and cholesterol by taking statins, which also have antioxidant, antiinflammatory, and other pleiotropic effects [22]. HMG-CoA reductase, the enzyme involved in the synthesis of cholesterol, is inhibited by this action. According to animal studies, fluvastatin pretreatment significantly lowered the toxic effects of anthracycline, reduced oxidative stress, and enhanced the expression of antioxidative enzyme mitochondrial superoxide dismutase-2 [23]. Retrospective study shows lower incidence of HF in 67 women who received statins during anthracycline-based chemotherapy for newly diagnosed breast cancer (HR, 0.3; 95% CI, 0.1 to 0.9; $P=0.03$)

compared to 134 women who did not receive statin treatment [24]. In a clinical trial of 40 patients who had normal LVEF before receiving chemotherapy (which included anthracyclines), the LVEF value at 6 months did not differ significantly among patients treated with atorvastatin, compared with an absolute decrease of 8% among the controls [25]. Several studies have identified statins as being associated with myopathies, possibly due to a decreased synthesis of mevalonic acid, which results in less energy generation and injury of the muscles. In addition to new-onset diabetes, statin usage can cause side effects [26]. Most of the benefits of statins can be attributed to their inhibition of heterotrimeric G-proteins, such as Ras, Rho, or Rac1 [27]. In view of this, the specific Rac and Rho inhibitors may be more appropriate targets for future chemopreventive strategies.

2.4 GPCRs

The class of GPCRs (G-protein [guanine nucleotide-binding protein]-coupled receptors) represents the largest and most versatile family of cell surface receptors and is also known as 7TMRs (7-transmembrane receptors) [28]. Agonists bind to GPCRs and can initiate both classical signaling cascades through heterotrimeric G-proteins (Gα, Gβ, and Gγ) as well as G-protein-independent pathways that involve β-arrestin [29]. It has been shown that β-arrestins act as scaffold proteins for MAP kinase cascades as well as serine/threonine kinase cascades [30]. There has been an effort to produce signal-biased drugs since it was discovered that some GPCRs prefer and activate G-protein- or arrestin-mediated pathways [31]. During the drug discovery process, a specific goal is to produce ligands with less adverse effects and/or allosteric forms without compromising their efficacy. In addition to their roles in a number of physiological processes, GPCRs have also become the major therapeutic targets. As an example, chronic activation of norepinephrine and Ang II, βARs (β-adrenergic receptors) and A1Rs (Ang II [angiotensin II] type 1 receptors) by their endogenous ligands, respectively, increases the workload on the heart, causing myocyte death and harmful remodeling of the heart [32]. In order to treat heart failure, drugs that block the activation of these receptors, including β-blockers, Ang II receptor blockers, and ACE (angiotensin-converting enzyme) inhibitors, are commonly used.

3 GPCR-targeted cardioprotective strategies against anticancer-induced cardiotoxicity

3.1 β-Blockers

The β-blockers, or β-adrenoceptor antagonists, are very effective in treating patients with cancer who may develop heart failure during treatment with potentially cardiotoxic therapy (which pose a high risk of developing heart failure). In addition to blocking the effects of epinephrine (adrenaline), β-blockers also slow the heart rate, cardiac contractility, and systolic pressure, which in turn decreases myocardial oxygen demand. Clinical and experimental evidence supports the use of β-blockers in ANT-induced CTX. ANTs and other types of dilated cardiomyopathies cause alterations to the signaling of the β-adrenergic receptor (β-AR) [33]. Additionally, experimental studies have demonstrated that β-AR blockage reduces cardiac oxidative stress and myocardial calcium overload [34]. The cardioprotective properties of new-generation β-blockers (i.e., carvedilol and nebivolol) have been investigated. Researchers compared carvedilol, a nonselective β- and α1-AR antagonist, with atenolol, a β-blocker with no antioxidative properties. Carvedilol provided the only protection from ANT-induced heart failure, and that effect is attributed to its antioxidant properties rather than to its ability to block the β-AR signal [35]. Apoptosis, ROS production, and mitochondrial alteration are inhibited by carvedilol in cardiomyocytes induced by ANT [36,37]. One small study of the cardioprotective effect of carvedilol in patients treated with ANTs showed a decreased incidence of LV dysfunction [38]. Nebivolol, a cardio-selective β-blocker with limited vasodilator properties, improved myocardial function in an experimental model of ANT-induced CTX, as well as enhancing NO levels and lowering oxidative stress [39]. There has been an association between β-blockers and a reduced risk of cardiovascular dysfunction in patients receiving trastuzumab and ANTs [40]. Recently, it has been shown that β-blockers such as bisoprolol and metoprolol do not provide a significant protection from trastuzumab-induced LV dysfunction, indicating that blocking β1 is not cardioprotective by itself [41,42]. Using nonselective β1- and β2-blockers is therefore supported [43].

3.2 Angiotensin-converting enzyme inhibitors and angiotensin receptor blockers

The renin-angiotensin-aldosterone (RAAS) system is primarily responsible for ANT-induced CTX [44]. In high-risk patients with heart failure, angiotensin-converting enzyme inhibitors (ACE-Is) and angiotensin II receptor blockers (ARBs) can help in slow progression of heart failure [45]. ACE-Is have been shown to be effective in fighting CTX caused by ANTs [46]. In addition to reducing oxidative stress, intracellular calcium overload, and fibrosis, ACE-Is also enhance mitochondrial respiration and cardiomyocyte metabolism, thereby protecting against ANT-related CTX [46]. Enalapril, captopril, and lisinopril can improve acute and chronic ANT-induced cardiotoxicity in experimental models. Patients treated with ANTs had a lower incidence of LV dysfunction compared to patients with placebo [47]. Candesartan modulates experimental cardiotoxicity induced by ANTs [48]. Doxorubicin-induced LV dysfunction in rats is prevented by pre and posttreatment with telmisartan [49]. Telmisartan affects the bioavailability of NO and inhibits the production of interleukin-6 (IL-6) and tumor necrosis factor-α (TNF-α) [50]. The cardiotoxic effects of epirubicin (EPI) were blunted by telmisartan in a small prospective study [51]. Telmisartan reversed myocardial dysfunction induced early by EPI and maintained normal systolic function for up to 18 months after treatment [52]. The cardioprotective effect of valsartan was evident in patients receiving ANTs [53]. Treatment of ANT-induced CTX may be improved with the combination of ACE-Is (enalapril) and β-blockers (carvedilol) [54]. ACE-Is and ARBs have been used as cardiopreventive agents in various clinical trials in patients receiving chemotherapy [55]. In the treatment of HF, ARBs, ACEIs, and β-blockers are necessary, but long-term studies are needed to assess whether ARBs have cardioprotective effects against the chronic aspects of cardiotoxicity caused by cancer treatments.

4 Newly identified GPCR agonist against anticancer-mediated cardiotoxicity

We discuss here several recently identified GPCR agonists that possess cardioprotective effects in both in vitro and in vivo preclinical models involving anticancer drugs (Fig. 1).

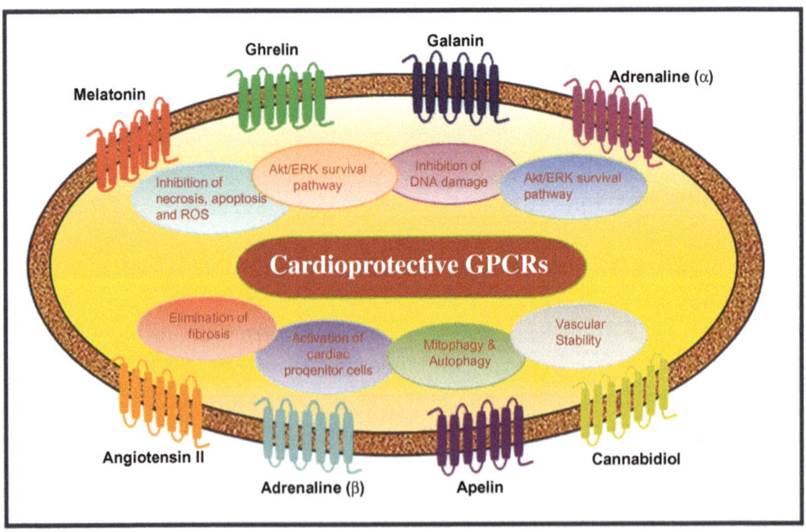

FIG. 1 Molecular mechanisms of cardioprotective GPCRs.

4.1 Alpha-adrenergic receptor

The alpha-adrenergic receptor (dabuzalgron) binds catecholamines (epinephrine and norepinephrine) through the receptors alpha 1 (α-AR1) and alpha 2 (α-AR2). When α-AR1 is coupled to Gαq type, inositol trisphosphate (IP3) and diacylglycerol (DAG) increase, and intracellular Ca^{2+} levels thus rise, causing smooth muscle contraction and glycogenolysis [56]. The α1-AR promotes cardiomyocyte survival, cardiac hypertrophy [57], and ischemic preconditioning [58] through activating phospholipase C and MAPK in cardiac tissues [59]. In mice, knockouts of α1A/α1B-adrenoceptors result in small hearts [60] and are more susceptible to pressure overload-induced HF. An analysis of a large-scale clinical trial supports the findings that doxazosin, an inhibitor of α-AR1 signaling, increased the risk of heart failure among hypertensive patients [61]. Inhibition of adenylyl cyclase by the α2-AR is accomplished through Gαi/o, thereby decreasing cAMP availability [62]. Additionally, it reduces neurotransmitter release and central vasodilation. Clinical trials of dabuzalgron against urinary incontinence have shown that it is a selective α1AR agonist [63]. In mice, dabuzalgron exhibited a strong cardioprotective effect against DOX-induced cardiotoxicity [64]. DOX-treated mice showed reduced ROS production, enhanced contractile function, and preserved the content of myocardial ATP by modulating mitochondrial function. It is not only MAPK1/2 pathways that are activated by α1-AR in order to produce cardioprotective effects [64], but this can also activate prosurvival pathways, including A kinase anchoring protein-Lbc (AKAP-Lbc) and its anchored protein kinase D1 (PKD1) in cardiotoxicity mouse models [65]. Hopefully, future studies will reveal if dabuzalgron can effectively treat HF caused by chemotherapy in cancer patients.

4.2 Adenosine receptor agonists

Degradation of ATP produces adenosine, a naturally occurring nucleoside. Adenosine receptors (ARs) A3, A2B, and A2A are endogenous agonists of four members of G-protein-coupled receptors [66]. Under circumstances of hypoxia and other stresses, the concentration of extracellular adenosine rises. In the context of cardiac ischemia, adenosine exhibits a significant cardioprotective effect by activating A1R and A3R [67]. A1R agonists are known to cause tachyphylaxis due to their off-target activation and desensitization of A1R, which produce cardiovascular adverse effects [68]. The selective partial agonist for A1AR thus promotes cardiac function without leading to atrioventricular blocks, bradycardia, or an adverse effect on blood pressure [69]. In response to pathological conditions, extracellular adenosine is increased along with local inflammatory cytokines, including interleukin-1β (IL-1β) and tumor necrosis factor (TNF), which results in a sustained increase in adenosine receptor expression on tissues and inflammatory cells [70]. A selective A3R agonist (Cl-IB-ME) is shown to alleviate bradycardia, raised serum creatine kinase levels, and cardiac histopathological abnormalities in mice treated with DOX. In vivo, Cl-IB-ME exerts cardioprotective effects by inhibiting ROS production and inflammation caused by DOX [71]. Activating A3ARs also decreases perioperative myocardial ischemia [72], protects ischemic cardiomyocytes by preconditioning [73], and induces ischemic tolerance by activating KATP channels [74]. It also demonstrated that A3AR agonists had cardioprotective effects in mouse cardiomyocytes lacking A3AR demonstrating an A3AR-mediated effect. The expression of A3AR in adult ventricular cardiomyocytes is very low compared to A1AR. Numerous clinical trials are being performed to assess the efficacy of two A3AR agonists [75].

4.3 Ghrelin receptor agonist

The 28-amino acid peptide, ghrelin, is synthesized as a preprohormone and proteolytically processed to create an endogenous ligand for the growth hormone secretagogue receptors (GHSRs). It was originally reported that this peptide was responsible for triggering growth hormone release, but recent studies have shown it to have many other physiological activities, such as regulating food intake and energy balance, as well as regulating lipid and glucose metabolism. There is substantial evidence both in animal models and in humans to support the wide expression of ghrelin receptors in cardiovascular tissues, indicating that ghrelin has beneficial effects on cardiovascular systems (Fig. 2). The effects of ghrelin on the heart were found in animal models to be beneficial in pathological conditions, to reduce the infarct size and to attenuate the

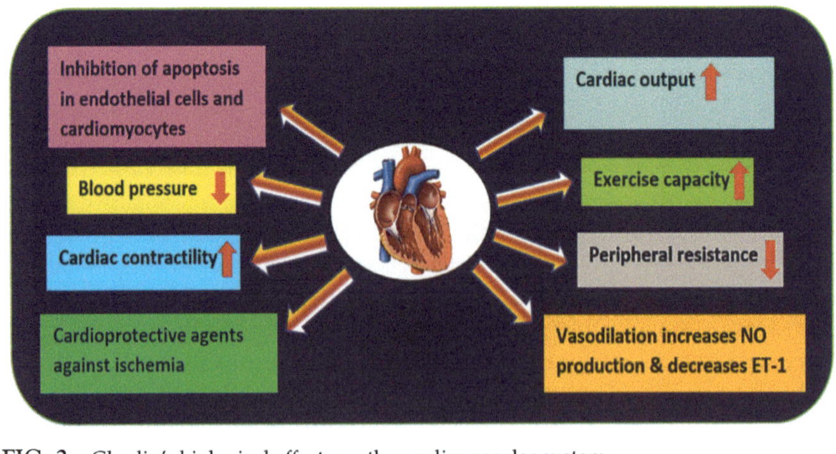

FIG. 2 Ghrelin's biological effects on the cardiovascular system.

ischemia-induced decrease in left ventricular function [76]. Specifically, Frascarelli et al. have found that the infarct size is significantly reduced with ghrelin administration according to triphenyltetrazolium chloride staining results in a rat ischemia-reperfusion model [77]. Studies in vitro indicated that ghrelin reduces endothelin-induced contractions by causing vasorelaxation in human mammary arteries [78]. The cardiovascular system has also been able to detect ghrelin as an antiapoptotic peptide. According to in vitro studies, ghrelin stimulates the proliferation of H9c2 cardiomyocytes and reduces doxorubicin-induced apoptosis on endothelial and cardiomyocyte cells.

Ghrelin also stimulates tyrosine phosphorylation and activates ERK-1/2 and Akt, signaling pathways that are necessary for its antiapoptotic effect [79]. According to Kola et al., ghrelin might modulate intracellular energy balance in an anaerobic state: Ghrelin might stimulate the 5′-AMP activated protein kinase (AMPK), which plays a key role in regulating energy balance in cells [80]. AMPK activity may be affected by ghrelin in the process of cardioprotection against cellular injury and ischemia-reperfusion injury [81].

4.4 Galanin receptor agonists

Galanin is an endogenous neuropeptide that regulates a variety of physiological processes in the central nervous system, including the central cardiovascular system. Recent experimental studies suggest that GalR1–3 receptor antagonists and agonists are potential therapeutic targets for myocardial ischemia and heart failure [82]. In addition to norepinephrine, this neuropeptide is part of the sympathetic family of co-transmitters, along with ATP and neuropeptide Y (NPY). The release of galanin from sympathetic nerves can decrease vagal neurotransmission [83]. Interestingly, galanin inhibits vagal bradycardia through GalR1 [84]. Naturally occurring N fragments of galanin [85] exhibit greater affinity for GalR2 than for GalR1 and GalR3, which limit acute myocardial infarction in rats [86]. Furthermore, natural galanin and GalR2 agonists have demonstrated to increase cell viability through suppressing caspase-3 and -9 activities in other cells when exposed to hypoxic conditions [87]. Rats exposed to GalR1–3 agonist [RAla14, His15]-galanin (2–15) show cardioprotective effects against DOX-mediated cardiac injury. When co-administered with DOX, this agonist inhibited the increase in plasma CK-MB activity, improved cardiac function parameters, and caused weight gain. In a study, a novel agonist of the galanin receptors GalR1–3 has been shown to attenuate DOX-induced cardiotoxicity [88]. Overall, galanin peptides alleviate cardiac dysfunctions induced by DOX through GalR1–3. Studying GalR1–3 agonist's impact on antitumor effect of DOX in cancer mice model is necessary.

4.5 Melatonin receptor agonists

The G-protein-coupled receptors (GPCRs) that bind melatonin are called melatonin receptors. In addition to being a neuroendocrine hormone, melatonin is recognized as an important component in controlling sleep-wake cycles. Melatonin affects the brain through two GPCRs, MT1 and MT2, which interact with receptors Gαi/Gαo and Gαq/Gα11 to cause antiadrenergic effects [89]. In addition to normalizing circadian rhythms and ameliorating nocturnal hypertension, melatonin directly affects blood vessels and the heart [90]. Additionally, they contribute to the functioning of the renin-angiotensin system [91] and the mitochondria [92]. A dose of melatonin enhances the antitumor activity of DOX and inhibits necrosis and apoptosis, as well as improving cardiac dysfunction induced by DOX in mice [93] and rats [94]. Several mechanisms have been implicated in its cardioprotective effect against DOX-induced cardiotoxicity, including antioxidant activity [95] and suppression of lipid peroxidation [96]. Studies have demonstrated that melatonin attenuates acute DOX cardiotoxicity by alleviating mitochondrial oxidative damage and apoptosis by activating AMPK, PGC1α [97], and sirtuins [98]. To exert its cardioprotective effect, melatonin must be taken in high doses to reach adequate subcellular concentrations. Ramelteon is a dual MT1 and MT2 melatonin receptor agonist used for insomnia. It exhibits strong cardioprotective properties in animal models of ischemic HF induced by coronary artery ligation [99], chronic intermittent hypoxia-induced heart failure [100], and isoproterenol-induced myocardial infarction [101]. However, no data exist on how ramelteon affects anticancer-mediated cardiotoxicity. Additionally, melatonin can enhance anthracycline's antitumor properties in animals [102]. As a result, anthracyclines and melatonin should be explored for their combined effects in cancer patients.

4.6 Apelin receptor agonists

An endogenous peptide called apelin binds to the G-protein-coupled receptor APJ. Due to its widespread distribution in tissues, apelin and its receptor are thought to play a role in a variety of physiological processes, including the regulation of blood pressure, body fluid balance, endocrine stress response, cardiomyocyte contractility, energy metabolism, and angiogenesis. Furthermore, this peptide contributes to pathological conditions, including heart failure, obesity, and diabetes. The mature apelin, apelin-36, and shorter forms (apelin-17, -12, and -13) arise from the cleavage of pre-pro-apelin. In vitro, apelin itself can also be cleaved by the angiotensin-converting enzyme 2 (ACE2) [103]. APJ and apelin are both expressed in cardiac myocytes during normal conditions. As a positive inotropic agent, apelin lowers arterial blood pressure [104], induces arterial vasodilation [105], and improves cardiac output in vitro [106].

Further, APJ-knockout mice demonstrated greater heart damage, including impaired contractility functions and lower survival rates when treated with DOX compared to wild-type mice [107]. Alternatively, APJ-overexpressing H9c2 cardiomyocytes are protected by apelin against DOX-mediated cell death. As a result of all these findings, it has been inferred that suppression of APJ expression may cause worsening of cardiac effects caused by DOX. DOX-treated hearts may exhibit partially suppressed protective signaling caused by impairments of the apelin-APJ system [107]. Activating PI3K/Akt signaling and MAPKs via ApeLIN-13 pretreatment suppresses cisplatin-induced cardiotoxicity in mice heart in vitro and in vivo by inhibiting apoptosis [108]. Cardioprotection occurs in vivo by attenuating ROS and superoxide anion accumulation, inhibiting DNA damage, suppressing PARP and caspases, and improving angiogenesis. Further, apelin concentration is altered in patients with various cardiovascular disorders, including diabetes, coronary artery disease, and left ventricular hypertrophy, demonstrating that apelin peptides are useful as biomarkers of cardiovascular disease.

4.7 Cannabidiol

In plants, CBD (cannabidiol) is by far the most prevalent and non-psychoactive cannabinoid (phytocannabinoid). The cannabinoid 1 receptor (CB1R) and the cannabinoid 2 receptor (CB2R) are antagonistic at nanomolar concentrations, while they are agonists and inverse agonists at micromolar concentrations [109]. In addition to activating the TRPV1 channel and several GPCRs, cannabidiol can also bind to the putative Abn-CBD receptor, α1-adrenoceptors, 5-HT1A receptors, and μ opioid receptors [110]. The cardioprotective effects of cannabidiol were demonstrated in several animal models of myocardial ischemic reperfusion injury [111] and acute myocardial infarction [112]. The drug is also effective in improving heart function in diabetics with cardiomyopathy. In rats and mice, CBD protects heart against damage caused by DOX [113]. Cardiomyopathy can be improved (i) by inhibiting ROS/RNS growth, (ii) by promoting mitochondrial function and biogenesis, (iii) by enhancing cell survival, and (iv) by decreasing myocardial inflammation. CB1 and CB2 signaling pathways were not clearly demonstrated in these studies. According to recently published data, the CB1R and CB2R receptors are oppositely influenced. Clozapine toxicity is effectively reduced by both CB1R antagonists and CB2R agonists [114]. Therefore, CB1R antagonists attenuate DOX-induced cardiotoxicity and cerebral infarction in the cortex [115]. On the other hand, CB2R agonists JWH-133 and AM 1241 were shown to reduce quetiapine cardiotoxicity [116]. Furthermore, cannabidiol has been shown to perform synergistic effects with anticancer drugs in the use of cardioprotective agents in multiple cancer cell lines and mice models [117]. Cannabinoid HU-331 has a greater potency than DOX and is less cardiotoxic [118]. Cannabidiol has acquired FDA orphan drug designation by Insys Therapeutics for the treatment of multiple gliomas and childhood epilepsy.

5 Conclusion

One of the most dreaded side effects of cancer chemotherapy, as well as cancer chemoprevention, is cardiotoxicity. To reduce cardiotoxicity-related morbidity and mortality, it is important to identify patients at higher risk. In addition, current screening techniques are time-consuming and ineffective. A standard cancer treatment protocol that takes cardiologic conditions into account is currently lacking and needs to be developed. In patients with reduced left ventricular function, it is important to carefully monitor their symptoms of heart failure throughout their treatment regime with the addition of angiotensin-converting enzyme inhibitors/ angiotensin receptor blockers or β-blockers. Additionally, there is a strong need to identify new compounds that can prevent cardiotoxicity because many chemotherapeutic agents commonly used in the treatment of cancer have debilitating effects on the cardiovascular system. Validating the potential cardioprotective properties of GPCR ligands will require identifying their signaling (e.g., G-protein- or arrestin-mediated pathways) and functional roles in both cancer cells and cardiac cells. It is also important to investigate whether these cardioprotective ligands interfere with the chemotherapeutic's antitumor effect. Induced pluripotent stem cell-derived cardiomyocytes (hiPSC-CMs), iPSC-CM-derived 3D cultures, and organoids offer human-based alternatives for studying cardiotoxicity and cardioprotection [119]. The systems could also serve as platforms for developing enhanced medications. The efficacy of GPCR ligands and their side effects can be evaluated in hiPSC-CMs and organoids, and thus, increased efficacy can be achieved. Furthermore, the majority of data on cardioprotective GPCR ligands have come from small animal models of cardiotoxicity and cancer cell-derived xenograft mouse models. Therefore, in order to apply these findings to human trials, further studies on larger animals are needed to discover whether or not they are effective and whether any adverse effects arise. It is noteworthy that certain cancer cell types may retain their ability to express GPCRs, which may serve as novel biomarkers or as potential therapeutic targets. As a final note, the development and promotion of clinical care models to improve long-term outcomes for cancer survivors and patients will greatly benefit from the multidisciplinary approach to cardio-oncology service that involves close collaboration among oncologists, cardiologists, and other healthcare professionals.

Conflict of interest

The authors state no conflict of interests.

Ethics approval

Not applicable.

Consent for publication

All authors agreed.

References

[1] J.V. McGowan, R. Chung, A. Maulik, I. Piotrowska, J.M. Walker, D.M. Yellon, Anthracycline chemotherapy and cardiotoxicity, Cardiovasc. Drugs Ther. 31 (1) (2017) 63–75.
[2] M.S. Ewer, S.M. Ewer, Troponin I provides insight into cardiotoxicity and the anthracycline-trastuzumab interaction, J. Clin. Oncol. 28 (25) (2010) 3901–3904.
[3] S.A. Crone, Y.Y. Zhao, L. Fan, et al., ErbB2 is essential in the prevention of dilated cardiomyopathy, Nat. Med. 8 (5) (2002) 459–465.
[4] T. Chen, T. Xu, Y. Li, et al., Risk of cardiac dysfunction with trastuzumab in breast cancer patients: a meta-analysis, Cancer Treat. Rev. 37 (4) (2011) 312–320.
[5] N.E. Papaioannou, O.V. Beniata, P. Vitsos, O. Tsitsilonis, P. Samara, Harnessing the immune system to improve cancer therapy, Ann. Transl. Med. 4 (14) (2016) 261.
[6] H. Ledford, Cocktails for cancer with a measure of immunotherapy, Nature 532 (7598) (2016) 162–164.
[7] K. Tajiri, M. Ieda, Cardiac complications in immune checkpoint inhibition therapy, Front. Cardiovasc. Med. 6 (2019) 3.
[8] Y.L. Lyu, J.E. Kerrigan, C.P. Lin, et al., Topoisomerase II beta mediated DNA double-strand breaks: implications in doxorubicin cardiotoxicity and prevention by dexrazoxane, Cancer Res. 67 (18) (2007) 8839–8846.
[9] US Food and Drug Administration, Orphan Drug Designations and Approvals, Available from: https://www.accessdata.fda.gov/scripts/opdlisting/oopd/detailedIndex.cfm?cfgridkey=441314.
[10] M.L. Hensley, L.M. Schuchter, C. Lindley, et al., American Society of Clinical Oncology clinical practice guidelines for the use of chemotherapy and radiotherapy protectants, J. Clin. Oncol. 17 (10) (1999) 3333–3355.
[11] S.E. Lipshultz, N. Rifai, V.M. Dalton, et al., The effect of dexrazoxane on myocardial injury in doxorubicin-treated children with acute lymphoblastic leukemia, N. Engl. J. Med. 351 (12) (2004) 145–153.
[12] J.H. Doroshow, G.Y. Locker, I. Ifrim, C.E. Myers, Prevention of doxorubicin cardiac toxicity in the mouse by N-acetylcysteine, J. Clin. Invest. 68 (4) (1981) 1053–1064.
[13] F. D'Agostini, M. Bagnasco, D. Giunciuglio, A. Albini, S. De Flora, Inhibition by oral N-acetylcysteine of doxorubicin-induced clastogenicity and alopecia, and prevention of primary tumors and lung micrometastases in mice, Int. J. Oncol. 13 (2) (1998) 217–224.
[14] A.R. Dresdale, L.H. Barr, R.O. Bonow, et al., Prospective randomized study of the role of N-acetyl cysteine in reversing doxorubicin-induced cardiomyopathy, Am. J. Clin. Oncol. 5 (6) (1982) 657–663.
[15] H.S. Shoukry, H.I. Ammar, L.A. Rashed, et al., Prophylactic supplementation of resveratrol is more effective than its therapeutic use against doxorubicin induced cardiotoxicity, PLoS One 12 (7) (2017) e0181535.

[16] B.D. Sahu, J.M. Kumar, M. Kuncha, R.M. Borkar, R. Srinivas, R. Sistla, Baicalein alleviates doxorubicin-induced cardiotoxicity via suppression of myocardial oxidative stress and apoptosis in mice, Life Sci. 144 (2016) 8–18.
[17] S. Cortassa, B. O'Rourke, M.A. Aon, Redox-optimized ROS balance and the relationship between mitochondrial respiration and ROS, Biochim. Biophys. Acta 1837 (2) (2014) 287–295.
[18] A.M. Agha, A.A. El-Fattah, H.H. Al-Zuhair, A.C. Al-Rikabi, Chemopreventive effect of Ginkgo biloba extract against benzo(a)pyrene-induced forestomach carcinogenesis in mice: amelioration of doxorubicin cardiotoxicity, J. Exp. Clin. Cancer Res. 20 (1) (2001) 39–50.
[19] S.D. Ray, D. Patel, V. Wong, D. Bagchi, In vivo protection of DNA damage associated apoptotic and necrotic cell deaths during acetaminophen-induced nephrotoxicity, amiodarone-induced lung toxicity and doxorubicin-induced cardiotoxicity by a novel IH636 grape seed proanthocyanidin extract, Res. Commun. Mol. Pathol. Pharmacol. 107 (1–2) (2000) 137–166.
[20] E.D. Danz, J. Skramsted, N. Henry, J.A. Bennett, R.S. Keller, Resveratrol prevents doxorubicin cardiotoxicity through mitochondrial stabilization and the Sirt1 pathway, Free Radic. Biol. Med. 46 (12) (2009) 1589–1597.
[21] E. Tatlidede, O. Sehirli, A. Velioğlu-Oğünc, et al., Resveratrol treatment protects against doxorubicin-induced cardiotoxicity by alleviating oxidative damage, Free Radic. Res. 43 (3) (2009) 195–205.
[22] R. Reddy, G. Chahoud, J.L. Mehta, Modulation of cardiovascular remodeling with statins: fact or fiction? Curr. Vasc. Pharmacol. 3 (1) (2005) 69–79.
[23] A. Riad, S. Bien, D. Westermann, et al., Pretreatment with statin attenuates the cardiotoxicity of doxorubicin in mice, Cancer Res. 69 (2) (2009) 695–699.
[24] S. Seicean, A. Seicean, J.C. Plana, G.T. Budd, T.H. Marwick, Effect of statin therapy on the risk for incident heart failure in patients with breast cancer receiving anthracycline chemotherapy: an observational clinical cohort study, J. Am. Coll. Cardiol. 60 (23) (2012) 2384–2390.
[25] Z. Acar, A. Kale, M. Turgut, et al., Efficiency of atorvastatin in the protection of anthracycline-induced cardiomyopathy, J. Am. Coll. Cardiol. 58 (9) (2011) 988–989.
[26] N. Sattar, D. Preiss, H.M. Murray, et al., Statins and risk of incident diabetes: a collaborative meta-analysis of randomised statin trials, Lancet 375 (9716) (2010) 735–742.
[27] C. Henninger, G. Fritz, Statins in anthracycline-induced cardiotoxicity: Rac and Rho, and the heartbreakers, Cell Death Dis. 8 (1) (2017) e2564.
[28] M.C. Lagerström, H.B. Schiöth, Structural diversity of G protein-coupled receptors and significance for drug discovery, Nat. Rev. Drug Discov. 7 (4) (2008) 339–357.
[29] R.J. Lefkowitz, S.K. Shenoy, Transduction of receptor signals by beta-arrestins, Science 308 (5721) (2005) 512–517.
[30] X.E. Zhou, K. Melcher, H.E. Xu, Understanding the GPCR biased signaling through G protein and arrestin complex structures, Curr. Opin. Struct. Biol. 45 (2017) 150–159.
[31] A.K. Shukla, K. Xiao, R.J. Lefkowitz, Emerging paradigms of β-arrestin-dependent seven transmembrane receptor signaling, Trends Biochem. Sci. 36 (9) (2011) 457–469.
[32] W. Zhu, A.Y. Woo, Y. Zhang, C.M. Cao, R.P. Xiao, β-Adrenergic receptor subtype signaling in the heart: from bench to the bedside, Curr. Top. Membr. 67 (2011) 191–204.
[33] M. Fu, M. Matoba, Q.M. Liang, K.G. Sjögren, A. Hjalmarson, Properties of G-protein modulated receptor-adenylyl cyclase system in myocardium of spontaneously hypertensive rats treated with adriamycin, Int. J. Cardiol. 44 (1994) 9–18.
[34] K. Nakamura, K. Kusano, Y. Nakamura, et al., Carvedilol decreases elevated oxidative stress in human failing myocardium, Circulation 105 (24) (2002) 2867–2871.
[35] H. Matsui, I. Morishima, Y. Numaguchi, Y. Toki, K. Okumura, T. Hayakawa, Protective effects of carvedilol against doxorubicin-induced cardiomyopathy in rats, Life Sci. 65 (12) (1999) 1265–1274.

[36] P. Spallarossa, S. Garibaldi, P. Altieri, et al., Carvedilol prevents doxorubicin-induced free radical release and apoptosis in cardiomyocytes in vitro, J. Mol. Cell. Cardiol. 37 (4) (2004) 837–846.
[37] D.L. Santos, A.J. Moreno, R.L. Leino, M.K. Froberg, K.B. Wallace, Carvedilol protects against doxorubicin-induced mitochondrial cardiomyopathy, Toxicol. Appl. Pharmacol. 185 (3) (2002) 218–227.
[38] N. Kalay, E. Basar, I. Ozdogru, et al., Protective effects of carvedilol against anthracycline-induced cardiomyopathy, J. Am. Coll. Cardiol. 48 (11) (2006) 2258–2262.
[39] C.G. Tocchetti, M. Molinaro, T. Angelone, V. Lionetti, R. Madonna, F. Mangiacapra, et al., Nitroso-redox balance and modulation of basal myocardial function: an update from the Italian society of cardiovascular research (SIRC), Curr. Drug Targets 16 (8) (2015) 895–903.
[40] S. Seicean, A. Seicean, N. Alan, J.C. Plana, G.T. Budd, T.H. Marwick, Cardioprotective effect of β-adrenoceptor blockade in patients with breast cancer undergoing chemotherapy: follow-up study of heart failure, Circ. Heart Fail. 6 (3) (2013) 420–426.
[41] E. Pituskin, J.R. Mackey, S. Koshman, et al., Multidisciplinary approach to novel therapies in cardio-oncology research (MANTICORE 101-breast): a randomized trial for the prevention of Trastuzumab-associated cardiotoxicity, J. Clin. Oncol. 35 (8) (2017) 870–877.
[42] G. Gulati, S.L. Heck, A.H. Ree, et al., Prevention of cardiac dysfunction during adjuvant breast cancer therapy (PRADA): a 2 × 2 factorial, randomized, placebo-controlled, double-blind clinical trial of candesartan and metoprolol, Eur. Heart J. 37 (21) (2016) 1671–1680.
[43] P. Sysa-Shah, C.G. Tocchetti, M. Gupta, et al., Bidirectional cross-regulation between ErbB2 and β-adrenergic signalling pathways, Cardiovasc. Res. 109 (3) (2016) 358–373.
[44] I. Andreadou, M. Papaefthimiou, A. Zira, et al., Metabonomic identification of novel biomarkers in doxorubicin cardiotoxicity and protective effect of the natural antioxidant oleuropein, NMR Biomed. 22 (6) (2009) 585–592.
[45] P. Ponikowski, A.A. Voors, S.D. Anker, et al., 2016 ESC guidelines for the diagnosis and treatment of acute and chronic heart failure: the task force for the diagnosis and treatment of acute and chronic heart failure of the European Society of Cardiology (ESC). Developed with the special contribution of the Heart Failure Association (HFA) of the ESC, Eur. J. Heart Fail. 18 (8) (2016) 891–975.
[46] M.A. Abd El-Aziz, A.I. Othman, M. Amer, M.A. El-Missiry, Potential protective role of angiotensin-converting enzyme inhibitors captopril and enalapril against adriamycin-induced acute cardiac and hepatic toxicity in rats, J. Appl. Toxicol. 21 (6) (2001) 469–473.
[47] D. Cardinale, A. Colombo, G. Bacchiani, et al., Early detection of anthracycline cardiotoxicity and improvement with heart failure therapy, Circulation 131 (22) (2015) 1981–1988.
[48] M. Soga, F.A. Kamal, K. Watanabe, M. Ma, S. Palaniyandi, P. Prakash, P. Veeraveedu, S. Mito, M. Kunisaki, H. Tachikawa, M. Kodama, Y. Aizawa, Effects of angiotensin II receptor blocker (candesartan) in daunorubicin-induced cardiomyopathic rats, Int. J. Cardiol. 110 (3) (2006) 378–385.
[49] M. Iqbal, K. Dubey, T. Anwer, A. Ashish, K.K. Pillai, Protective effects of telmisartan against acute doxorubicin-induced cardiotoxicity in rats, Pharmacol. Rep. 60 (2008) 382–390.
[50] S. Yamagishi, M. Takeuchi, Telmisartan is a promising cardiometabolic sartan due to its unique PPAR-gamma-inducing property, Med. Hypotheses 64 (3) (2005) 476–478.
[51] C. Cadeddu, A. Piras, G. Mantovani, et al., Protective effects of the angiotensin II receptor blocker telmisartan on epirubicin-induced inflammation, oxidative stress, and early ventricular impairment, Am. Heart J. 160 (3) (2010) 487.e1–487.e4877.
[52] M. Dessì, A. Piras, C. Madeddu, et al., Long-term protective effects of the angiotensin receptor blocker telmisartan on epirubicin-induced inflammation, oxidative stress and myocardial dysfunction, Exp. Ther. Med. 2 (5) (2011) 1003–1009.

[53] H. Nakamae, K. Tsumura, Y. Terada, et al., Notable effects of angiotensin II receptor blocker, valsartan, on acute cardiotoxic changes after standard chemotherapy with cyclophosphamide, doxorubicin, vincristine, and prednisolone, Cancer 104 (11) (2005) 2492–2498.

[54] X. Bosch, M. Rovira, M. Sitges, et al., Enalapril and carvedilol for preventing chemotherapy-induced left ventricular systolic dysfunction in patients with malignant hemopathies: the OVERCOME trial (preventiOn of left ventricular dysfunction with Enalapril and carvedilol in patients submitted to intensive ChemOtherapy for the treatment of malignant hEmopathies), J. Am. Coll. Cardiol. 61 (23) (2013) 2355–2362.

[55] S.L. Lim, C.S. Lam, V.F. Segers, D.L. Brutsaert, G.W. De Keulenaer, Cardiac endothelium-myocyte interaction: clinical opportunities for new heart failure therapies regardless of ejection fraction, Eur. Heart J. 36 (2015) 2050–2060.

[56] G.R. Biazi, I.G. Frasson, D.R. Miksza, et al., Decreased hepatic response to glucagon, adrenergic agonists, and cAMP in glycogenolysis, gluconeogenesis, and glycolysis in tumor-bearing rats, J. Cell. Biochem. 119 (9) (2018) 7300–7309.

[57] A. Sabri, E. Pak, S.A. Alcott, B.A. Wilson, S.F. Steinberg, Coupling function of endogenous alpha (1)- and beta-adrenergic receptors in mouse cardiomyocytes, Circ. Res. 86 (2000) 1047–1053.

[58] R.K. Kudej, Y.T. Shen, A.P. Peppas, et al., Obligatory role of cardiac nerves and alpha1-adrenergic receptors for the second window of ischemic preconditioning in conscious pigs, Circ. Res. 99 (11) (2006) 1270–1276.

[59] C.D. Wright, Q. Chen, N.L. Baye, et al., Nuclear alpha1-adrenergic receptors signal activated ERK localization to caveolae in adult cardiac myocytes, Circ. Res. 103 (9) (2008) 992–1000.

[60] T.D. O'Connell, S. Ishizaka, A. Nakamura, et al., The alpha(1A/C)- and alpha(1B)-adrenergic receptors are required for physiological cardiac hypertrophy in the double-knockout mouse, J. Clin. Invest. 111 (11) (2003) 1783–1791.

[61] T.D. O'Connell, R.C. Block, S.P. Huang, G.C. Shearer, ω3-polyunsaturated fatty acids for heart failure: effects of dose on efficacy and novel signaling through free fatty acid receptor 4, J. Mol. Cell. Cardiol. 103 (2017) 74–92.

[62] C.A. Durkee, A. Covelo, J. Lines, P. Kofuji, J. Aguilar, A. Araque, $G_{i/o}$ protein-coupled receptors inhibit neurons but activate astrocytes and stimulate gliotransmission, Glia 67 (6) (2019) 1076–1093.

[63] D.M. Musselman, A.P. Ford, D.J. Gennevois, M.L. Harbison, A.L. Laurent, A.S. Mokatrin, et al., A randomized crossover study to evaluate Ro 115-1240, a selective alpha1A/1L-adrenoceptor partial agonist in women with stress urinary incontinence, BJU Int. 93 (2004) 78–83.

[64] J. Beak, W. Huang, J.S. Parker, et al., An oral selective alpha-1A adrenergic receptor agonist prevents doxorubicin cardiotoxicity, JACC Basic Transl. Sci. 2 (1) (2017) 39–53.

[65] S. Caso, D. Maric, M. Arambasic, S. Cotecchia, D. Diviani, AKAP-Lbc mediates protection against doxorubicin-induced cardiomyocyte toxicity, Biochim. Biophys. Acta, Mol. Cell Res. 1864 (12) (2017) 2336–2346.

[66] Z.G. Gao, K.A. Jacobson, Emerging adenosine receptor agonists: an update, Expert Opin. Emerg. Drugs 16 (4) (2011) 597–602.

[67] V.J. McIntosh, R.D. Lasley, Adenosine receptor-mediated cardioprotection: are all 4 subtypes required or redundant? J. Cardiovasc. Pharmacol. Ther. 17 (1) (2012) 21–33.

[68] P.K. Deb, Therapeutic potentials of adenosine receptors: the state of the art, Curr. Pharm. Des. 25 (26) (2019) 2789–2791.

[69] D. Meibom, B. Albrecht-Küpper, N. Diedrichs, et al., Neladenoson bialanate hydrochloride: a prodrug of a partial adenosine A_1 receptor agonist for the chronic treatment of heart diseases, ChemMedChem 12 (10) (2017) 728–737.

[70] U. Schingnitz, K. Hartmann, C.F. Macmanus, et al., Signaling through the A2B adenosine receptor dampens endotoxin-induced acute lung injury, J. Immunol. 184 (9) (2010) 5271–5279.

[71] A. Galal, W.M. El-Bakly, E.N.A. Al Haleem, E. El-Demerdash, Selective A3 adenosine receptor agonist protects against doxorubicin-induced cardiotoxicity, Cancer Chemother. Pharmacol. 77 (2016) 309–322.
[72] M.P. DeNinno, H. Masamune, L.K. Chenard, K.J. DiRico, C. Eller, J.B. Etienne, J.E. Tickner, S.P. Kennedy, D.R. Knight, J. Kong, J.J. Oleynek, W.R. Tracey, R.J. Hill, 3′-Aminoadenosine-5′-uronamides: discovery of the first highly selective agonist at the human adenosine A_3 receptor, J. Med. Chem. 46 (2003) 353–355.
[73] R.D. Lasley, Adenosine receptor-mediated Cardioprotection-current limitations and future directions, Front. Pharmacol. 9 (2018) 310.
[74] T.C. Wan, A. Tampo, W.M. Kwok, J.A. Auchampach, Ability of CP-532,903 to protect mouse hearts from ischemia/reperfusion injury is dependent on expression of A_3 adenosine receptors in cardiomyocytes, Biochem. Pharmacol. 163 (2019) 21–31.
[75] K.A. Jacobson, D.K. Tosh, S. Jain, Z.G. Gao, Historical and current adenosine receptor agonists in preclinical and clinical development, Front. Cell. Neurosci. 13 (2019) 124.
[76] N. Nagaya, M. Uematsu, M. Kojima, et al., Elevated circulating level of ghrelin in cachexia associated with chronic heart failure: relationships between ghrelin and anabolic/catabolic factors, Circulation 104 (17) (2001) 2034–2038.
[77] S. Frascarelli, S. Ghelardoni, S. Ronca-Testoni, R. Zucchi, Effect of ghrelin and synthetic growth hormone secretagogues in normal and ischemic rat heart, Basic Res. Cardiol. 98 (6) (2003) 401–405.
[78] K.E. Wiley, A.P. Davenport, Comparison of vasodilators in human internal mammary artery: ghrelin is a potent physiological antagonist of endothelin-1, Br. J. Pharmacol. 136 (8) (2002) 1146–1152.
[79] G. Baldanzi, N. Filigheddu, S. Cutrupi, et al., Ghrelin and des-acyl ghrelin inhibit cell death in cardiomyocytes and endothelial cells through ERK1/2 and PI 3-kinase/AKT, J. Cell Biol. 159 (6) (2002) 1029–1037.
[80] B. Kola, E. Hubina, S.A. Tucci, et al., Cannabinoids and ghrelin have both central and peripheral metabolic and cardiac effects via AMP-activated protein kinase, J. Biol. Chem. 280 (26) (2005) 25196–25201.
[81] E.A. García, M. Korbonits, Ghrelin and cardiovascular health, Curr. Opin. Pharmacol. 6 (2) (2006) 142–147.
[82] A. Chen, M. Li, L. Song, et al., Effects of the galanin receptor antagonist M40 on cardiac function and remodeling in rats with heart failure, Cardiovasc. Ther. 33 (5) (2015) 288–293.
[83] E.K. Potter, M.A. Smith-White, Galanin modulates cholinergic neurotransmission in the heart, Neuropeptides 39 (3) (2005) 345–348.
[84] N. Herring, J. Cranley, M.N. Lokale, et al., The cardiac sympathetic co-transmitter galanin reduces acetylcholine release and vagal bradycardia: implications for neural control of cardiac excitability, J. Mol. Cell. Cardiol. 52 (3) (2012) 667–676.
[85] K.E. Webling, J. Runesson, T. Bartfai, U. Langel, Galanin receptors and ligands, Front. Endocrinol. (Lausanne) 3 (2012) 146.
[86] L.I. Serebryakova, I.M. Studneva, M.V. Ovchinnikov, et al., Kardiometabolicheskaia éffektivnost' i toksikologicheskaia kharakteristika farmakologicheskogo agonista retseptorov galanina [Cardiometabolic efficacy and toxicological evaluation of a pharmacological galanin receptor agonist], Biomed Khim. 65 (3) (2019) 231–238.
[87] J. Sun, S. Xu, H. Li, L. Li, Z.D. Xu, Galanin protects rat cortical astrocyte from oxidative stress: involvement of GalR2 and pERK1/2 signal pathway, Mediat. Inflamm. 2019 (2019) 2716028.
[88] I.M. Studneva, M.E. Palkeeva, O.M. Veselova, et al., Zashchitnoe deĭstvie modifitsirovannogo fragmenta galanina pri serdechnoĭ nedostatochnosti u krys, vyzvannoĭ doksorubitsinom (Protective action of a modified fragment of galanine in rats with doxorubicin-induced heart failure), Biomed Khim. 65 (1) (2019) 51–56.

[89] O.C. Baltatu, S. Senar, L.A. Campos, J. Cipolla-Neto, Cardioprotective melatonin: translating from proof-of-concept studies to therapeutic use, Int. J. Mol. Sci. 20 (18) (2019) 4342.
[90] M. Stroethoff, I. Christoph, F. Behmenburg, et al., Melatonin receptor agonist Ramelteon reduces ischemia-reperfusion injury through activation of mitochondrial potassium channels, J. Cardiovasc. Pharmacol. 72 (2) (2018) 106–111.
[91] L.A. Campos, J. Cipolla-Neto, F.G. Amaral, L.C. Michelini, M. Bader, O.C. Baltatu, The angiotensin-melatonin axis, Int. J. Hypertens. 2013 (2013) 521783.
[92] O.C. Baltatu, F.G. Amaral, L.A. Campos, J. Cipolla-Neto, Melatonin, mitochondria and hypertension, Cell. Mol. Life Sci. 74 (2017) 3955–3964.
[93] X. Liu, Z. Chen, C.C. Chua, et al., Melatonin as an effective protector against doxorubicin-induced cardiotoxicity, Am. J. Physiol. Heart Circ. Physiol. 283 (1) (2002) H254–H263.
[94] H. Aygun, S.S. Gul, Cardioprotective effect of melatonin and agomelatine on doxorubicin-induced cardiotoxicity in a rat model: an electrocardiographic, scintigraphic and biochemical study, Bratisl. Lek. Listy 120 (4) (2019) 249–255.
[95] M.F. Xu, S. Ho, Z.M. Qian, P.L. Tang, Melatonin protects against cardiac toxicity of doxorubicin in rat, J. Pineal Res. 31 (4) (2001) 301–307.
[96] E. Balli, U.O. Mete, A. Tuli, O. Tap, M. Kaya, Effect of melatonin on the cardiotoxicity of doxorubicin, Histol. Histopathol. 19 (4) (2004) 1101–1108.
[97] D. Liu, Z. Ma, S. Di, et al., AMPK/PGC1α activation by melatonin attenuates acute doxorubicin cardiotoxicity via alleviating mitochondrial oxidative damage and apoptosis, Free Radic. Biol. Med. 129 (2018) 59–72.
[98] J. Govender, B. Loos, E. Marais, A.M. Engelbrecht, Melatonin improves cardiac and mitochondrial function during doxorubicin-induced cardiotoxicity: a possible role for peroxisome proliferator-activated receptor gamma coactivator 1-alpha and sirtuin activity? Toxicol. Appl. Pharmacol. 358 (2018) 86–101.
[99] H. Pei, J. Du, X. Song, et al., Melatonin prevents adverse myocardial infarction remodeling via Notch1/Mfn2 pathway, Free Radic. Biol. Med. 97 (2016) 408–417.
[100] H.M. Yeung, M.W. Hung, C.F. Lau, M.L. Fung, Cardioprotective effects of melatonin against myocardial injuries induced by chronic intermittent hypoxia in rats, J. Pineal Res. 58 (1) (2015) 12–25.
[101] D. Mukherjee, S.G. Roy, A. Bandyopadhyay, et al., Melatonin protects against isoproterenol-induced myocardial injury in the rat: antioxidative mechanisms, J. Pineal Res. 48 (3) (2010) 251–262.
[102] C. Kim, N. Kim, H. Joo, et al., Modulation by melatonin of the cardiotoxic and antitumor activities of adriamycin, J. Cardiovasc. Pharmacol. 46 (2) (2005) 200–210.
[103] T. Sato, T. Suzuki, H. Watanabe, et al., Apelin is a positive regulator of ACE2 in failing hearts, J. Clin. Invest. 123 (12) (2013) 5203–5211.
[104] I. Szokodi, P. Tavi, G. Földes, et al., Apelin, the novel endogenous ligand of the orphan receptor APJ, regulates cardiac contractility, Circ. Res. 91 (5) (2002) 434–440.
[105] K. Tatemoto, K. Takayama, M.X. Zou, et al., The novel peptide apelin lowers blood pressure via a nitric oxide-dependent mechanism, Regul. Pept. 99 (2–3) (2001) 87–92.
[106] A.G. Japp, N.L. Cruden, D.A. Amer, et al., Vascular effects of apelin in vivo in man, J. Am. Coll. Cardiol. 52 (11) (2008) 908–913.
[107] J. Hamada, A. Baasanjav, N. Ono, et al., Possible involvement of downregulation of the apelin-APJ system in doxorubicin-induced cardiotoxicity, Am. J. Physiol. Heart Circ. Physiol. 308 (8) (2015) H931–H941.
[108] P. Zhang, L.H. Yi, G.Y. Meng, H.Y. Zhang, H.H. Sun, L.Q. Cui, Apelin-13 attenuates cisplatin-induced cardiotoxicity through inhibition of ROS-mediated DNA damage and regulation of MAPKs and AKT pathways, Free Radic. Res. 51 (5) (2017) 449–459.
[109] M. Rajesh, P. Mukhopadhyay, G. Haskó, L. Liaudet, K. Mackie, P. Pacher, Cannabinoid-1 receptor activation induces reactive oxygen species-dependent and -independent

mitogen-activated protein kinase activation and cell death in human coronary artery endothelial cells, Br. J. Pharmacol. 160 (3) (2010) 688–700.
[110] R.G. Pertwee, Ligands that target cannabinoid receptors in the brain: from THC to anandamide and beyond, Addict. Biol. 13 (2) (2008) 147–159.
[111] R. Durst, H. Danenberg, R. Gallily, et al., Cannabidiol, a nonpsychoactive Cannabis constituent, protects against myocardial ischemic reperfusion injury, Am. J. Physiol. Heart Circ. Physiol. 293 (6) (2007) H3602–H3607.
[112] S.K. Walsh, C.Y. Hepburn, K.A. Kane, C.L. Wainwright, Acute administration of cannabidiol in vivo suppresses ischaemia-induced cardiac arrhythmias and reduces infarct size when given at reperfusion, Br. J. Pharmacol. 160 (5) (2010) 1234–1242.
[113] E. Hao, P. Mukhopadhyay, Z. Cao, et al., Cannabidiol protects against doxorubicin-induced cardiomyopathy by modulating mitochondrial function and biogenesis, Mol. Med. 21 (1) (2015) 38–45.
[114] D.W. Williams, L.C. Askew, E. Jones, J.E. Clements, CCR2 signaling selectively regulates IFN-α: role of β-arrestin 2 in IFNAR1 internalization, J. Immunol. 202 (1) (2019) 105–118.
[115] S.Y. Lim, S.M. Davidson, D.M. Yellon, C.C. Smith, The cannabinoid CB1 receptor antagonist, rimonabant, protects against acute myocardial infarction, Basic Res. Cardiol. 104 (6) (2009) 781–792.
[116] X. Li, Z. Peng, Y. Zhou, et al., Quetiapine induces myocardial necroptotic cell death through bidirectional regulation of cannabinoid receptors, Toxicol. Lett. 313 (2019) 77–90.
[117] P. Massi, M. Solinas, V. Cinquina, D. Parolaro, Cannabidiol as potential anticancer drug, Br. J. Clin. Pharmacol. 75 (2) (2013) 303–312.
[118] N.M. Kogan, M. Schlesinger, M. Peters, G. Marincheva, R. Beeri, R. Mechoulam, A cannabinoid anticancer quinone, HU-331, is more potent and less cardiotoxic than doxorubicin: a comparative in vivo study, J. Pharmacol. Exp. Ther. 322 (2) (2007) 646–653.
[119] P. Nair, M. Prado, I. Perea-Gil, I. Karakikes, Concise review: precision matchmaking: induced pluripotent stem cells meet cardio-oncology, Stem Cells Transl. Med. 8 (8) (2019) 758–767.

CHAPTER

12

Principles of cardiovascular rehabilitation

Preeti Sharma[a], Poornima Pandey[a], Tikam Chand Dakal[b], and Narendra Kumar Sharma[a]

[a]Department of Bioscience and Biotechnology, Banasthali Vidyapith, Banasthali Tonk, Rajasthan, India, [b]Genome & Computational Biology Lab, Department of Biotechnology, Mohanlal Sukhadia University, Udaipur, Rajasthan, India

1 Introduction

Cardiovascular disease is the leading cause of death in the world. According to the data, cardiovascular disease is responsible for an estimated 17 million deaths (30% of all deaths) each year around the world. Cardiac rehabilitation is an important factor in the prevention of cardiovascular disease and can help to minimize age-related cardiovascular mortality, which is a national health priority [1]. Cardiovascular rehabilitation recommends exercise, cardiac risk factors, therapy, and modification in educating people, according to the medical assessment. These services are explicitly designed to minimize the risk of premature death by limiting the physical and physiological effects of cardiac disease [2]. The overall principle is the identification of the patient first, then enrollment in cardiac rehabilitation programs. The ideas of family medicine are aligned with the modern concept of medical home, which includes complete risk factor reduction and longitudinal treatment [1].

Cardiovascular rehabilitation is a program that focuses on physical activity. It is a patient-specific program aimed at enhancing the patient's physical, behavioral, social, and occupational conditions. Via physical activity and educational techniques, cardiovascular rehabilitation services foster a number of advantages [3,4]. Although, the increased metabolic demand required for physical activity causes a number of adaptations in the

Cardiovascular Toxicity and Therapeutic Modalities Targeting Cardio-oncology
https://doi.org/10.1016/B978-0-323-90461-2.00011-0

cardiopulmonary and autonomic systems, which can be linked to the appearance of signs and symptoms. Cardiovascular rehabilitation comprises a multidisciplinary intercede, which includes psychological counseling, smoking cessation, patient education, exercise training risk factor control, and vocational reintegration [4].

2 Cardiopulmonary stress testing in prehabilitation, habilitation, and rehabilitation

Cardiopulmonary stress testing is a way to measure how well a person's heart and lungs are working during exercise. The measurements are made of the amount of oxygen taken by the body while riding a bicycle or walking on a treadmill. Cardiopulmonary exercise testing (CPET), also known as cardiopulmonary stress testing, has emerged as a valuable clinical method for evaluating exercise ability and projecting outcomes in patients with cardiac arrest and other cardiac issues. It demonstrates the importance of the integrative exercise response involving the pulmonary, cardiovascular, hematopoietic, neuropsychological, and skeletal muscle systems, which is not adequately expressed by individual organ system function measurements [5]. During a symptom-limited exercise test, CPET includes assessing respiratory oxygen absorption (VO_2), carbon dioxide emission (VCO_2), and ventilator steps.

3 Cardiopulmonary exercise testing

Several methods for calculating ventilation and respiratory gas parameters during exercise still exist in cardiopulmonary exercise testing. Breath-to-breath analysis methods are used in the most advanced clinical systems because they provide the most accurate indicator of metabolic response to exercise. A nonrebreathing valve is attached to a mouthpiece in the cardiopulmonary exercise testing system to avoid the mixing of inspired and expired air. A metabolic cart specifically designed for practical testing is normally built-in cart in the cardiopulmonary exercise testing machine for this oxygen, and carbon dioxide gas analyzers are usually built-in cart. The air flow signals are integrated over the time of inspiration and expiration to calculate respiratory volumes. The average minute volumes can be determined by multiplying the breath-to-breath data by the respiratory rate [6].

4 CPET protocol

Many different procedures are used for cardiopulmonary exercise testing. In addition, the procedure chosen is determined by the test's intent

TABLE 1 Comparative exercise equipment [5].

Variable	Cycle	Treadmill
(PVO_2) level	Low	High
Blood gas connection	Easier	More difficult
Noise	Comparatively less	Moderately high
In hand on safety	Safe	Moderately less safe
Load bearing in large subject	A lesser amount	Moderately high
Scale of leg strength exercise	A smaller amount	More
Suitable for	Patients	Energetic common subject

and the patient's functional ability. Both bicycle and treadmill procedures have been used in the treatment of patients with cardiopulmonary issues. Ergometry treadmill fitness testing has many benefits over cycling (Table 1), and for the majority of people, treadmill walking is a more familiar practice than cycling. It necessitates more muscle mass to function against gravity. As a result, PVO_2 on the treadmill is 5%–10% higher than on a cycle ergometer [5,6]. Treadmill walking with your hands on the handrails lowers the metabolic cost and should be avoided if at all necessary. With improved metabolic cost quantification, a cycle ergometer is less likely to cause noise. It is generally less expensive and takes up less space than a treadmill.

During exercise testing, patients with heart disease are subjected to constant ECG monitoring and repeated blood pressure tests. Since oral contact is typically difficult with the mouthpiece apparatus, the patient usually communicates with hand signals during exercise [5,6].

During an exercise-based stress test (CPET tests), respiratory oxygen uptake (VO_2), peak oxygen assimilation, respiratory switch ratio, carbon dioxide production (VCO_2), and ventilator aerobic ratio are measured for this state.

5 Oxygen assimilation

VO_2 (oxygen assimilation) is regulated by cellular oxygen demand up to a point where it equals the maximum rate of oxygen transport. As external work increases, one or more of the determinants of VO_2 approach limitation, and the relationship between VO_2 and work rate can begin to plateau. The strongest proof of VO_2max has traditionally been this VO_2 plateau. Genetic factors, age, sex, body size, and the amount of exercising muscle are the key determinants of PVO_2 [6]. The patient's preparation and encouragement may also have an effect on PVO_2. PVO_2 should be expressed as a percentage of the expected value and in absolute values (1/min). PVO_2 values can range from 15 times the resting value (30–50 mL/kg/min) [6]. VO_2 can rise from a resting value of about 3.5 mL/kg/min to PVO_2 values of about 15 times the resting value (30–50 mL/kg/min).

6 Respiratory switch ratio

The gas exchange ratio, also known as the respiratory exchange ratio, is the ratio of carbon dioxide emission to oxygen absorption (VCO_2/VO_2) (RER). The RER equals the RQ (respiratory quotient) in a steady-state situation. The RQ value is determined by the fuels used in metabolic processes.

An RQ of 1 indicates metabolisms primarily of carbohydrates.

RQ < 1 indicates a mixture of carbohydrate with fat (an RQ of about 0.7) and protein (an RQ of about 0.8).

RER is a rough indicator of metabolic events. During exercise, the RER rises due to either buffered lactic acid or hyperventilation (usually at the end of the workout) [7].

7 Ventilator anaerobic ratio

The anaerobic ratio, formerly known as the ventilator anaerobic threshold, is used to estimate the exercise ability. Expired ventilation rises linearly with respect to VO_2 and represents aerobically generated CO_2 in the muscles during CPET, which lasts until 50%–60% VO_2max is reached. Since oxygen supply cannot keep up with the increased metabolic demands of exercising muscles, anaerobic metabolism occurs during the second half of the exercise. A substantial rise in blood lactate concentration and lactic acid production in the muscles occurs at this time. The lactate threshold corresponds to the VO_2 level at the start of blood accumulation. The VAR is a valuable metric since work at this stage covers the majority of everyday activities. Since patients who fatigue before hitting VAR are more likely to have a noncardiac problem, the ability to achieve the VAR may help distinguish between cardiac and noncardiac causes of exercise limitation. There are a number of strategies for addressing the VAR, which can be categorized as invasive or noninvasive. For the intrusive estimate of the VAR, direct blood sampling may be used. Since the function of the point of growing lactate necessitates multiple blood samples, it is rarely used in clinical applications. When this approach is used, the VAR is described as the VO_2 at which blood lactate levels begin to rise or, finally, when bicarbonate levels begin to fall. On the other hand, the exact meaning is still up for debate. The pattern of change in VCO_2 and

VE is measured using this noninvasive approach. As workout strength increases, this is proportional to VO_2.

While exercise-based intervention programs are recommended for cardiovascular prevention, different exercise parameters, such as intensity, duration, and frequency, may have a variety of effects on cardiovascular metabolic endpoints. Furthermore, there are interindividual differences in response to different forms or intensities of exercise, which may explain why some studies show similar findings with different types of exercise modalities. Exercise training research is being done now and in the future as well [6]. The effects of exercise-based CR on the progression of cardiovascular disease have been consistently positive. Recent research has shown that training impact varies greatly at the individual level, with up to a third of patients failing to display a substantial improvement in exercise ability despite sufficient compliance. An understanding of the variables and processes controlling the reaction to activity in bodies apart from the skeletal muscles, the heart, and the spinal cord, as well as their interaction at a systemic level, is required to enhance the effectiveness of fitness CR. Presently, cardiac recovery methods are used in cardiac prehabilitation, but they are used proactively rather than reactively. According to comprehensive evidence, such treatments before cardiac surgery may enhance result in a number of domains, including aerobic conditioning, respiratory muscle preparation, lifestyle improvements, diabetic regulation, sleep, and psychoeducation [6] (Table 2).

8 Positive outcomes of cardiac rehabilitation

Cardiac rehabilitation has been shown to have a number of positive results in properly chosen patients. Hypertension, hypotension, and nausea are all signs that have been diminished. Both fitness efficiency and physical activity engagement have increased. The standard of living has also improved. The number of people admitted to the hospital has decreased. About half of the decline in cardiac mortality is believed to be attributed to dietary improvements in stroke risk profiles. Exercise-only cardiac rehabilitation services resulted in a 27% decrease across all fatalities as compared to control program. The results of cardiac recovery in patients who have had a bypass and pericardial effusion surgery are poorly understood due to a lack of trial data. According to one study, cardiac recovery improves VO2max by 50% after a transplant. Following valve replacement surgery, patients who underwent cardiac rehabilitation showed a 25% improvement in VO2max and a 25% increase in functional ability. In the general population, it is unclear if cardiac recovery improves survival [9].

Prehabilitation is the process of improving an individual's functional ability so that he or she can survive a traumatic incident. Individuals are

TABLE 2 Measurement of performance from a cardiac rehabilitation program based on the following criteria: [1,8].

Performance measure	Goal	Intervention
Blood pressure control	< 140/90 mmHg or < 130 mmHg (diabetic patient)	Lifestyle moderation; drug
Diabetes control	AIC < 7%	Patient education, diet planning, BMI, and physical activity programming
Depression	Assess for the presence of depression	Use of a suitable screening method in combination with the required interference
Exercise capacity	Determine if you have a low patience for exercise	Each patient's exercise prescription is special
Lipid control	For patients with a heart condition LDL cholesterol level of 100 mg/dL	Lifestyle moderation; drug
Physical activity	30 min/day 5 days/week	Self-directed resistance training activity is customized to each person's unique needs and desires
Prevention medication	Taking preventive medications as prescribed	Preventive counseling and intervention for people of all ages and gender
Tobacco intervention	Cession of prohibition	Tobacco cession strategy
BMI	A BMI of 18.5–24.9 kg per m^2 and a waist circumference of 40 in. for men and 35 in. for women are recommended	Dietary modification, physical activity programming, and psychological support are all part of a multidisciplinary program

prepared to weather the storm of their activities and to stop or resolve applications by prehabilitation. The preparatory time is regarded as a teaching opportunity. Prehabilitation models should typically include three phases of screening, intervention, tracking, and postintervention evaluation for the benefit, as well as a high level of patient encouragement to improve behavior for preoperative benefits. All patients undergoing major surgery will benefit from prehabilitation, but the duration and severity may differ. To promote behavior change prior to surgery, the multidisciplinary team makes each touch count. Physical activity, inspiratory muscle conditioning, smoking, alcohol, diet, and psychological factors are all factors that are linked to prehabilitation strategies. The aim of inspiratory muscle training is to strengthen breathing muscles and lower the risk of preoperative pulmonary complications. It is preoperative smoking cessation, preoperative alcohol-adjusted consumption of nonhazardous alcohol, preoperative management of anxiety and depression, and development of self-effectiveness for prehabilitation and surgery, and preoperative nutrition.

Physiological factor may affect rehabilitation as follows:

8.1 Aerobic capacity

In terms of physical activity, the purpose of prehabilitation is to increase aerobic ability. Increased frequency physical activity rehabilitation systems are used in patients with stable coronary artery disease and heart disease. Improved exercise efficiency and hemodynamic gain more than moderate-intensity exercise training, with no substantial increase in complications. This modality's exercise protocols differ. In one study, a 10-min warm-up at 50%–60% VO_2max was presented in the form of 4-min periods at 90%–95% maximum cardiac output (Rate of Perceived Exertion 171), with periods split by three-minute cycles of walking at 50%–70% peak cardiac output [8,10].

8.1.1 Expected outcomes

- A 5%–10% increase in cardiorespiratory fitness, flexibility, muscular endurance, and strength.
- Symptoms are lessened, and physiological responses are lessened.
- An increase in psychological well-being.

8.2 Nutrient

Prior to surgery, the main objectives of nutrient prehabilitation are to optimize nutrient stores and aerobic reserve, as well as to provide an enough reserve to substitute for the aerobic response of serious trauma or injury. Acutely replacing dietary deficiencies is not the same as nutritional prehabilitation. The nutritional intervention involves the timeline that begins with the consideration of surgery and continues through the preoperative and postoperative phases to ensure early patient involvement. The importance of moving away from reactive to proactive nutrition evaluations and action must be stressed. Nutritional therapies are less well studied, and when they are, it is usually in conjunction with exercise as part of a prehabilitation program. These promising preliminary findings warrant further investigation, with a focus on fine-tuning approaches, elucidating mechanisms, and assessing correlations between therapies and large-scale clinical efficacy trials [10].

8.2.1 Expected outcomes

Changes in risk factors are linked to an unhealthy or insufficient diet.

8.3 Smoking

Smoking status is strongly linked to how people use CR. Smokers seem to be less interested in attending even a regular CR training, and those that do participate are often more likely to break out. Addition to stopping smoking should be a priority. Priority is given to patients while they are in hospital. To provide extra assistance in the hospital, a variety of techniques may be used. For stop smoking, medications drugs and nicotine pharmacological intervention referrals to smoking cessation services may be given to patients. Smoking cessation assistance should be available during the prehabilitation program. This could be delivered in the form of community, consultations in learning, and focusing on cessation assistance, including specific targeting. In the case of a relapse, daily monitoring and deep obey and assistance is in place. Intensive ongoing counseling for smoking cessation should be available in the prehabilitation services. This could be achieved through group health teaching focused on cessation support, with personalized therapy available if appropriate, pharmacotherapy assistance if needed, and regular contact with long-term accept and assistance in the event of a repeat. The availability of an intensive cessation support network could entice more smokers to enroll in the program, and the increased support for cessation they receive could enable them to stay in the program longer [8].

8.3.1 Expected outcomes

It has the potential to improve smoking cessation for a long time.

8.4 Alcohol drinking

When compared to abstinence and heavy consumption, reasonable alcohol intake (1–2 drinks/day) is linked to lower cardiovascular and all-cause mortality. According to a collective estimation from approaching longitudinal studies of patients with coronary heart disease, patients who consumed the modest improvement quantities of alcohol daily had a 20% reduction in heart failure (relative risk 0.80, 95% confidence interval [CI] 0.78–0.83) compared to nondrinkers. When we look up the data regarding alcohol intake, it is represented by the J-shaped curve with a considerable added safety against cardiovascular mortality at approximately 25 g/d and significant maximum protection against all-cause mortality at 6–11 g/d. The quantity and pattern of alcohol consumption tends to be further significant than the form of alcohol consumed. HDL is enhanced by promoting the production of apo A-I and A-II in the liver, lowering fibrinogen levels, increasing fibrinolysis, reducing inflammation, and suppressing platelet activation, which are all possible reasons for the benefits of moderate alcohol intake [11].

8.5 Behavioral and technical problems management

Personal and behavioral problems are common among heart patients, and they can have an effect on both morbidity and mortality. Following a myocardial infarction, up to 20% of patients experience depression, anxiety, and denial. Patients are routinely screened for nervousness, insomnia, and drug addiction, and family or other social justice issues are just some of the issues that people face during cardiac rehabilitation follow-up. In cardiac rehabilitation center, social workers and other experts on the multidisciplinary patients are given the tools and help they need in order to prepare for their workplace and daily routine. Clinical, behavioral, and cultural therapies tailored to the individual are available; they have been revealed to improve health.

The Interheart Research clearly showed that stress, after lipids and smoking, is one of the most significant risk factors for cardiac incidents. Social anxiety is responsible for 30% of the population's risk of myocardial infarction, which affects the progression of cardiovascular disease by increasing blood glucose, cholesterol, lipid profile, and body mass [11,12]. Atherosclerosis, inflammation, and endothelial dysfunction are all accelerated by it. Exercise training has been linked to lower stress levels and, as a result, lowers mortality rates. Several cardiac treatment services also have workshops on stress control to assist patients in recognizing, avoiding, and dealing with traumatic conditions. Cardiac rehabilitation is thus an effective therapeutic intervention for cardiac patients in distress. Numerous cardiac rehabilitation centers provide further anxiety control technique instructions, such as consideration, relaxation inhalation, and yoga, in addition to exercise [11].

8.5.1 Expected outcomes

The psychological sense of enough and a sense of well-being or the quality of life can be improved by reducing psychological difficulties or by acquiring stress management skills.

Prehabilitation aims to enhance these health metrics, such as endurance, in order to improve postoperative recovery and preserve physical function. During the preoperative screening cycle for elective procedures, it is the best time to create a prehabilitation program. At this stage, a multidisciplinary team composed of medicine, gerontology, anesthesia, surgery, dietetics, kinesiology/physiotherapy, and nursing will recommend a risk stratification model and determine the form and length of prehabilitation required to balance the potential benefit of such intervention against the potential harm of delaying surgery [13].

9 Risk factor for prehabilitation

All patients undergoing major surgery will benefit from prehabilitation, but the duration and severity may differ. A tiring system has been proposed [9]. Educational elements underpin cardiac and pulmonary recovery, with the aim of improving patient awareness of their conditions and the benefits of lifestyle changes. This is one way to improve a patient's self-efficacy. The transition from this to other major-surgery "surgery school" is now a key component of many prehabilitation programs. Patients are better equipped to participate in prehabilitation programs and improved recovery pathways if they are informed and perceptions of the preoperative period are managed [13].

10 Phases of cardiac rehabilitation

Cardiac recovery is divided into four stages. After your cardiac event, the first phase takes place in the hospital, and the remaining three stages take place in the home or in a cardiac treatment facility.

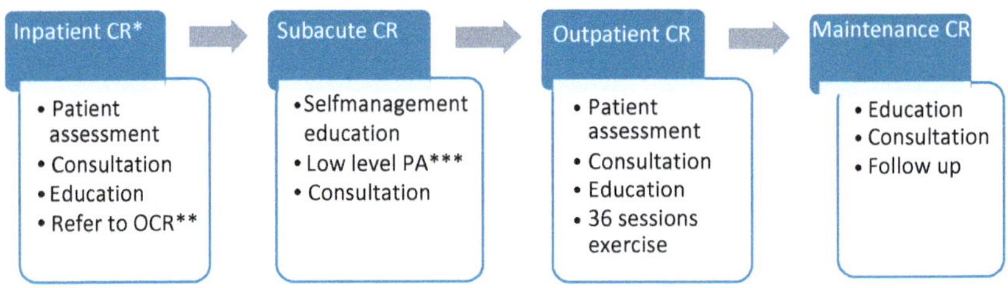

* Cardiac Rehabilitation

** Outpatient Cardiac Rehabilitation

*** Physical Activity

10.1 Phase 1

Acute phase Step 1 occurs when a patient is admitted to the hospital for an acute MI or another type of coronary incident. Throughout that time, the patient is subjected to appropriate regular exercise that is monitored and planned. However, hospitalization, clinical practice, potential risk adjustment, and treatment planning using close to zero multistage fitness tolerance testing are common. This is the first step that occurs during a cardiac incident. This process primarily aids at the beginning of mobility, determining the resilience and the effect of simple functional flexibility on the heart muscle, which are among the first targets of phase 1 cardiac rehabilitation. Prescribe healthy exercises to aid in the improvement of mobility and cardiac health. If an individual has had open-heart surgery, medical personnel should be consulted. Trigger factors for heart events are being monitored. Patients and their families should be informed about their condition as well as the potential advantages and risks of a cardiac rehabilitation program. An individual may enter the second phase after a sufficient amount of time has passed and substantial healing has occurred.

10.2 Phase 2

The early outpatient phase of cardiac recovery is known as the subacute phase. The patient participates in a controlled regimen of physical exercise, diet therapy, behavioral support, and relative risk change during this period after being discharged from the hospital. To improve patient adherence, a program located in specialty clinic, a part of a hospital, or a cultural center can be constructed in the residential home. Cardiopulmonary recovery can take anywhere from 2 to 7 weeks and includes ongoing monitoring of the heart's reaction to exercise and activity. This step focuses on educating participants on correct workout techniques and how and when to consciously your heart rate and level of exertion when exercising. The individual ought to be able to start to do more individual research and operation at the end of phase 2.

10.3 Phase 3

Phase 3 of intensive outpatient therapy is aimed at maintaining lifestyle improvements made in phases 1 and 2. During step 3, the patient's symptoms, risk factor drug usage, and psychosocial support are all evaluated on a regular basis. The individual must be able to monitor his or her own pulse rate, simply a symptom stress response, and estimated exhaustion level (RPE). During this process of cardiac therapy, a personal trainer would be available to assist in improving exercise capacity and controlling any detrimental changes that might arise. The physical therapist helps customize a regimen of exercises, including agility, mobility, and strength training, as the individual becomes more confident during stage 3 of cardiovascular recovery.

10.4 Phase 4

Stage 4 is a self-governing upholding stage. Independent fitness and conditioning are critical for maintaining good health and avoiding any future heart complications. If a person has completed all three stages, he or she should have a thorough understanding of the disease, risk factors, and strategies for maintaining optimal health. Family physicians play an important role in all three stages of cardiac recovery by ensuring patient continuity; they also help to enhance the quality of life [14].

11 Is rehabilitation too late?

The effect on the result per day in CR remain time was determined using CR timing (i.e., the time among recommendation and begin of core CR) as a nonstop variable. A separate study looked at the effect of CR time as per the current guidelines for "early on" or "late" CR, which is described as starting within 4 weeks of recommendation.

For this, the following CR timing parameters were formed:

- early CR (0–27 days);
- late CR (27–365 days).

In certain circumstances, some patients began the CR in later age, mostly older persons tended to do so. Female, with a lower BMI, at minimum one comorbidity, high systolic blood pressure low diastolic blood pressure, and a low amount of exercise at the moment (less than 150 min per week), has a shorter baseline ISWT distance (all $p < 0.05$). Patients with MI-PCI (myocardial infarction-percutaneous coronary intervention) made up the majority of the participants in both the CR grades (early and late). In terms of patient improvement after CR, late CR attendees saw a smaller benefit for all fitness-related outcomes. Out of 100 people, 63% are identified as slow starter in CR; the average time it takes for a CR applicant to become a CR as per guidelines, it is 39 days. The percentage of MI, MI-PCI, PCI, and CABG patients who were delayed was 69% for MI, 64% for MI-PCI, 56% for PCI, and 63% for CABG patients.

The median wait times (days) for each patient group were compared with the recommended wait times (28 or 42 days for CABG). The median wait time exceeded the recommended maximum waiting time in each patient group; moreover, the post-MI population's median wait time exceeded the highest optional wait time, which was 12 days. Following several analyses, it was discovered that the duration of delayed cardiovascular recovery was a significant independent reason for lower fitness levels as compared to early CR. CR timing was also an important indicator when used as a variable. Latest cardiac care guidelines and articles suggest starting CR as early as possible whenever feasible. Comorbidity burden could be one reason because of the higher percentage of late comers. Compared to 69% of early CR attendees, 73% of late CR attendees had one comorbidity. The complexity of the case can cause a delay in the start of recovery. Difference in time interval among classes is concerning.

Post-MI patients belonged to another patient category. The median minimum wait time was well surpassed by the largest number of days (12 days). This may appear to be counterintuitive, given that post-MI patients do not undergo invasive revascularization operations, which may postpone the start of CR due to the recovery time. Pre-CR to post-CR changes in terms of fitness-linked outcomes were observed regardless of

CR timing. However, in late CR attendees, the level of change was limited. There are two methods used to investigate the effect of cardiovascular rehabilitation timing: for this continuous measure to investigate the connection between escalating wait time and outcome, and timing as a categorical measure (early/late) to investigate the relationship in the form of current guidelines. CR timing was found to be an important predictor of patient outcome in terms of fitness level in each strategy. According to these findings, delaying CR reduces the probability of reporting a positive physical activity level and health outcome. According to the recent study of CR patients, it is found that delayed enrolment is linked to improved patient outcomes [14].

The protection of early enrollment in CR was recently investigated in cardiovascular patients, with no significant difference in main incident rates between early and late enrolls. Aside from clinical results, additional research indicates that the timing of CR enrollment can have an effect on initial CR enrollment. A second study using routine patient records found a correlation between wait time and enrollment. The observed connection between CR timing and treatment outcomes in results of these findings supports the need for timely CR as prompted by current guidance. Since there is a clear correlation between exercise-based CR and lower mortality, avoiding any potential causes of suboptimal fitness shift is important. While it is understood that CR timing should be determined on a one-to-one basis, caution should be exercised to avoid unnecessary delays, such as long waiting lists [14].

12 A paradigm shift in cardiopulmonary rehabilitation

This will be linked to the patient, payer, physician, hospital, and societal health-care-financing preferences. Physical therapists and other rehabilitation providers will play an important role in the incorporation of health promotion in the changing paradigm change from illness and impairment prevention to secondary condition prevention in people with disabilities. Excess weight, as measured by body mass index, is linked to a higher risk of repeated coronary events after a heart attack, particularly in obese people. Jayawardena et al. describe how they used the "plate model" as a dietary intervention for cardiac rehabilitation [15,16]. The "plate model" is a practical way to defy the latest dietary pattern by through standard serving sizes of staple foods in daily courses while still having enough veggies and proteins. The intervention group lost substantially more weight than the control group ($P = 0.029$) over the 12-week follow-up period (intervention group: 1.27 3.58 kg; control group: 0.26 2.42 kg). Additionally, the intervention community saw a nonsignificant reduction in blood pressure and lipid levels. Just 15%–50% of CR patients continue to exercise 6 months after CR, and even fewer do so after 12 months. About half of patients who smoked prior to a coronary incident continue to smoke 6 months later, and less than half of obese patients adhere to dietary guidelines. As a result, there are a few methods to obtaining the beneficial effects of healing. Patients will further integrate what else they have experienced throughout CR when they have a similar obey [15,16].

References

[1] M.B. Stephens, Cardiac rehabilitation, Am. Fam. Physician 80 (9) (2009) 955–959.

[2] C. Takahashi, F. Ribeiro, L.M. Vanzella, I.M. Lima, A.L. Ricci-Vitor, D.G.D. Christofaro, L.C.M. Vanderlei, Are signs and symptoms in cardiovascular rehabilitation correlated with heart rate variability? An observational longitudinal study, Geriatr. Gerontol. Int. 20 (10) (2020) 853–859.

[3] A.B. Gevaert, V. Adams, M. Bahls, T.S. Bowen, V. Cornelissen, M.D. Dorr, H.M.C. Kemps, P. Leeson, K. Craenenbroeck, Toward a personalized approach in exercise based cardiovascular rehabilitation: how can translation research help A call to action from the section on secondary prevention and cardiac rehabilitation of the European Association of Prevention Cardiology, Eur. J. Prev. Cardiol. 27 (13) (2019) 1369–1385.

[4] H.J.N. Bethell, R.J.P. Lewin, H.M. Dalal, Article on cardiac rehabilitation: so why it works so why is it not it done, Br. J. Gen. 58 (555) (2008) 677–679.

[5] R.M. Ross, ATS/ACCP statement on cardiopulmonary exercise testing, Am. J. Respir. Crit. Care Med. 167 (10) (2003) 1451.

[6] K. Albouaini, M. Egred, A. Alahmar, D.J. Wright, Cardiopulmonary exercise testing and its application, Postgrad. Med. J. 83 (985) (2007) 675–682.

[7] M. McCann, N. Stamp, A. Ngui, E. Litton, Cardiac prehabilitation, J. Cardiothorac. Vasc. Anesth. 33 (8) (2019) 2255–2265.

[8] G.J. Balady, M.A. Williams, P.A. Ades, et al., Core component of cardiac rehabilitation secondary prevention program updated: a scientific statement from American Heart Association Exercise, Cardiac Rehabilitation and Prevention Committee, the Counseling on Clinical Cardiology; the Councils on Cardiovascular Nursing, Epidemiology and Prevention and Nutrition Physical Activity and Metabolism; and the American Association of Cardiovascular Rehabilitation, Circulation 115 (20) (2007) 2675–2682.

[9] D. James, J.S. Sally, G. Dangous, Prehabilation, Clin. Med. (Lond.) 19 (6) (2019) 458–464.

[10] M.A. West, P.E. Wischmeyer, M.P.W. Grocott, Prehabilitation and nutritional support to improve perioperative outcomes, Anesthesiol. Rep. 7 (4) (2017) 340–349.

[11] M. Warner, Mampuya., Cardiac rehabilitation past, present and future: an overview, Cardiovasc. Diagn. Ther. 2 (1) (2012) 38–49.

[12] E. Diann, A.Y. Gaalema, S.T. Cutler, P.A. Higgins, Smoking and cardiac rehabilitation participation. Associations with referral, attendance and adherence, Prev. Med. 80 (2015) 67–74.

[13] F. Carli, C.S. Bergdahl, Prehabilitation to enhance perioperative care, Anesthesiol. Clin. 33 (1) (2015) 17–33.

[14] J. Fell, V. Dale, Does the timing of cardiac rehabilitation impact fitness outcomes? An observational analysis, Open Heart 3 (1) (2016), e000369.

[15] B. Barbara, L. Tina, G. Kathrin, R. Laura, R. Andres, The beneficial effects of cardiac rehabilitation, Cardiol. Ther. 9 (1) (2020) 35–44.

[16] D.W. Schopfer, E.F. Daniel, Cardiac rehabilitation in older adults, Can. J. Cardiol. 32 (9) (2016) 1088–1096.

CHAPTER

13

Hematopoietic stem cell transplantation: Sources of cells, cryopreservation, and applications

Kumar Pranay[a], *Deepak Kumar Jha*[b], *Rakesh Bhaskar*[c], *and Mukesh Kumar Gupta*[c]

[a]Department of Biochemistry, Indira Gandhi Institute of Medical Sciences, Patna, Bihar, India, [b]P.C. Vigyan Mahavidyala, Chapra, Bihar, India, [c]Department of Biotechnology and Medical Engineering, National Institute of Technology Rourkela, Rourkela, Odisha, India

1 Introduction

Hematopoietic stem cells (HSCs) are the self-renewing population of multipotent stem cells found in bone marrow (BM) and umbilical cord blood (UCB) and are capable of generating all the formed elements of blood. They are generally identified as Lin^-, $cKit^+$, $Sca1^+$, $Thy1^+$, and $CD34^{-/low}$ cells and are capable of restoring complete hematopoiesis in myeloablated mice [1,2]. Consequently, HSC transplantation (HSCT) has emerged as a promising curative treatment for hematological malignancies such as lymphomas, leukemias, and myelomas, and various other nonmalignant diseases such as severe combined immunodeficiency, sickle cell anemia, and immune deficiencies. The HSCs and their precursors (HPCs) can be isolated from a healthy donor and intravenously infused into the recipient to replace mutant HSCs or restore impaired BM functions in hematolymphoid diseases. Alternatively, the HSCs and HPCs can be frozen and stored in liquid nitrogen until required (Fig. 1). The latter approach is more practicable with the use of UCB-derived cells and has led to the

Cardiovascular Toxicity and Therapeutic Modalities Targeting Cardio-oncology
https://doi.org/10.1016/B978-0-323-90461-2.00007-9

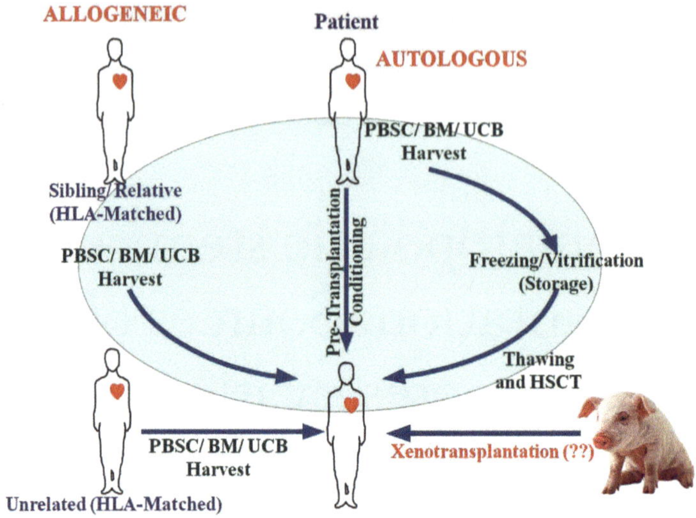

FIG. 1 Types of HSCT based on the source of HSCs. The HSCs can be harvested from patients (autologous HSCT) or from an HLA-matched donor (allogeneic HSCT). Use of HSCs from humanized pig (xenotransplantation) is also envisaged and under extensive research. The harvested HSCs can either be transplanted directly into the preconditioned patient or be cryopreserved for future use.

establishment of several public and private cord blood banks worldwide. However, the clinical practice of HSCT is hindered by several procedural limitations, including the lack of HLA-typed donor, poor cryosurvival of frozen cells, immunological complications, etc., and thus, HSCT is chosen only in those life-threatening diseases wherein other methods fail to justify the posttransplant mortality rate associated with HSCT [3]. A yet another major limitation of allogeneic HSCT is the graft-versus-host disease (GVHD) reaction, which is a potentially life-threatening condition. An allogeneic transplant may also cause delayed immune reconstitution that can lead to enhanced infection rates and chronic GVHD [4]. This chapter provides an overview of sources of HSCs and their cryopreservation for long-term storage. Various indications and applications of HSCT are also discussed.

2 History and evolution of HSC and HSCT

Hematopoiesis in mammals is a well-organized and highly orchestrated process initiated by HSCs. The HSCs undergo progressive differentiation to produce committed progenitors that eventually form terminally differentiated blood cells (Fig. 2). Defective hematopoiesis due to mutation or

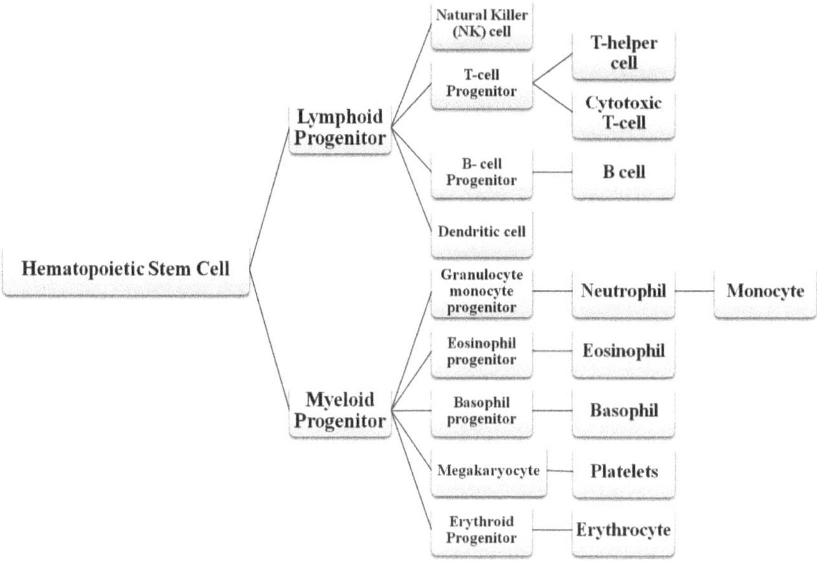

FIG. 2 HSC lineage during hematopoiesis. The multipotent HSCs give rise to lymphoid and myeloid progenitors. Lymphoid cells are derived from lymphoid progenitor cells, while myeloid progenitor gives rise to myeloid cells.

malignancy is associated with high mortality, and hence, several attempts have been made for their restoration or reconstitution by HSCT. The concept of HSCT is not new and dates back to World War II. The atom and plutonium bombs used to end World War II had devastating consequences on human life and had persuaded the scientific fraternity to explore ways of repairing or preventing radiation-based tissue damage in humans [5]. In 1949, a study conducted by Jacobson and his colleagues found that exteriorization and protection of spleen in mice enabled them to survive otherwise lethal irradiation [6]. In addition, animals injected with spleen cells or infused with BM cells exhibited a similar level of protection against irradiation [6]. Donor-derived BM cells were also successfully engrafted in lethally irradiated dogs and mice [7,8].

The first successful BM transplantation was performed by Thomas and colleagues for acute leukemia patients [9]. However, major impediments like GVHD, graft rejection, and/or death due to infections led to extremely poor results of transplants. All patients receiving BM transplantation during this period died. With the discovery of human leukocyte antigen (HLA) genetics [10,11] in graft immunology, the importance of HLA-matched donor was realized, and the success of BM transplantation improved. An advanced HSCT protocol was successfully developed in 1968 for monkeys by Balner and his team. In the same year, HSCs, derived from an HLA-identical sibling donor, were successfully transplanted in patients suffering from congenital immune deficiency [12]. The first successful allografting was reported in 1972 for severe aplastic anemia [13]. Subsequently, successful allogeneic HSCT was achieved in patients suffering from hematologic malignancy, including acute leukemia. Till the 1970s, allogeneic grafting was limited to HLA-identical sibling pairs. The year 1979 saw the first grafting from an unrelated donor for a leukemic patient with some success [14]. The period of 1990s saw tremendous changes in the transplantation methodology, including significant advances in the prevention and treatment of infectious diseases.

The breakthrough in HSC was seen in the year 1988 when it was shown that HSCs could be purified from BM by using surface markers [15]. These cells could proliferate and differentiated into myelomonocytic cells, B cells, and T cells and could reconstitute all blood cell types in irradiated mice [15]. This finding has led to significant developments such as marker-based isolation of HSCs and their progenitors (Table 1). Newer methods such as long-term culture-initiating cell assay were developed for identifying HSCs capable of hematopoietic cells [16]. Subsequently, a very large number of studies have characterized various cell types and intermediary progenitors of HSCs and specific cell-surface markers for their identification [1,2]. Mouse models such as SCID-hu, SCID/beige/XID, NOD-SCID, NOD-SCID/IL-2R$^{\gamma}$-null strain (NOG), etc. have been developed that were suitable for the identification of human HSCs and their long-term hematopoietic upon HSCT in recipient mouse models [17].

3 Sources of HSCs

Peripheral blood, BM, and UCB have been used as sources of HSCs in various studies. They can be distinguished from their progenitors by using specific surface protein markers (Table 1), although it is not

TABLE 1 Marker speciation during commitment of HSCs lineage in human and mice.

	Species	
Phenotype	**Human**	**Mice**
Multipotent HSCs	Lin^-, $Sca1^+$, $CD38^{-/low}$, $CD48RA^-$, $CD49f^-$, $CD90^+$	Lin^-, $Sca1^+$, $cKit^+$, $CD34_{-/low}$, $CD201^+$, $CD48^-$, $CD150^+$
Multipotent progenitors	Lin^-, $Sca1^{low}$, $CD38^+$, $CD48RA^-$, $CD49f^-$, $CD90^-$	Lin^-, $Sca1^{low}$, $cKit^+$, $CD34_+$, $CD201^+$, $CD48^-$, $CD150^-$
Lineage-committed progenitors	Lin^-, $CD34^+$, $CD38^+$	Lin^-, $Sca1^-$, $cKit^+$, $CD201^{+/-}$, $CD48^+$, $CD150^-$
Precursor cells	$Lin^{+/-}$, $CD34^-$,	$Lin^{+/-}$
Mature blood cells	Lin^+	Lin^+, $CD34^-$

always easy to categorically distinguish the functionally different progenitors. The selection of stem cell sources is influenced by donor and recipients' age, disease stage, and associated comorbidities [5]. Attempts have also been made to derive HPCs and their progenies from pluripotent stem cells such as embryonic stem cells (ESCs) and induced pluripotent stem cells (iPSCs). The use of iPSCs offers the advantage of being patient specific as they are derived from the nuclear reprogramming of the patient's own somatic cells (Fig. 3). Further, iPSCs also allow the genetic correction of mutations prior to their transplantation. Studies have shown that autologous transplantation of mutation-corrected iPSC-derived hematopoietic progenitors could correct sickle cell anemia in humanized mouse models [18]. Similar results have also been obtained for genetic correction of Fanconi anemia in human wherein iPSC-derived hematopoietic progenitors were found to be disease free [19]. More recently, CRISPR/Cas9-mediated genetic correction of iPSC-derived HSCs has emerged as a newer tool for gene therapy via HSCT [20].

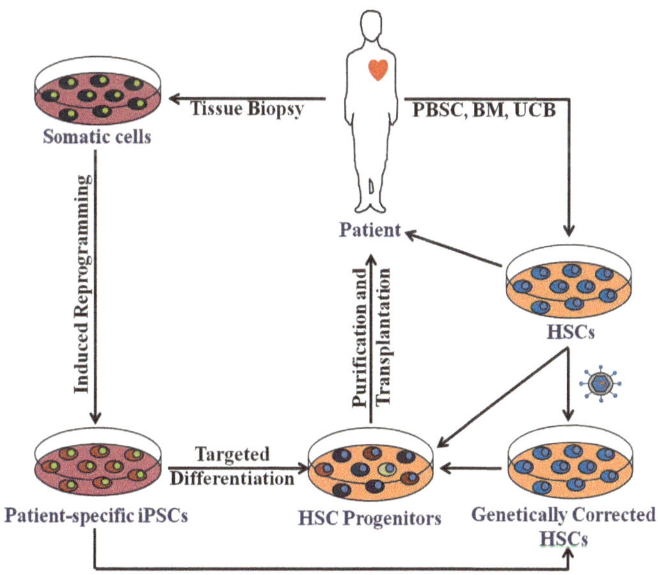

FIG. 3 Sources of HSCs and their genetic correction for HSCT. HSCs can be purified from peripheral blood (PBSC), bone marrow (BM), or umbilical cord blood (UCB). Alternatively, it is also possible to differentiate HSC and HSC progenitors by targeted differentiation of induced pluripotent stem cells (iPSCs). Since HSCs and iPSCs are proliferating cells, they can be genetically modified for correction of genetic mutations prior to their differentiation and transplantation. In case of malignancies, the HSCs and their progenitors can be purified to exclude neoplastic cells prior to their transplantation.

3.1 Peripheral blood

The presence of circulating HSCs and HPCs in peripheral blood was first reported in the 1960s. However, most attempts of peripheral blood HSCT failed due to the low number of HSCs and HPCs until the 1980s when it was found that the administration of granulocyte colony-stimulating factor (GCSF) can mobilize HSCs from BM [21]. Subsequently, GCSF-mobilized $CD34^+$ cells in the blood have been employed as autologous as well as allogeneic HSCT. The peripheral blood-derived stem cells (PBSCs), stimulated by granulocyte colony-stimulating factor (GCSF), can be collected in outpatient settings and spare patients from general anesthesia and have the advantage of being less invasive. Several devices and techniques such as CEPRATE stem cell concentrator system (CellPro Inc., Bothell, WA), Isolex Magnetic Cell Separation System (Baxter Health Corporation, Irvine, CA), AmCell Selection Device (Miltenyi Biotech GmbH, Bergisch Gladbach, Germany), and high-speed fluorescence-activated cell sorting (FACS) have been used to separate $CD34^+$ progenitors that were capable of rapid and stable hematopoietic engraftment upon autologous or allogeneic transplantation [22]. The autologous PBSCs separated from patients affected with neoplasm such as breast cancer, multiple myeloma, lymphoma, etc., may cause disease recurrence or secondary malignancies [23]. However, studies have shown that marker-mediated selection of PBSCs can better screen the cells to reduce the malignancies and hence, show a lower incidence of secondary cancer in recipients than in those from BM donors [24]. Studies have also shown that PBSC-mediated HSCT leads to faster hematopoietic engraftment and immune reconstitution than those of BM-mediated transplants [25]. Unfortunately, the concentration of PBSCs in peripheral blood is very low and may require developing a cytokine-supported ex vivo culture system for their expansion in sufficient numbers. The ex vivo expansion of PBSCs is still under development [26]. Another concern with transplantation of PBSCs has been the potential risk of GVHD due to a higher number of T cells, NK cells, and monocytes in peripheral blood [27].

3.2 Bone marrow

The BM represents the primary niche of HSCs wherein they can self-renew and proliferate by symmetric cell division to maintain their population. Upon cell signaling cues, they can undergo asymmetric cell division to proliferate, differentiate, and mature to generate more committed progenitors and terminally differentiated blood cells that are mobilized into the peripheral blood. Thus, BM has been the first and the obvious source of cells for HSCT, although they are less preferred over

PBSCs due to invasiveness in the procedure, difficulty in finding donors, higher graft complicacies, and high cost. They represent a rich source of HSCs and have been extensively used in clinics to augment or restore hematopoietic function in hematologic and lymphoproliferative diseases wherein PB-derived HSCT is not practical. The BM-HSCT is generally preferred in cases such as children's aplastic anemia and haploidentical transplantation for reducing nonrelapse mortality and GVHD, whereas PB-HSCT is usually preferred in hematological malignancies such as acute myeloid leukemia, lymphoma, etc. The benefit/risk of BM-HSCT versus PB-HSCT has been compared in several studies and reviewed in Ref. [27].

The marrow is generally aspirated from iliac crests using ACD anticoagulants. It can be collected from the patient's own body (autologous) or a twin sibling (syngeneic), or an HLA-matched donor (allogeneic) and transplanted with or without separation of HSCs. The concentration of HSCs from BM can be increased by administering GCSF [28]. Studies have shown that BM-derived HSCs can form all types of HPCs and achieve long-term donor-derived multilineage reconstitution upon clinical HSCT [29]. Upon intravenous injection into recipients, BM-derived HSCs can migrate from peripheral blood to home into the host's BM and start proliferating to produce donor-derived hematopoiesis [30]. However, graft failure and high mortality remain major concerns in allogeneic HSCT. The BM-HSCT is also associated with GVHD, infection, relapse of disease, and cardiotoxic events (reviewed in Ref. [31]).

3.3 Umbilical cord blood

The UCB contains a mixed population of HSCs, HPCs, and their progenies, including terminally differentiated blood cells. They can be easily harvested from the cord blood at the time of birth and cryopreserved for later use in life. Studies have shown that UCB is an excellent alternative to BM for HSCT in patients lacking HLA-matched donors. Cord blood also offers advantages such as the lower risk of GVHD and relapse and higher tolerable HLA disparity over BM and PBSCs [5]. Thus, the cryobanking of UCB has become common, and many public and private cord blood banks have been established worldwide. However, a lower number of HSCs and HPCs, prolonged time to engraftment, delayed or incomplete immune reconstitution, risk of infections, and treatment-related mortalities remain major hurdles [32]. Novel strategies such as intra-bone injection of cells, modification in homing, and co-injection with mesenchymal stromal cells are thus being explored to improve UCB-mediated transplant [33]. Attempts have also been made for ex vivo expansion of UCB-derived HSCs and HPCs but have met with limited success (reviewed in Ref. [34]).

4 Preservation and storage of HSCs

Freshly collected HSCs remain viable only for few hours to few days, which occasionally limits their long-distance transportation or long-term storage until use (Fig. 1). Cryopreservation offers an effective means to overcome these difficulties. Frozen cells/blood can be easily transported to a clinic and increases the shelf life of the cells. Cryopreservation also permits the enforcement of more strict quality controls through extensive testing, thereby enhancing the safety of HSCT.

A purified population of HSCs can be cryopreserved by slow freezing using a protocol similar to those of somatic cells. However, to comply with regulatory agencies, which limit extensive processing of cells before HSCT, the HSCs are typically cryopreserved in the form of peripheral blood or UCB. Typically, the UCB are cryopreserved by mixing with cryoprotective agents (CPA) such as dimethyl sulfoxide (DMSO) and slow freezing in a controlled-rate freezer. The slow freezing is generally achieved by cooling to (−) 40–(−) 50°C at the rate of (−) 1–(−) 2°C/min followed by cooling at the rate of (−) 5–(−) 10°C/min to (−) 80–(−) 90°C and plunging in liquid nitrogen. Uncontrolled slow freezing (USF) in a −80°C freezer has also been tried for cryopreservation of HSCs [35]. The USF does not require an expensive programmable freezer and is a less time-consuming protocol [36]. However, it is met with lower postthaw viability of cells and is not commonly practiced.

The frozen cells can be stored in liquid nitrogen or liquid nitrogen vapor. At the time of use, the frozen cells can be thawed in a water bath set at 37°C and used for infusion. Unfortunately, most cryopreservation protocols significantly reduce the postthaw viability of cells, which might necessitate the pooling of two or more samples for HSCT. Moreover, the CPAs themselves can cause cytotoxicity and adverse reactions upon their introduction into patients. High concentrations of DMSO in the blood have been reported to cause cardiac arrest, respiratory stress, and epileptic seizures [37]. Studies have shown that DMSO-cryopreserved stem cell infusion may lead to nausea, vomiting, and cardiac arrhythmia [38–40]. Thus, attempts are made to optimize CPA concentration or reduce its requirement by combining with nonpermeating CPAs such as trehalose, sucrose, and hydroxyethyl starch (HES) [41]. Currently, the most freezing protocol uses DMSO at the final concentration to ~ 5% but a lower concentration of 2.2%–3.5% has also been reported [42,43]. A class of small molecules known as ice recrystallization inhibitors (IRIs) has also shown promise as deep freezing agents [44,45]. Alternatively, it is also possible to remove CPAs from cryopreserved HSCs before their transplantation to reduce their adverse effects upon HSCT. Devices such as Cytomate and Sepax are available for the removal of DMSO from frozen-thawed HSCs [46]. A detailed discussion on various considerations for cryopreservation of HSCs can be found elsewhere [47].

5 Applications of HSCT

The HSCT has potential applications in a wide variety of life-threatening diseases that are nonresponsive to medical treatment. It has been proven useful in malignant hematological disorders such as leukemia, lymphoma, and multiple myeloma, and nonmalignancies such as autoimmune and metabolic diseases. It can also be combined with immunotherapies such as monoclonal antibodies and chimeric antigen receptor T cell (CAR-T cell) therapies. Some of the common applications of HSCT are described below.

5.1 HSCT in primary immunodeficiency diseases

Primary immunodeficiencies (PID) are disorders of monogenic nature originating from mutations involved in immune regulation and immune host defense. Major characteristics of these disorders are autoimmunity, recurrent infusions, inflammatory manifestations, lymphoproliferation, and malignancy. The majority of PIDs are due to genetic defects in hematopoietic cells, and hence, HSCT is a rational therapeutic approach [48].

5.1.1 X-linked SCID and JAK3 deficiency

X-linked SCID is an X-linked recessive immunodeficiency disorder characterized by chronic diarrhea, periodic respiratory infection, and acquired maternal T cell-mediated GVHD. Mutations in interleukin-2 receptor gamma (IL2RG) can result in X-SCID. In addition, Janus kinase 3 (JAK3) deficiency also shares similar clinical and phenotypic manifestations that are indistinguishable from X-linked SCID. In a study conducted at Duke from 1992 to 1998, a large cohort of patients suffering from SCID were evaluated for the beneficial effects of HSCT [49]. Among the patients, 13 showed mutations of adenosine deaminase (ADA)-SCID, six were JAK3 mutants, three showed a defect in the IL-7 receptor chain, and 21 had autosomal recessive SCID of unclassified nature. T cell-depleted HSCTs from parents were performed in 77 infants, while 12 underwent matched related donor transplants. Results showed patient survival rates that varied between 3 months and 16.5 years. Although no chemotherapy or GVHD prophylaxis was given to any patient, only 28 patients developed mild GVHD. Among long-term surviving patients, less than 50% showed normal B cell reconstitution and required immunoglobulin replacement. In another study, after successful T cell engraftment to patients having X-SCID and JAK3 deficiencies, no antibodies were produced by the B cells [50]. In addition, several patients in long-term follow-up showed a weakening of T cell-mediated immunity with both the deficiencies. Similarly, ~ 72% survival rate was observed in patients having IL2RG/JAK3 SCID deficiency [51].

5.1.2 *IL-7 receptor deficiency*

IL-receptor deficiency is an autosomal recessive form of SCID wherein B and NK lymphocytes population are normal, but there is a selective absence of T cells [52]. In this deficiency, B cells show normal function as long as helper T cells are provided. This observation led to performing HSCT in the case of IL7R-deficient patients [52]. Maternal T cell engraftment was performed in one of the two infants showing IL7R-deficient SCID. Unrelated donor with alemtuzumab-dependent conditioning was used for HSCT, which led to the temporary removal of T cells in the infant having maternal engraftment. T cell engraftment was leisurely in both the patients in the absence of B cell donor. Reconstituted donor T cells enabled humoral immune reconstitution. These findings suggest that in patients having IL7R deficiency, chemotherapeutic treatment is not a prerequisite for the induction of humoral and cellular immune reconstitution.

5.1.3 *Recombination-activating genes deficiency*

The deficiency associated with the recombination-activating genes 1 and 2 (RAG 1 and RAG 2) causes impaired VDJ recombination resulting in defective B and T cell development. Hypomorphic mutations in these genes lead to a diverse range of clinical phenotypic manifestations such as Omenn syndrome, combined immune deficiency with granulomas, and atypical SCID [53]. For a long time, HSCT for RAG deficiency was not promising compared to other forms of SCID, particularly in the absence of matched related donors. In a study conducted on 76 patients with RAG deficiency having undergone HSCT, it was observed that 25 patients who underwent HSCT from HLA-matched donors with myeloablative conditioning had a survival rate of about 80% [54]. T cell reconstitution was observed in most patients, but the $CD4^+$ T cell count was below the normal value. In addition, less consistency was observed in the reconstitution of B cell immunity, and more than 50% of patients required regular immunoglobulin therapy.

5.2 HSCT in autoimmune diseases

Autoimmune diseases (ADs) are a class of heterogeneous diseases affecting about 5%–10% population in developed countries [55]. The clinical manifestations associated with ADs include a wide range of symptoms depending on the immunologic components and tissue under consideration, but a common feature is the chronic inflammation of target organs and decreased self-tolerance [56]. The increased understanding of molecular mechanism of ADs has led to the development of many novel therapies that block pathological inflammatory responses [57]. However, these therapies are prolonged in nature and can have long-term side effects, and often, the organ damage is not reversed. Thus, HSCT is increasingly employed in the treatment of AD since the 1990s [58,59]. According to the Autoimmune Disease Working Party (ADWP) of the European Society for Blood and Marrow Transplantation (EBMT) database, approximately 3300 patients have registered in 25 years [60]. EBNT registry has reported 3320 HSCT procedures, with most patients ($n = 3103$) receiving autologous HSCT, while only a small number of patients ($n = 217$) undergoing the allogeneic transplant. Multiple sclerosis (MS), systemic sclerosis (SS), Crohn's disease, inflammatory arthritis, and systemic lupus erythematosus (SLE) are major indications of the autologous HSCT. Improved HSCT outcomes have created a firm ground for the use of HSCT in resetting severe ADs [61].

5.2.1 *Multiple sclerosis*

Multiple sclerosis (MS) is a terminal disorder with inflammatory and degenerative phases. These phases follow different clinical courses, which include malignant rapidly evolving severe (RESMS), relapsing-remitting (RRMS), secondary progressive MS (SPMS), and primary progressive MS (PPMS) [62]. Many immunomodulatory and immunosuppressive therapies are currently in use, which can reduce the relapse rate but have little effect on disability progression [63,64]. Autologous HSCT as a treatment for MS has been used since 1995. Initially, autologous HSCT was confined mainly for patients with advanced disability for primary and secondary progressive MS in the form of retrospective or case studies with phase I/II trials [65,66]. Subsequently, a decline in transplant-related death (TDR) resulted in a profound shift as patients with relapsing-remitting (RRMS) and aggressive form disease were included in HSCT [67–69]. This trend resulted in benefiting patients having the inflammatory form of the disease [70]. Presently, HSCT is mainly being used for RRMS patients [60].

At least eight retrospective studies, six single-arm clinical trials on autologous HSCT, and only two randomized clinical trials involving a comparison of HSCT versus disease-modifying therapies for MS have been reported [71]. These studies advocate the efficacy of HSCT in patients having the relapsing form of MS. According to the EBMT register, the most extensive retrospective study involved 281 patients [72]. The progression-free survival rate was about 46%. Survival of patients with RRMS or PRMS was 82% at 3 years and 73% at 5 years. Single-arm clinical trials of HSCT have been highly effective for RRMS and PRMS patients [69,73,74]. Another study conducted on 21 RRMS patients with a conditioning regimen of cyclophosphamide plus alemtuzumab showed 100% survival at 3 years with no progression on Expanded Disability Status Scale (EDSS) [65]. In another report, a group of 35 patients with RRMS and SPMS with the conditioned regimen of BEAM plus ATG showed an 82% survival rate at 1 year [75]. Randomized controlled trials involving HSCT versus disease-modifying studies have also given better results for the former with regard to disease progression and associated mortality [76]. According to the exhaustive literature review conducted by the American Society of Bone Marrow Transplantation, autologous HSCT has been recommended as the gold standard for the treatment of refractory relapsing MS [77].

5.2.2 *Systemic sclerosis and systemic lupus erythematosus*

The major indications of HSCT have changed several times in the field of rheumatology in the past 25 years due to the ever-changing landscape of available biologic therapies. Initially, autologous HSCT was confined to diseases like rheumatoid arthritis and juvenile idiopathic arthritis [60]. The high rate of relapse in conjunction with the success of drugs like TNF-α inhibitors did not hold a rationale for the use of HSCT in these indications. As a result, HSCT was primarily used for the treatment of SLE. Studies involving Phase I/II trials showed treatment-free remission rates of more than 50% after 5 years [73,78]. EBMT guidelines recommend HSCT as a mode of treatment in SLE patients [60]. However, effective use of drugs like rituximab and belimumab in B-cell therapies has restricted HSCT use to cases of poor prognosis.

Systemic sclerosis (SS), a chronic autoimmune disease having characteristics like vasculopathy and tissue fibrosis [79,80], has become the main indication for HSCT. Initial Phase I/II trials showed skin improvement and partial stabilization of pulmonary condition [81,82]. A vital aspect of these studies was to stress the importance of proper heart assessment as a tool to reduce transplant-related mortality. Recently randomized trials have shown that HSCT promotes more prolonged disease-free survival, longer overall survival, and better life quality [83]. Consequently, European League Against Rheumatism (ELULAR) guidelines recommend autologous HSCT for SSc patients at risk of organ failures [84].

Various transplant centers have adopted well-defined protocols. However, few contentious issues remain. There is no agreement among centers on using graft selection or adopt nonselected grafts for transplantation. CD34 + graft selection prevents reinfusion of autoreactive cells and possible disease reactivation, but the process is costly and enhances the chance of contamination and cell death. A retrospective study carried out by European Blood and Marrow transplant Group (EBMT) showed that SS patients undergoing HSCT, with or without graft selection, led to similar clinical results [71]. In contrast, a prospective observational study reported a higher rate of disease improvement in patients with CD34 + graft selection [85]. In addition, myeloablative conditioning led to a better outcome in patients undergoing transplant [86].

5.2.3 *Crohn's disease*

Crohn's disease has multifactorial features that affect the gastrointestinal tract and causes inflammation and ulceration [87]. It is an immunologically mediated disease and heterogeneous among the patients. The chronic nature of disease and absence of therapeutic tools in refractory patients has made the HSCT the choice of treatment for the affected patients. The first successful experience involving the treatment of Crohn's disease with HSCT was published in 2003 [88]. Studies from European Countries reported similar findings of low toxicity and increased rates of short-term disease remission [89]. Another randomized study stated that autologous HSCT promotes endoscopic and clinical benefits, despite the burden of adverse events [90]. Crohn's disease is one of the major indications for autologous HSCT, according to the EBMT registry [61].

5.3 HSCT in malignancies and cancer

Cancer is one of the deadliest diseases responsible for the death of millions worldwide [91]. The fast growth in research and technology has caused a reduction in the death rate (about 1.5%) by cancer. HSCT is a crucial standard procedure for treating leukemia, multiple myeloma, and lymphomas. Additionally, HSCT is a widely used therapy to treat other kinds of cancer such as neuroblastoma, sarcomas (NCT01807468), brain tumors (NCT00528437), and breast cancer (NCT00003927). Recipients of allogeneic HSCT are highly susceptible to GVHD [92].

The HSCT has been used to treat patients with acute leukemia since 1979 [93]. The current clinical guidelines recommend allogeneic HSCT for patients with AML when the chance of disease recurrence is higher than the probability of transplant-associated death [94,95]. However, remission of leukemic patients before allogeneic HSCT is critical for the overall success of treatment. Targeted therapies like tyrosine kinase inhibitors for acute lymphoblastic leukemia (ALL) patients following allogeneic transplant have successfully decreased the recurrence rate [96,97]. The AML therapy has witnessed a dramatic shift with the approval of drugs like CPX-351, gemtuzumab ozogamicin, midostaurin, gilteritinib, and venetoclax (Table 1) [98]. These target-based drugs help patients achieve remission and improve the outcome of allogeneic HSCT [99]. However, they do not replace allogeneic HSCT. Clinical trials of drugs like enasidenib (NCT03515512 and NCT03728334) and ivosidenib (NCT03564821) against isocitrate dehydrogenase type 2 mutant AML have been approved [100,101].

For lymphomas, autologous HSCT is the most accepted method of care for recurrent/refractory (R/R) patients with diffused large B-cell lymphoma (DLBCL) [98]. At present, novel and emerging therapies have no clear role in the treatment of patients with DLBCL. Studies in patients with R/R settings are mixed but promising. Drugs like ibrutinib, lenalidomide, venetoclax, and two CART-cell therapy products have been approved [98]. However, their optimal role remains unclear and requires further investigation. Similarly, the management of multiple myeloma also involves using autologous HSCT following induction therapy, which can be either early or delayed as the first choice method. Studies are ongoing in various laboratories worldwide to explore the role of cellular therapies and immunotherapies in conjunction with consolidative HSCT. Allogeneic HSCT is still an important consideration for patients with an increased risk of disease [98].

6 Complications of HSCT in recipients

Patients undergoing HSCT may face many complications such as GVHD, primary disease relapse, chemotherapy- and radiation-mediated toxicities, and infection [102]. Some of the important complications of HSCT are discussed below.

6.1 Immune-mediated complications

The HSCT often requires immunosuppression in autologous as well as allogeneic HSCT. The GVHD, an inflammatory condition due to invasion of donor T cells into recipient tissue, is particularly common in patients undergoing allogeneic HSCT. It can have acute and chronic manifestations. The clinicopathologic features associated with GVHD are well understood and involve the liver, gastrointestinal tract, skin, and other organs [103]. In its chronic form, GVHD is mainly associated with tissues such as the eyes, pulmonary tract, mouth, gastrointestinal tract, and genital tract. These tissues provide the anatomical and physical barriers between the host and the pathogens. GVHD involves the reactivation of donor lymphocytes against host-specific tissue markers after allogeneic HSCT [104]. Animal model studies have shown that there is probably a shift in T helper cell phenotype through tumor growth factor β and interleukin-6. Activation of quiescent autoreactive donor T cells possibly directs T cells along an autoreactive pathway [105]. Additionally, many of the immunosuppressive drugs inhibit thymic-dependent clonal deletion and thus disrupt immune reconstitution [106]. Chronic GVHD has been linked with several autoimmune-like manifestations, such as scleroderma-like syndrome, myasthenia gravis (MG), and chronic inflammatory liver disease.

The allogeneic HSCT recipients are highly prone to autoimmune reactions, and most commonly suffer from hematologic, i.e., autoimmune cytopenia (AICs) [107]. AICs include autoimmune hemolytic anemia (AIHA), autoimmune neutropenia (AIN), immune thrombocytopenic purpura (ITP), and others [108,109]. AIHA is the most common AIC after allogeneic HSCT. AICs occur at high frequency in certain allogeneic HSCT settings. The highest rate of AICs has been observed in infants with metabolic disorders receiving unrelated UCB [106]. The important risk factors associated with the manifestation of AICs among children have been documented recently and include the use of an unrelated donor, nonmalignant primary diagnosis, absence of total body irradiation (TBI) in the conditioning regimen, and chronic GVHD [107].

6.2 Infections

The HSCT requires conditioning regimens, which may induce mucosal injuries, affect hematopoiesis, and cause severe immunodeficiency, particularly in those who have received myeloablative regimens [110]. Consequently, commensal microflora of the gut and opportunistic pathogens that enter the bloodstream through the catheter-related infections are the primary cause of mortality and morbidity in patients receiving HSCT. Longer engraftment median time for BM-HSCT accounts for a higher infection rate in comparison with PB-HSCT [111].

The major risks of infection in HSCT are neutropenia, central venous catheter, and mucositis [112]. Preengraftment-phase infections are mainly due to Gram-positive bacterium such as *Clostridium difficile* [113], while fungal infections are mediated by *Candida* spp. and *Aspergillus* spp. During the postengraftment phase, pneumonia and gastrointestinal infections are the major cause of trouble [114]. Encapsulated bacterial infections in the late postengraftment phase primarily result from immunosuppressive drug treatment used to reduce GVHD. This phase also witnesses invasive fungal infections and viral infections.

6.3 Secondary malignancies

Allogeneic HSCT is highly prone to secondary malignancies. A meta-analysis, which included cohort studies to evaluate the risk of cancer development upon HSCT [115], reported 9233 cases of malignancies after transplantation. After HSCT, the risk of cancer was high, with a standardized incidence rate (SIR) of 1.66. The most common forms of malignancies were head and neck cancers, bone cancer, and melanoma with SIRs of 6.35, 10.04, and 3.52, respectively.

6.4 Medical and procedural risks

6.4.1 Developing nontoxic conditioning regimens

For the expanded futuristic use of autologous HSCT, new methods need to be developed to reduce the associated toxicity. One promising approach relies on using antibodies against CD117 to disrupt its binding with ligand stem cell factors (SCF), resulting in HSC exhaustion [116]. In another approach, a single dose of an antibody-drug conjugate of immunotoxin CD45-saporin led to better engraftment of donor cells in mice [117].

6.4.2 Immune reconstitution

One of the major limitations of HSCT is the slow reconstitution of the adaptive immune system, which results in increased risks of bacterial, fungal, and viral infections in recipients. Many novel approaches are currently under investigation, including co-transplantation with ex vivo generated lymphocyte populations and cytokine administration [118]. A yet another approach describes the use of tyrosine phosphatase-sigma (PTPσ), which stimulates hematological recuperation and enhances survival in mice models irradiated with 5-fluorouracil [119].

6.4.3 Tissue mismatch

Tissue mismatch between the donor and recipient is one of the major concerns associated with HSCT. It leads to the development of GVHD and other difficulties. Histocompatible family donor transplant is possible for less than 20 patients, and not many few can depend on matched, unrelated donor. One strategy to confront this has been the development of haploid-identical HSCT [120], HLA donor cell bank creation, and graft engineering, a method in which immune cells specific to GVHD are removed ex vivo [121].

6.4.4 Immune evasion

Many other strategies are being used to bypass the adaptive immune system, but a new and more detailed investigation is required. One approach aims to develop "universal donor" cells, which would theoretically be tolerated by the recipient's immune system [122]. An alternate strategy is to create conditions, which would make transplanted cells "inconspicuous" to the host. These methods comprise HLA gene engineering and/or inducing the expression of immunosuppressive genes [122].

7 Summary

The HSCT is a powerful tool for the treatment of many hematological malignancies and autoimmune diseases. Last decade saw a rapid development in the field of allogeneic transplant. However, few major worries still remain. Newer methods need to be developed in order to mitigate problems like toxicities associated with conditioning regimens, HSCT-associated relapse, and chronic GVHD. Studies have found that a decrease in the systemic regimen intensity reduces the chances of toxicities and relapse of malignancies. In this context, an enhanced understanding of minor HLAs may help in developing the methods for increased graft acceptance. In addition to it, researchers are also trying to develop novel cellular and immunotherapies, which can potentially lead to better HSCT outcomes. Lastly, limited availability of HSC sources represents a major hurdle in diversifying the use of HSCT. Differentiated HSCs, genetically engineered HSCs, and genome editing are promising technologies that may revolutionize the applications of HSCT in the future.

References

[1] R. Majeti, C.Y. Park, I.L. Weissman, Identification of a hierarchy of multipotent hematopoietic progenitors in human cord blood, Cell Stem Cell 1 (6) (2007) 635–645, https://doi.org/10.1016/j.stem.2007.10.001.

[2] Y. Sonoda, Human CD34-negative hematopoietic stem cells: the current understanding of their biological nature, Exp. Hematol. (2021), https://doi.org/10.1016/j.exphem.2021.02.004. S0301-472X(21)00038-2. Advance online publication.

[3] A. Czechowicz, I.L. Weissman, Purified hematopoietic stem cell transplantation: the next generation of blood and immune replacement, Immunol. Allergy Clin. N. Am. 30 (2) (2010) 159–171, https://doi.org/10.1016/j.iac.2010.03.003.

[4] N. Bejanyan, C.G. Brunstein, Q. Cao, A. Lazaryan, C. Ustun, E.D. Warlick, M. Arora, J.E. Wagner, D.J. Weisdorf, Predictive value of disease risk comorbidity index for overall survival after allogeneic hematopoietic transplantation, Blood Adv. 3 (3) (2019) 230–236, https://doi.org/10.1182/bloodadvances.2018018549.

[5] M.K. Juric, S. Ghimire, J. Ogonek, E.M. Weissinger, E. Holler, J.J. van Rood, M. Oudshoorn, A. Dickinson, H.T. Greinix, Milestones of hematopoietic stem cell transplantation—from first human studies to current developments, Front. Immunol. 7 (November) (2016) 1–16, https://doi.org/10.3389/fimmu.2016.00470.

[6] L.O. Jacobson, E.K. Marks, M. Robson, R.E. Zirkle, The effect of spleen protection on mortality following X-irradiation, J. Lab. Clin. Med. 34 (1949) 1538–1543.

[7] J.W. Ferrebee, H.L. Lochte Jr., A. Jaretzki 3rd, O.D. Sahler, E.D. Thomas, Successful marrow homograft in the dog after radiation, Surgery 43 (3) (1958) 516–520.

[8] C.E. Ford, J.L. Hamerton, D.W. Barnes, J.F. Loutit, 177-4506.Indd, Nature 177 (2005) 1–3.

[9] E. Thomas, L. Harry, J. Lochte, W. Ching Lu, J. Ferrebee, 1957_Intravenousinfusion_Thomas1957, N. Engl. J. Med. 12 (257 (11)) (1957) 491–496.

[10] J.J. Van Rood, A. Van Leeuwen, Leukocyte Grouping, A method and its application, J. Clin. Invest. 42 (9) (1963) 1382–1390, https://doi.org/10.1172/JCI104822. 14060982. PMC289412.

[11] J.J. Van Rood, J.G. Eernisse, A. Van Leeuwen, Leucocyte antibodies in sera from pregnant women, Nature 181 (4625) (1958) 1735–1736, https://doi.org/10.1038/1811735a0.

[12] M.M. Bortin, F.H. Bach, D.W. van Bekkum, R.A. Good, J.J. van Rood, 25th anniversary of the first successful allogeneic bone marrow transplants, Bone Marrow Transplant. 14 (2) (1994) 211–212.

[13] E.D. Thomas, R. Storb, A. Fefer, S.J. Slichter, J.I. Bryant, C.D. Buckner, P.E. Neiman, R.A. Clift, D.D. Funk, K.E. Lerner, Aplastic anaemia treated by marrow transplantation, Lancet 1 (7745) (1972) 284–289, https://doi.org/10.1016/s0140-6736(72)90292-9.

[14] J.A. Hansen, R.A. Clift, E.D. Thomas, C.D. Buckner, R. Storb, E.R. Giblett, Transplantation of marrow from an unrelated donor to a patient with acute leukemia, N. Engl. J. Med. 303 (10) (1980) 565–567, https://doi.org/10.1056/NEJM198009043031007.

[15] G.J. Spangrude, S. Heimfeld, I.L. Weissman, Purification and characterization of mouse hematopoietic stem cells, Science 241 (4861) (1988) 58–62, https://doi.org/10.1126/science.2898810.
[16] H.J. Sutherland, C.J. Eaves, A.C. Eaves, W. Dragowska, P.M. Lansdorp, Characterization and partial purification of human marrow cells capable of initiating long-term hematopoiesis in vitro, Blood 74 (5) (1989) 1563–1570.
[17] M. Michallet, T. Philip, I. Philip, H. Godinot, C. Sebban, G. Salles, A. Thiebaut, P. Biron, F. Lopez, P. Mazars, N. Roubi, T. Leemhuis, E. Hanania, C. Reading, G. Fine, K. Atkinson, C. Juttner, B. Coiffier, D. Fière, E. Archimbaud, Transplantation with selected autologous peripheral blood CD34+Thy1+ hematopoietic stem cells (HSCs) in multiple myeloma: impact of HSC dose on engraftment, safety, and immune reconstitution, Exp. Hematol. 28 (7) (2000) 858–870, https://doi.org/10.1016/s0301-472x(00)00169-7.
[18] J. Hanna, A.W. Cheng, K. Saha, J. Kim, C.J. Lengner, F. Soldner, J.P. Cassady, J. Muffat, B.W. Carey, R. Jaenisch, Human embryonic stem cells with biological and epigenetic characteristics similar to those of mouse ESCs, Proc. Natl. Acad. Sci. USA 107 (20) (2010) 9222–9227, https://doi.org/10.1073/pnas.1004584107.
[19] A. Raya, I. Rodríguez-Pizà, G. Guenechea, R. Vassena, S. Navarro, M.J. Barrero, A. Consiglio, M. Castellà, P. Río, E. Sleep, F. González, G. Tiscornia, E. Garreta, T. Aasen, A. Veiga, I.M. Verma, J. Surrallés, J. Bueren, J.C. Izpisúa Belmonte, Disease-corrected haematopoietic progenitors from Fanconi anaemia induced pluripotent stem cells, Nature 460 (7251) (2009) 53–59, https://doi.org/10.1038/nature08129.
[20] A.C. Wilkinson, D.P. Dever, R. Baik, J. Camarena, I. Hsu, C.T. Charlesworth, C. Morita, H. Nakauchi, M.H. Porteus, Cas9-AAV6 gene correction of beta-globin in autologous HSCs improves sickle cell disease erythropoiesis in mice, Nat. Commun. 12 (1) (2021) 686, https://doi.org/10.1038/s41467-021-20909-x.
[21] U. Dührsen, J.L. Villeval, J. Boyd, G. Kannourakis, G. Morstyn, D. Metcalf, Effects of recombinant human granulocyte colony-stimulating factor on hematopoietic progenitor cells in cancer patients, Blood 72 (6) (1988) 2074–2081.
[22] W. Vogel, S. Scheding, L. Kanz, W. Brugger, Clinical applications of CD34(+) peripheral blood progenitor cells (PBPC), Stem Cells (Dayton, Ohio) 18 (2) (2000) 87–92, https://doi.org/10.1634/stemcells.18-2-87.
[23] F. Olm, A. Urbansky, J.H. Dykes, T. Laurell, S. Scheding, Label-free neuroblastoma cell separation from hematopoietic progenitor cell products using acoustophoresis—towards cell processing of complex biological samples, Sci. Rep. 9 (1) (2019) 8777, https://doi.org/10.1038/s41598-019-45182-3.
[24] W.I. Bensinger, R. Clift, P. Martin, F.R. Appelbaum, T. Demirer, T. Gooley, K. Lilleby, S. Rowley, J. Sanders, R. Storb, C.D. Buckner, Allogeneic peripheral blood stem cell transplantation in patients with advanced hematologic malignancies: a retrospective comparison with marrow transplantation, Blood 88 (7) (1996) 2794–2800.
[25] S. Pagliuca, P.H. Prata, A. Xhaard, C. Frieri, L. Giannoni, A. Sutra Del Galy, A. Brignier, F. Sicre de Fontbrune, D. Michonneau, N. Dhedin, R. Peffault de Latour, G. Socié, M. Robin, Long-term outcomes and risk factor analysis of steroid-refractory graft versus host disease after hematopoietic stem cell transplantation, Bone Marrow Transplant. 56 (1) (2021) 38–49, https://doi.org/10.1038/s41409-020-0977-3.
[26] I. McNiece, R. Jones, S.I. Bearman, P. Cagnoni, Y. Nieto, W. Franklin, J. Ryder, A. Steele, J. Stoltz, P. Russell, J. McDermitt, C. Hogan, J. Murphy, E.J. Shpall, Ex vivo expanded peripheral blood progenitor cells provide rapid neutrophil recovery after high-dose chemotherapy in patients with breast cancer, Blood 96 (9) (2000) 3001–3007.
[27] A. Amouzegar, B.R. Dey, T.R. Spitzer, Peripheral blood or bone marrow stem cells? Practical considerations in hematopoietic stem cell transplantation, Transfus. Med. Rev. 33 (1) (2019) 43–50, https://doi.org/10.1016/j.tmrv.2018.11.003.
[28] S.A. Grupp, H. Frangoul, D. Wall, M.A. Pulsipher, J.E. Levine, K.R. Schultz, Use of G-CSF in matched sibling donor pediatric allogeneic transplantation: a consen-

sus statement from the Children's Oncology Group (COG) Transplant Discipline Committee and Pediatric Blood and Marrow Transplant Consortium (PBMTC) Executive Committee, Pediatr. Blood Cancer 46 (4) (2006) 414–421, https://doi.org/10.1002/pbc.20800.

[29] F. Dong, S. Hao, S. Zhang, C. Zhu, H. Cheng, Z. Yang, F.K. Hamey, X. Wang, A. Gao, F. Wang, Y. Gao, J. Dong, C. Wang, J. Wang, Y. Lan, B. Liu, H. Ema, F. Tang, B. Göttgens, P. Zhu, T. Cheng, Differentiation of transplanted haematopoietic stem cells tracked by single-cell transcriptomic analysis, Nat. Cell Biol. 22 (6) (2020) 630–639, https://doi.org/10.1038/s41556-020-0512-1.

[30] G. Caocci, M. Greco, G. La Nasa, Bone marrow homing and engraftment defects of human hematopoietic stem and progenitor cells, Mediterr. J. Hematol. Infect Dis. 9 (1) (2017) e2017032, https://doi.org/10.4084/MJHID.2017.032.

[31] M. Tuzovic, M. Mead, P.A. Young, G. Schiller, E.H. Yang, Cardiac complications in the adult bone marrow transplant patient, Curr. Oncol. Rep. 21 (3) (2019) 28, https://doi.org/10.1007/s11912-019-0774-6.

[32] J. Montoro, J.L. Piñana, F. Moscardó, J. Sanz, Infectious complications after umbilical cord-blood transplantation from unrelated donors, Mediterr. J. Hematol. Infect Dis. 8 (1) (2016) e2016051, https://doi.org/10.4084/MJHID.2016.051.

[33] T.C. Lund, A.E. Boitano, C.S. Delaney, E.J. Shpall, J.E. Wagner, Advances in umbilical cord blood manipulation-from niche to bedside, Nat. Rev. Clin. Oncol. 12 (3) (2015) 163–174, https://doi.org/10.1038/nrclinonc.2014.215.

[34] L. Papa, M. Djedaini, R. Hoffman, Ex vivo HSC expansion challenges the paradigm of unidirectional human hematopoiesis, Ann. NY Acad. Sci. 1466 (1) (2020) 39–50, https://doi.org/10.1111/nyas.14133.

[35] S. Makino, M. Harada, K. Akashi, S. Taniguchi, T. Shibuya, S. Inaba, Y. Niho, A simplified method for cryopreservation of peripheral blood stem cells at −80 degrees C without rate-controlled freezing, Bone Marrow Transplant. 8 (4) (1991) 239–244.

[36] M. Montanari, D. Capelli, A. Poloni, D. Massidda, M. Brunori, L. Spitaleri, M. Offidani, M. Lucesole, M.C. Masia, F. Balducci, C. Refe, M. Piani, P. Leoni, A. Olivieri, Long-term hematologic reconstitution after autologous peripheral blood progenitor cell transplantation: a comparison between controlled-rate freezing and uncontrolled-rate freezing at 80 degrees C, Transfusion 43 (1) (2003) 42–49, https://doi.org/10.1046/j.1537-2995.2003.00271.x.

[37] M.A. Cox, J. Kastrup, M. Hrubiško, Historical perspectives and the future of adverse reactions associated with haemopoietic stem cells cryopreserved with dimethyl sulfoxide, Cell Tissue Bank. 13 (2) (2012) 203–215, https://doi.org/10.1007/s10561-011-9248-2.

[38] R. Cordoba, R. Arrieta, A. Kerguelen, F. Hernandez-Navarro, The occurrence of adverse events during the infusion of autologous peripheral blood stem cells is related to the number of granulocytes in the leukapheresis product, Bone Marrow Transplant. 40 (11) (2007) 1063–1067, https://doi.org/10.1038/sj.bmt.1705861.

[39] A. Donmez, M. Tombuloglu, A. Gungor, N. Soyer, G. Saydam, S. Cagirgan, Clinical side effects during peripheral blood progenitor cell infusion, Transfus. Apher. Sci. 36 (1) (2007) 95–101, https://doi.org/10.1016/j.transci.2006.05.019.

[40] R. Zenhäusern, A. Tobler, L. Leoncini, O.M. Hess, P. Ferrari, Fatal cardiac arrhythmia after infusion of dimethyl sulfoxide-cryopreserved hematopoietic stem cells in a patient with severe primary cardiac amyloidosis and end-stage renal failure, Ann. Hematol. 79 (9) (2000) 523–526, https://doi.org/10.1007/s002770000186.

[41] S.D. Rowley, Z. Feng, L. Chen, L. Holmberg, S. Heimfeld, B. MacLeod, W.I. Bensinger, A randomized phase III clinical trial of autologous blood stem cell transplantation comparing cryopreservation using dimethylsulfoxide vs dimethylsulfoxide with hydroxyethylstarch, Bone Marrow Transplant. 31 (11) (2003) 1043–1051, https://doi.org/10.1038/sj.bmt.1704030.

[42] P. Halle, O. Tournilhac, W. Knopinska-Posluszny, J. Kanold, P. Gembara, N. Boiret, C. Rapatel, M. Berger, P. Travade, S. Angielski, J. Bonhomme, F. Deméocq, Uncontrolled-rate freezing and storage at −80 degrees C, with only 3.5-percent DMSO in cryoprotective solution for 109 autologous peripheral blood progenitor cell transplantations, Transfusion 41 (5) (2001) 667–673, https://doi.org/10.1046/j.1537-2995.2001.41050667.x.

[43] P. Windrum, T.C. Morris, M.B. Drake, D. Niederwieser, T. Ruutu, EBMT Chronic Leukaemia Working Party Complications Subcommittee, Variation in dimethyl sulfoxide use in stem cell transplantation: a survey of EBMT centres, Bone Marrow Transplant. 36 (7) (2005) 601–603, https://doi.org/10.1038/sj.bmt.1705100.

[44] J.G. Briard, J.S. Poisson, T.R. Turner, C.J. Capicciotti, J.P. Acker, R.N. Ben, Small molecule ice recrystallization inhibitors mitigate red blood cell lysis during freezing, transient warming and thawing, Sci. Rep. 6 (March) (2016) 2–11, https://doi.org/10.1038/srep23619.

[45] L.K. Wu, J.M. Tokarew, J.L. Chaytor, E. Von Moos, Y. Li, C. Palii, R.N. Ben, D.S. Allan, Carbohydrate-mediated inhibition of ice recrystallization in cryopreserved human umbilical cord blood, Carbohydr. Res. 346 (1) (2011) 86–93, https://doi.org/10.1016/j.carres.2010.10.016.

[46] K. Hornberger, G. Yu, D. McKenna, A. Hubel, Cryopreservation of hematopoietic stem cells: emerging assays, cryoprotectant agents, and technology to improve outcomes, Transfus. Med. Hemother. 46 (3) (2019) 188–196, https://doi.org/10.1159/000496068.

[47] D. Berz, E.M. McCormack, E.S. Winer, G.A. Colvin, P.J. Quesenberry, Cryopreservation of hematopoietic stem cells, Am. J. Hematol. 82 (6) (2007) 463–472, https://doi.org/10.1002/ajh.20707.

[48] R. Castagnoli, O.M. Delmonte, E. Calzoni, L.D. Notarangelo, Hematopoietic stem cell transplantation in primary immunodeficiency diseases: current status and future perspectives, Front. Pediatr. 7 (2019) 295, https://doi.org/10.3389/fped.2019.00295.

[49] R.H. Buckley, S.E. Schiff, R.I. Schiff, L. Markert, L.W. Williams, J.L. Roberts, L.A. Myers, F.E. Ward, Hematopoietic stem-cell transplantation for the treatment of severe combined immunodeficiency, N. Engl. J. Med. 340 (7) (1999) 508–516, https://doi.org/10.1056/NEJM199902183400703.

[50] A.M. Miggelbrink, B.R. Logan, R.H. Buckley, R.E. Parrott, C.C. Dvorak, N. Kapoor, H. Abdel-Azim, S.E. Prockop, D. Shyr, H. Decaluwe, I.C. Hanson, A. Gillio, B.J. Dávila Saldaña, H. Eibel, G. Hopkins, J.E. Walter, J.S. Whangbo, D.B. Kohn, J.M. Puck, S.Y. Pai, B-cell differentiation and IL-21 response in IL2RG/JAK3 SCID patients after hematopoietic stem cell transplantation, Blood 131 (26) (2018) 2967–2977, https://doi.org/10.1182/blood-2017-10-809822.

[51] I.J. Abd Hamid, M.A. Slatter, F. McKendrick, M.S. Pearce, A.R. Gennery, Long-term outcome of hematopoietic stem cell transplantation for IL2RG/JAK3 SCID: a cohort report, Blood 129 (15) (2017) 2198–2201, https://doi.org/10.1182/blood-2016-11-748616.

[52] A. Puel, S.F. Ziegler, R.H. Buckley, W.J. Leonard, Defective IL7R expression in T-B+NK+ severe combined immunodeficiency, Nat. Genet. 20 (4) (1998) 394–397, https://doi.org/10.1038/3877.

[53] L.D. Notarangelo, M.S. Kim, J.E. Walter, Y.N. Lee, Human RAG mutations: biochemistry and clinical implications, Nat. Rev. Immunol. 16 (4) (2016) 234–246, https://doi.org/10.1038/nri.2016.28.

[54] C. Schuetz, B. Neven, C.C. Dvorak, S. Leroy, M.J. Ege, U. Pannicke, K. Schwarz, A.S. Schulz, M. Hoenig, M. Sparber-Sauer, S.A. Gatz, C. Denzer, S. Blanche, D. Moshous, C. Picard, B.N. Horn, J.P. De Villartay, M. Cavazzana, K.M. Debatin, M.J. Cowan, SCID patients with ARTEMIS vs RAG deficiencies following HCT: increased risk of late toxicity in ARTEMIS-deficient SCID, Blood 123 (2) (2014) 281–289, https://doi.org/10.1182/blood-2013-01-476432.

[55] N.R. Rose, Prediction and prevention of autoimmune disease in the 21st century: a review and preview, Am. J. Epidemiol. 183 (5) (2016) 403–406, https://doi.org/10.1093/aje/kwv292.
[56] L. Wang, F.S. Wang, M.E. Gershwin, Human autoimmune diseases: a comprehensive update, J. Intern. Med. 278 (4) (2015) 369–395, https://doi.org/10.1111/joim.12395.
[57] T. Alexander, A. Thiel, O. Rosen, G. Massenkeil, A. Sattler, S. Kohler, H. Mei, H. Radtke, E. Gromnica-Ihle, G.R. Burmester, R. Arnold, A. Radbruch, F. Hiepe, Depletion of autoreactive immunologic memory followed by autologous hematopoietic stem cell transplantation in patients with refractory SLE induces long-term remission through de novo generation of a juvenile and tolerant immune system, Blood 113 (1) (2020) 214–223, https://doi.org/10.1182/blood-2008-07-168286.
[58] J.R. Passweg, H. Baldomero, P. Bader, G.W. Basak, C. Bonini, R. Duarte, C. Dufour, N. Kröger, J. Kuball, A. Lankester, S. Montoto, A. Nagler, J.A. Snowden, J. Styczynski, M. Mohty, Is the use of unrelated donor transplantation leveling off in Europe? The 2016 European Society for Blood and Marrow Transplant activity survey report, Bone Marrow Transplant. 53 (9) (2018) 1139–1148, https://doi.org/10.1038/s41409-018-0153-1.
[59] J.A. Snowden, R. Saccardi, M. Allez, S. Ardizzone, R. Arnold, R. Cervera, C. Denton, C. Hawkey, M. Labopin, G. Mancardi, R. Martin, J.J. Moore, J. Passweg, C. Peters, M. Rabusin, M. Rovira, J.M. Van Laar, D. Farge, Haematopoietic SCT in severe autoimmune diseases: updated guidelines of the European group for blood and marrow transplantation, Bone Marrow Transplant. 47 (6) (2012) 770–790, https://doi.org/10.1038/bmt.2011.185.
[60] J.A. Snowden, M. Badoglio, M. Labopin, S. Giebel, E. McGrath, Z. Marjanovic, J. Burman, J. Moore, M. Rovira, N.M. Wulffraat, M. Kazmi, R. Greco, E. Snarski, T. Kozak, K. Kirgizov, T. Alexander, P. Bader, R. Saccardi, D. Farge, Evolution, trends, outcomes, and economics of hematopoietic stem cell transplantation in severe autoimmune diseases, Blood Adv. 1 (27) (2017) 2742–2755, https://doi.org/10.1182/bloodadvances.2017010041.
[61] J.A. Snowden, J. Panés, T. Alexander, M. Allez, S. Ardizzone, D. Dierickx, J. Finke, P. Hasselblatt, C. Hawkey, M. Kazmi, J.O. Lindsay, F. Onida, A. Salas, R. Saccardi, S. Vermeire, M. Rovira, E. Ricart, Autologous haematopoietic stem cell transplantation (AHSCT) in severe Crohn's disease: a review on behalf of ECCO and EBMT, J. Crohns Colitis 12 (4) (2018) 1–13, https://doi.org/10.1093/ecco-jcc/jjx184.
[62] P.J. Kelsey, M.C. Oliveira, M. Badoglio, B. Sharrack, D. Farge, J.A. Snowden, Haematopoietic stem cell transplantation in autoimmune diseases: from basic science to clinical practice, Curr Res. Transl. Med. 64 (2) (2016) 71–82, https://doi.org/10.1016/j.retram.2016.03.003.
[63] S. Meyer-Moock, Y.S. Feng, M. Maeurer, F.W. Dippel, T. Kohlmann, Systematic literature review and validity evaluation of the Expanded Disability Status Scale (EDSS) and the Multiple Sclerosis Functional Composite (MSFC) in patients with multiple sclerosis, BMC Neurol. 14 (1) (2014) 1–10, https://doi.org/10.1186/1471-2377-14-58.
[64] R. Saccardi, M.S. Freedman, M.P. Sormani, H. Atkins, D. Farge, L.M. Griffith, G. Kraft, G.L. Mancardi, R. Nash, M. Pasquini, R. Martin, P.A. Muraro, A prospective, randomized, controlled trial of autologous haematopoietic stem cell transplantation for aggressive multiple sclerosis: a position paper, Mult. Scler. J. 18 (6) (2012) 825–834, https://doi.org/10.1177/1352458512438454.
[65] R.K. Burt, Y. Loh, B. Cohen, D. Stefosky, R. Balabanov, G. Katsamakis, Y. Oyama, E.J. Russell, J. Stern, P. Muraro, J. Rose, A. Testori, J. Bucha, B. Jovanovic, F. Milanetti, J. Storek, J.C. Voltarelli, W.H. Burns, Autologous non-myeloablative haemopoietic stem cell transplantation in relapsing-remitting multiple sclerosis: a phase I/II study, Lancet Neurol. 8 (3) (2009) 244–253, https://doi.org/10.1016/S1474-4422(09)70017-1.

[66] E. Krasulová, M. Trněný, T. Kozák, B. Vacková, D. Pohlreich, D. Kemlink, P. Kobylka, I. Kovářová, P. Lhotáková, E. Havrdová, High-dose immunoablation with autologous haematopoietic stem cell transplantation in aggressive multiple sclerosis: a single centre 10-year experience, Mult. Scler. 16 (6) (2010) 685–693, https://doi.org/10.1177/1352458510364538.

[67] J. Burman, E. Iacobaeus, A. Svenningsson, J. Lycke, M. Gunnarsson, P. Nilsson, M. Vrethem, S. Fredrikson, C. Martin, A. Sandstedt, B. Uggla, S. Lenhoff, J.E. Johansson, C. Isaksson, H. Hägglund, K. Carlson, J. Fagius, Autologous haematopoietic stem cell transplantation for aggressive multiple sclerosis: the Swedish experience, J. Neurol. Neurosurg. Psychiatry 85 (10) (2014) 1116–1121, https://doi.org/10.1136/jnnp-2013-307207.

[68] R.K. Burt, R. Balabanov, X. Han, B. Sharrack, A. Morgan, K. Quigley, K. Yaung, I.B. Helenowski, B. Jovanovic, D. Spahovic, I. Arnautovic, D.C. Lee, B.C. Benefield, S. Futterer, M.C. Oliveira, J. Burman, Association of nonmyeloablative hematopoietic stem cell transplantation with neurological disability in patients with relapsing-remitting multiple sclerosis, JAMA, J. Am. Med. Assoc. 313 (3) (2015) 275–284, https://doi.org/10.1001/jama.2014.17986.

[69] R.A. Nash, G.J. Hutton, M.K. Racke, U. Popat, S.M. Devine, K.C. Steinmiller, L.M. Griffith, P.A. Muraro, H. Openshaw, P.H. Sayre, O. Stuve, D.L. Arnold, M.H. Wener, G.E. Georges, A. Wundes, G.H. Kraft, J.D. Bowen, High-dose immunosuppressive therapy and autologous HCT for relapsing-remitting MS, Neurology 88 (9) (2017) 842–852, https://doi.org/10.1212/WNL.0000000000003660.

[70] S.M. Bell, B. Sharrack, J.A. Snowden, Autologous hematopoietic cell transplantation in multiple sclerosis, Expert. Opin. Biol. Ther. 17 (1) (2017) 77–86, https://doi.org/10.1080/14712598.2017.1239706.

[71] M.C. Oliveira, M. Labopin, J. Henes, J. Moore, N.D. Papa, A. Cras, I. Sakellari, R. Schroers, H.U. Scherer, A. Cuneo, S. Kyrcz-Krzemien, T. Daikeler, T. Alexander, J. Finke, M. Badoglio, B. Simões, J.A. Snowden, D. Farge, Does ex vivo CD34+ positive selection influence outcome after autologous hematopoietic stem cell transplantation in systemic sclerosis patients? Bone Marrow Transplant. 51 (4) (2016) 501–505, https://doi.org/10.1038/bmt.2015.299.

[72] P.A. Muraro, M. Pasquini, H.L. Atkins, J.D. Bowen, D. Farge, A. Fassas, M.S. Freedman, G.E. Georges, F. Gualandi, N. Hamerschlak, E. Havrdova, V.K. Kimiskidis, T. Kozak, G.L. Mancardi, L. Massacesi, D.A. Moraes, R.A. Nash, S. Pavletic, J. Ouyang, R. Saccardi, Long-term outcomes after autologous hematopoietic stem cell transplantation for multiple sclerosis, JAMA Neurol. 74 (4) (2017) 459–469, https://doi.org/10.1001/jamaneurol.2016.5867.

[73] H.L. Atkins, M. Bowman, D. Allan, G. Anstee, D.L. Arnold, A. Bar-Or, I. Bence-Bruckler, P. Birch, C. Bredeson, J. Chen, D. Fergusson, M. Halpenny, L. Hamelin, L. Huebsch, B. Hutton, P. Laneuville, Y. Lapierre, H. Lee, L. Martin, S. McDiarmid, M.S. Freedman, Immunoablation and autologous haemopoietic stem-cell transplantation for aggressive multiple sclerosis: a multicentre single-group phase 2 trial, Lancet (London, England) 388 (10044) (2016) 576–585, https://doi.org/10.1016/S0140-6736(16)30169-6.

[74] J.J. Moore, J.C. Massey, C.D. Ford, M.L. Khoo, J.J. Zaunders, K. Hendrawan, Y. Barnett, M.H. Barnett, K.A. Kyle, R. Zivadinov, K.C. Ma, S.T. Milliken, I.J. Sutton, D.D.F. Ma, Prospective phase II clinical trial of autologous haematopoietic stem cell transplant for treatment refractory multiple sclerosis, J. Neurol. Neurosurg. Psychiatry 90 (5) (2019) 514–521, https://doi.org/10.1136/jnnp-2018-319446.

[75] J.J. Moore, P. Brooks, S. Milliken, J. Biggs, D. Ma, M. Handel, P. Cannell, R. Will, S. Rule, D. Joske, B. Langlands, K. Taylor, J. O'Callaghan, J. Szer, I. Wicks, G. McColl, F. Passeullo, J. Snowden, A pilot randomized trial comparing CD34-selected versus unmanipulated hemopoietic stem cell transplantation for severe, refractory rheumatoid arthritis, Arthritis Rheum. 46 (9) (2002) 2301–2309, https://doi.org/10.1002/art.10495.

[76] G.L. Mancardi, M.P. Sormani, F. Gualandi, A. Saiz, E. Carreras, E. Merelli, A. Donelli, A. Lugaresi, P. Di Bartolomeo, M.R. Rottoli, A. Rambaldi, M.P. Amato, L. Massacesi, M. Di Gioia, L. Vuolo, D. Currò, L. Roccatagliata, M. Filippi, U. Aguglia, P. Iacopino, ASTIMS Haemato-Neurological Collaborative Group On behalf of the Autoimmune Disease Working Party ADWP of the European Group for Blood and Marrow Transplantation EBMT, Autologous hematopoietic stem cell transplantation in multiple sclerosis: a phase II trial, Neurology 84 (10) (2015) 981–988, https://doi.org/10.1212/WNL.0000000000001329.

[77] J.A. Cohen, L.E. Baldassari, H.L. Atkins, J.D. Bowen, C. Bredeson, P.A. Carpenter, J.R. Corboy, M.S. Freedman, L.M. Griffith, R. Lowsky, N.S. Majhail, P.A. Muraro, R.A. Nash, M.C. Pasquini, S. Sarantopoulos, B.N. Savani, J. Storek, K.M. Sullivan, G.E. Georges, Autologous hematopoietic cell transplantation for treatment-refractory relapsing multiple sclerosis: position statement from the American Society for Blood and Marrow Transplantation, Biol. Blood Marrow Transplant. 25 (5) (2019) 845–854, https://doi.org/10.1016/j.bbmt.2019.02.014.

[78] R.K. Burt, X. Han, P. Gozdziak, K. Yaung, A. Morgan, A.M. Clendenan, J. Henry, M.A. Calvario, S.K. Datta, I. Helenowski, J. Schroeder, Five year follow-up after autologous peripheral blood hematopoietic stem cell transplantation for refractory, chronic, corticosteroid-dependent systemic lupus erythematosus: effect of conditioning regimen on outcome, Bone Marrow Transplant. 53 (6) (2018) 692–700, https://doi.org/10.1038/s41409-018-0173-x.

[79] C.P. Denton, D. Khanna, Systemic sclerosis, Lancet 390 (10103) (2017) 1685–1699, https://doi.org/10.1016/S0140-6736(17)30933-9.

[80] C.P. Denton, A.U. Wells, J.G. Coghlan, Major lung complications of systemic sclerosis, Nat. Rev. Rheumatol. 14 (9) (2018) 511–527, https://doi.org/10.1038/s41584-018-0062-0.

[81] N. Del Papa, F. Onida, E. Zaccara, G. Saporiti, W. Maglione, E. Tagliaferri, R. Andracco, D. Vincenti, T. Montemurro, L. Mircoli, C. Vitali, A. Cortelezzi, Autologous hematopoietic stem cell transplantation has better outcomes than conventional therapies in patients with rapidly progressive systemic sclerosis, Bone Marrow Transplant. 52 (1) (2017) 53–58, https://doi.org/10.1038/bmt.2016.211.

[82] J.C. Henes, M. Schmalzing, W. Vogel, G. Riemekasten, F. Fend, L. Kanz, I. Koetter, Optimization of autologous stem cell transplantation for systemic sclerosis—a single-center longterm experience in 26 patients with severe organ manifestations, J. Rheumatol. 39 (2) (2012) 269–275, https://doi.org/10.3899/jrheum.110868.

[83] R.K. Burt, M.C. Oliveira, S.J. Shah, D.A. Moraes, B. Simoes, M. Gheorghiade, J. Schroeder, E. Ruderman, D. Farge, Z.J. Chai, Z. Marjanovic, S. Jain, A. Morgan, F. Milanetti, X. Han, B. Jovanovic, I.B. Helenowski, J. Voltarelli, Cardiac involvement and treatment-related mortality after non-myeloablative haemopoietic stem-cell transplantation with unselected autologous peripheral blood for patients with systemic sclerosis: a retrospective analysis, Lancet 381 (9872) (2013) 1116–1124, https://doi.org/10.1016/S0140-6736(12)62114-X.

[84] O. Kowal-Bielecka, J. Fransen, J. Avouac, M. Becker, A. Kulak, Y. Allanore, O. Distler, P. Clements, M. Cutolo, L. Czirjak, N. Damjanov, F. Del Galdo, C.P. Denton, J.H.W. Distler, I. Foeldvari, K. Figelstone, M. Frerix, D.E. Furst, S. Guiducci, B. Imbert, Update of EULAR recommendations for the treatment of systemic sclerosis, Ann. Rheum. Dis. 76 (8) (2017) 1327–1339, https://doi.org/10.1136/annrheumdis-2016-209909.

[85] J. Henes, M.C. Oliveira, M. Labopin, M. Badoglio, H.U. Scherer, N. Del Papa, T. Daikeler, M. Schmalzing, R. Schroers, T. Martin, G. Pugnet, B. Simoes, D. Michonneau, E.W.A. Marijt, B. Lioure, J.O. Bay, J.A. Snowden, M. Rovira, A. Huynh, D. Farge, Autologous stem cell transplantation for progressive systemic sclerosis: a prospective non-interventional study from the European Society for Blood and Marrow Transplantation Autoimmune Disease Working Party, Haematologica 106 (2) (2021) 375–383, https://doi.org/10.3324/haematol.2019.230128.

[86] K.M. Sullivan, E.A. Goldmuntz, L. Keyes-Elstein, P.A. McSweeney, A. Pinckney, B. Welch, M.D. Mayes, R.A. Nash, L.J. Crofford, B. Eggleston, S. Castina, L.M. Griffith, J.S. Goldstein, D. Wallace, O. Craciunescu, D. Khanna, R.J. Folz, J. Goldin, E.W. St. Clair, D.E. Furst, Myeloablative autologous stem-cell transplantation for severe scleroderma, N. Engl. J. Med. 378 (1) (2018) 35–47, https://doi.org/10.1056/nejmoa1703327.

[87] D.C. Baumgart, W.J. Sandborn, Crohn's disease, Lancet 380 (9853) (2012) 1590–1605, https://doi.org/10.1016/S0140-6736(12)60026-9.

[88] R.M. Craig, A. Traynor, Y. Oyama, R.K. Burt, Hematopoietic stem cell transplantation for severe Crohn's disease, Bone Marrow Transplant. 32 (Suppl. 1) (2003) S57–S59, https://doi.org/10.1038/sj.bmt.1703945.

[89] A. Cassinotti, C. Annaloro, S. Ardizzone, F. Onida, A. Della Volpe, M. Clerici, P. Usardi, S. Greco, G. Maconi, G. Bianchi Porro, G. Lambertenghi Deliliers, Autologous haematopoietic stem cell transplantation without CD34 + cell selection in refractory Crohn's disease, Gut 57 (2) (2008) 211–217, https://doi.org/10.1136/gut.2007.128694.

[90] J.O. Lindsay, M. Allez, M. Clark, M. Labopin, E. Ricart, G. Rogler, M. Rovira, J. Satsangi, D. Farge, C.J. Hawkey, Autologous stem-cell transplantation in treatment-refractory Crohn's disease: an analysis of pooled data from the ASTIC trial, Lancet Gastroenterol. Hepatol. 2 (6) (2017) 399–406, https://doi.org/10.1016/S2468-1253(17)30056-0.

[91] R.L. Siegel, K.D. Miller, A. Jemal, Cancer statistics, 2016, CA Cancer J. Clin. 66 (1) (2016) 7–30, https://doi.org/10.3322/caac.21332.

[92] J. Casper, D. Wolff, W. Knauf, I.W. Blau, T. Ruutu, L. Volin, H. Wandt, K. Schäfer-Eckart, J. Holowiecki, S. Giebel, J. Aschan, A.R. Zander, N. Kröger, I. Hilgendorf, J. Baumgart, H.A. Mylius, U. Pichlmeier, M. Freund, Allogeneic hematopoietic stem-cell transplantation in patients with hematologic malignancies after dose-escalated treosulfan/fludarabine conditioning, J. Clin. Oncol. 28 (20) (2010) 3344–3351, https://doi.org/10.1200/JCO.2009.23.3429.

[93] P.L. Weiden, N. Flournoy, E.D. Thomas, R. Prentice, A. Fefer, C.D. Buckner, R. Storb, Antileukemic effect of graft-versus-host disease in human recipients of allogeneic-marrow grafts, N. Engl. J. Med. 300 (19) (1979) 1068–1073, https://doi.org/10.1056/NEJM197905103001902.

[94] H. Döhner, E. Estey, D. Grimwade, S. Amadori, F.R. Appelbaum, T. Büchner, H. Dombret, B.L. Ebert, P. Fenaux, R.A. Larson, R.L. Levine, F. Lo-Coco, T. Naoe, D. Niederwieser, G.J. Ossenkoppele, M. Sanz, J. Sierra, M.S. Tallman, H.F. Tien, C.D. Bloomfield, Diagnosis and management of AML in adults: 2017 ELN recommendations from an international expert panel, Blood 129 (4) (2017) 424–447, https://doi.org/10.1182/blood-2016-08-733196.

[95] M.S. Tallman, E.S. Wang, J.K. Altman, F.R. Appelbaum, V.R. Bhatt, D. Bixby, S.E. Coutre, M. De Lima, A.T. Fathi, M. Fiorella, J.M. Foran, A.C. Hall, M. Jacoby, J. Lancet, T.W. LeBlanc, G. Mannis, G. Marcucci, M.G. Martin, A. Mims, N. Ogba, Acute myeloid leukemia, version 3.2019, J. Natl. Compr. Cancer Netw. 17 (6) (2019) 721–749, https://doi.org/10.6004/jnccn.2019.0028.

[96] S. Giebel, A. Czyz, O. Ottmann, F. Baron, E. Brissot, F. Ciceri, J.J. Cornelissen, J. Esteve, N.C. Gorin, B. Savani, C. Schmid, M. Mohty, A. Nagler, Use of tyrosine kinase inhibitors to prevent relapse after allogeneic hematopoietic stem cell transplantation for patients with Philadelphia chromosome-positive acute lymphoblastic leukemia: a position statement of the acute leukemia working party of the, Cancer 122 (19) (2016) 2941–2951, https://doi.org/10.1002/cncr.30130.

[97] A. Shimoni, Y. Volchek, M. Koren-Michowitz, N. Varda-Bloom, R. Somech, N. Shem-Tov, R. Yerushalmi, A. Nagler, Phase 1/2 study of nilotinib prophylaxis after allogeneic stem cell transplantation in patients with advanced chronic myeloid leukemia or Philadelphia chromosome-positive acute lymphoblastic leukemia, Cancer 121 (6) (2015) 863–871, https://doi.org/10.1002/cncr.29141.

[98] S.M. Bair, J.D. Brandstadter, E.C. Ayers, E.A. Stadtmauer, Hematopoietic stem cell transplantation for blood cancers in the era of precision medicine and immunotherapy, Cancer 126 (9) (2020) 1837–1855, https://doi.org/10.1002/cncr.32659.

[99] T.L. Lin, J.E. Cortes, L.F. Newell, E.K. Ritchie, D.E. Hogge, S.R. Solomon, G.J. Schiller, M.J. Wieduwilt, D.H. Ryan, R.J. Ryan, A.C. Louie, The impact of hematopoietic cell transplantation (HCT) on survival: an exploratory analysis of a phase 3 study of CPX-351 versus 7+3 in older patients with newly diagnosed, high-risk/secondary AML (sAML), Biol. Blood Marrow Transplant. 25 (3) (2019) S125–S126, https://doi.org/10.1016/j.bbmt.2018.12.408.

[100] C.D. DiNardo, A.S. Stein, A.T. Fathi, P. Montesinos, O. Odenike, H.M. Kantarjian, R.M. Stone, D.O. Koralek, J. Van Oostendorp, J. Gong, I. Gupta, P. Vyas, Mutant isocitrate dehydrogenase (mIDH) inhibitors, enasidenib or ivosidenib, in combination with azacitidine (AZA): preliminary results of a phase 1b/2 study in patients with newly diagnosed acute myeloid leukemia (AML), Blood 130 (Suppl. 1) (2017) 639. http://www.bloodjournal.org/content/130/Suppl_1/639.abstract.

[101] E.M. Stein, C.D. DiNardo, A.T. Fathi, A.S. Mims, K.W. Pratz, M.R. Savona, A.S. Stein, R.M. Stone, E.S. Winer, C.S. Seet, H. Döhner, D.A. Pollyea, J. McCloskey, O. Odenike, B. Löwenberg, G.J. Ossenkoppele, P.A. Patel, M. Roshal, M.G. Frattini, F. Lersch, M.S. Tallman, Ivosidenib or enasidenib combined with intensive chemotherapy in patients with newly diagnosed AML: a phase 1 study, Blood (2020), https://doi.org/10.1182/blood.2020007233Fassas. blood.2020007233. Advance online publication. A., Kimiskidis, V. K., Sakellari, I., Kapinas, K., Anagnostopoulos, A., Tsimourtou, V., Sotirakoglou, K., & Kazis, A. (2018). Long-term results of stem cell transplantation for MS: A single-center experience. Neurology, 76(12), 1066–1070. doi:10.1212/WNL.0b013e318211c537.

[102] Z. Li, S.M. Rubinstein, R. Thota, M. Savani, E. Brissot, B.E. Shaw, N.S. Majhail, M. Mohty, B.N. Savani, Immune-mediated complications after hematopoietic stem cell transplantation, Biol. Blood Marrow Transplant. 22 (8) (2016) 1368–1375, https://doi.org/10.1016/j.bbmt.2016.04.005.

[103] J.L. Ferrara, J.E. Levine, P. Reddy, E. Holler, Graft-versus-host disease, Lancet (London, England) 373 (9674) (2009) 1550–1561, https://doi.org/10.1016/S0140-6736(09)60237-3.

[104] T. Daikeler, M. Labopin, M. Di Gioia, M. Abinun, T. Alexander, I. Miniati, F. Gualandi, A. Fassas, T. Martin, C.P. Schwarze, N. Wulffraat, M. Buch, A. Sampol, E. Carreras, B. Dubois, B. Gruhn, T. Güngör, D. Pohlreich, A. Schuerwegh, E. Snarski, EBMT Autoimmune Disease Working Party, Secondary autoimmune diseases occurring after HSCT for an autoimmune disease: a retrospective study of the EBMT Autoimmune Disease Working Party, Blood 118 (6) (2011) 1693–1698, https://doi.org/10.1182/blood-2011-02-336156.

[105] P. Sagoo, K. Ratnasothy, Y. Tsang, L.D. Barber, A. Noble, R.I. Lechler, G. Lombardi, Alloantigen-specific regulatory T cells prevent experimental chronic graft-versus-host disease by simultaneous control of allo- and autoreactivity, Eur. J. Immunol. 42 (12) (2012) 3322–3333, https://doi.org/10.1002/eji.201242770.

[106] K.M. Page, A.M. Mendizabal, V.K. Prasad, P.L. Martin, S. Parikh, S. Wood, G.D. Sempowski, P. Szabolcs, J. Kurtzberg, Posttransplant autoimmune hemolytic anemia and other autoimmune cytopenias are increased in very young infants undergoing unrelated donor umbilical cord blood transplantation, Biol. Blood Marrow Transplant. 14 (10) (2018) 1108–1117, https://doi.org/10.1016/j.bbmt.2008.07.006.Posttransplant. NIH Public Access.

[107] C.E. Neunert, J.M. Despotovic, Autoimmune hemolytic anemia and immune thrombocytopenia following hematopoietic stem cell transplant: a critical review of the literature, Pediatr. Blood Cancer 66 (4) (2019) 1–7, https://doi.org/10.1002/pbc.27569.

[108] M.D. Kruizinga, M.J.D. van Tol, V. Bekker, T. Netelenbos, F.J. Smiers, D. Bresters, A.M. Jansen-Hoogendijk, M.M. van Ostaijen-ten Dam, W.J.W. Kollen, J.J. Zwaginga, A.C. Lankester, R.G.M. Bredius, Risk factors, treatment, and immune dysregulation in autoimmune cytopenia after allogeneic hematopoietic stem cell transplantation in pediatric patients, Biol. Blood Marrow Transplant. 24 (4) (2018) 772–778, https://doi.org/10.1016/j.bbmt.2017.12.782.

[109] P.D.E. Miller, J.A. Snowden, R.P. De Latour, S. Iacobelli, D.J. Eikema, C. Knol, J.C.W. Marsh, C. Rice, M. Koh, F. Fagioli, S. Chaganti, J. Finke, R.F. Duarte, P. Bader, D. Farge, J.R. Passweg, J.A. Madrigal, C. Dufour, Autoimmune cytopenias (AIC) following allogeneic haematopoietic stem cell transplant for acquired aplastic anaemia: a joint study of the Autoimmune Diseases and Severe Aplastic Anaemia Working Parties (ADWP/SAAWP) of the European Society for Blood and Marrow Transplantation (EBMT), Bone Marrow Transplant. 55 (2) (2020) 441–451, https://doi.org/10.1038/s41409-019-0680-4.

[110] D. Zama, E. Biagi, R. Masetti, P. Gasperini, A. Prete, M. Candela, P. Brigidi, A. Pession, Gut microbiota and hematopoietic stem cell transplantation: where do we stand? Bone Marrow Transplant. 52 (1) (2017) 7–14, https://doi.org/10.1038/bmt.2016.173.

[111] J.H. Young, B.R. Logan, J. Wu, J.R. Wingard, D.J. Weisdorf, C. Mudrick, K. Knust, M.M. Horowitz, D.L. Confer, E.R. Dubberke, S.A. Pergam, F.M. Marty, L.M. Strasfeld, J. Brown, A.A. Langston, M.G. Schuster, D.R. Kaul, S.I. Martin, C. Anasetti, Blood and Marrow Transplant Clinical Trials Network Trial 0201, Infections after transplantation of bone marrow or peripheral blood stem cells from unrelated donors, Biol. Blood Marrow Transplant. 22 (2) (2016) 359–370, https://doi.org/10.1016/j.bbmt.2015.09.013.

[112] C. Gudiol, C. Garcia-Vidal, M. Arnan, I. Sánchez-Ortega, B. Patiño, R. Duarte, J. Carratalà, Etiology, clinical features and outcomes of pre-engraftment and post-engraftment bloodstream infection in hematopoietic SCT recipients, Bone Marrow Transplant. 49 (6) (2014) 824–830, https://doi.org/10.1038/bmt.2014.37.

[113] M.G. Schuster, A.A. Cleveland, E.R. Dubberke, C.A. Kauffman, R.K. Avery, S. Husain, D.L. Paterson, F.P. Silveira, T.M. Chiller, K. Benedict, K. Murphy, P.G. Pappas, Infections in hematopoietic cell transplant recipients: results from the organ transplant infection project, a multicenter, prospective, cohort study, Open Forum Infect. Dis. 4 (2) (2018) 1–7, https://doi.org/10.1093/ofid/ofx050.

[114] S.Y. Cho, H.J. Lee, D.G. Lee, Infectious complications after hematopoietic stem cell transplantation: current status and future perspectives in Korea, Korean J. Intern. Med. 33 (2) (2018) 256–276, https://doi.org/10.3904/kjim.2018.036.

[115] K. Heydari, A. Shamshirian, P. Lotfi-Foroushani, A. Aref, A. Hedayatizadeh-Omran, M. Ahmadi, G. Janbabei, S. Keyhanian, E. Zaboli, S.M. Ghasemzadeh, R. Alizadeh-Navaei, The risk of malignancies in patients receiving hematopoietic stem cell transplantation: a systematic review and meta-analysis, Clin. Transl. Oncol. 22 (10) (2020) 1825–1837, https://doi.org/10.1007/s12094-020-02322-w.

[116] W.W. Pang, A. Czechowicz, A.C. Logan, R. Bhardwaj, J. Poyser, C.Y. Park, I.L. Weissman, J.A. Shizuru, Anti-CD117 antibody depletes normal and myelodysplastic syndrome human hematopoietic stem cells in xenografted mice, Blood 133 (19) (2019) 2069–2078, https://doi.org/10.1182/blood-2018-06-858159.

[117] R. Palchaudhuri, B. Saez, J. Hoggatt, A. Schajnovitz, D.B. Sykes, T.A. Tate, A. Czechowicz, Y. Kfoury, D.J. Rossi, G.L. Verdine, M.K. Mansour, D.T. Scadden, R. Biology, M.G. Hospital, M.G. Hospital, Non-genotoxic conditioning for hematopoietic stem cell transplantation using a hematopoietic-cell-specific internalizing immunotoxin, Nat. Biotechnol. 34 (7) (2016) 738–745, https://doi.org/10.1038/nbt.3584.Non-genotoxic. HHS Public Access.

[118] D. Wolf, H. Barreras, C.S. Bader, S. Copsel, C.O. Lightbourn, B.J. Pfeiffer, N.H. Altman, E.R. Podack, K.V. Komanduri, R.B. Levy, Marked in vivo donor regulatory T cell expansion via interleukin-2 and TL1A-Ig stimulation ameliorates graft-versus-host dis-

ease but preserves graft-versus-leukemia in recipients after hematopoietic stem cell transplantation, Biol. Blood Marrow Transplant. 23 (5) (2017) 757–766, https://doi.org/10.1016/j.bbmt.2017.02.013.

[119] M. Quarmyne, P.L. Doan, H.A. Himburg, X. Yan, M. Nakamura, L. Zhao, N.J. Chao, J.P. Chute, Protein tyrosine phosphatase-σ regulates hematopoietic stem cell-repopulating capacity, J. Clin. Investig. 125 (1) (2015) 177–182, https://doi.org/10.1172/JCI77866.

[120] J.R. Passweg, H. Baldomero, P. Bader, C. Bonini, R.F. Duarte, C. Dufour, A. Gennery, N. Kröger, J. Kuball, F. Lanza, S. Montoto, A. Nagler, J.A. Snowden, J. Styczynski, M. Mohty, Use of haploidentical stem cell transplantation continues to increase: the 2015 European Society for Blood and Marrow Transplant activity survey report, Bone Marrow Transplant. 52 (6) (2017) 811–817, https://doi.org/10.1038/bmt.2017.34.

[121] A. Bertaina, M.G. Roncarolo, Graft engineering and adoptive immunotherapy: new approaches to promote immune tolerance after hematopoietic stem cell transplantation, Front. Immunol. 10 (July) (2019), https://doi.org/10.3389/fimmu.2019.01342.

[122] R. Lanza, D.W. Russell, A. Nagy, Engineering universal cells that evade immune detection, Nat. Rev. Immunol. 19 (12) (2019) 723–733, https://doi.org/10.1038/s41577-019-0200-1.

Index

Note: Page numbers followed by *f* indicate figures and *t* indicate tables.

Kolibri 360 GmbH
Ferdinand-Jühlke-Straße 7
99095 Erfurt, Deutschland
produktsicherheit@kolibri360.de